SYMMETRY PROPERTIES OF NUCLEI

SYMMETRY PROPERTIES OF NUCLEI

Proceedings
of the
Fifteenth Solvay
Conference on Physics
September 28-October 3, 1970

GORDON AND BREACH SCIENCE PUBLISHERS
London New York Paris

Gordon and Breach Science Publishers Ltd.
42 William IV Street
London W.C.2.

Editorial office for the United States of America
Gordon and Breach Science Publishers Inc.
One Park Avenue
New York, N.Y. 10016.

Editorial office for France
Gordon & Breach
7–9 rue Emile Dubois
Paris 14^e.

Library of Congress catalog card number 73–84467.
ISBN 0 677 14450 4. Printed in German Democratic Republic by Offizin Andersen Nexö.

ADMINISTRATIVE COMMISSION OF THE INTERNATIONAL INSTITUTES OF PHYSICS AND CHEMISTRY *founded by* ERNEST SOLVAY

Scientific Committee of the International Institute of Physics

A. Amaldi (*President*)	Istituto di Fisica "Guglielmo Marconi", Universita degli Studi, Piazzale delle Scienze, 5, Roma Italy
A. Abragam	CEN/Saclay, Boîte Postale n^0 2, F-91 Gif-sur-Yvette, France
L. Artsimovitch	Academy of Sciences of the U.S.S.R., Moscow, U.S.S.R.
L. Bragg	The Royal Institution, 21, Albemarle Street, London, W. 1, England
C. J. Gorter	Kamerlingh Onnes Laboratory, 18, Nieuwsteeg, Leiden, Nederland
W. Heisenberg	Max-Planck Institut für Physik und Astrophysik, Aumeisterstrasse, 6, München 23, Deutsche Bundesrepubliek
C. Møller	Nordita, Blegdamsvej, 17, DK-2100 København, Ø, Denmark
F. Perrin	Commissariat à l'Energie Atomique, 20–33, rue de la Fédération F-75 Paris, France
A. B. Pippard	University of Cambridge, Department of Physics, Cavendish Laboratory, Free School Lane, Cambridge, England

Fifteenth Conference of Physics
Chairman **E. Amaldi**
Invited Speakers

A. Bohr	U.I.T.F., Blegdamsvej, 17, DK-2100 København, Ø, Denmark
D. M. Brink	Department of Theoretical Physics, Oxford University, Oxford, England
J. P. Elliott	Department of Physics, University of Sussex, Brighton, England
K. T. Hecht	The University of Michigan, Department of Physics, Ann Arbor, Michigan 48104, U.S.A.
P. Kramer	Institut für Theoretische Physik der Universität Tübingen, D-74 Tübingen, Köstlinstrasse, 6, Deutsche Bundesrepubliek
M. Moshinsky	Universidad Nacional Autonoma de Mexico, APDO Post. 20-364, Mexico 20, D.F., Mexico
B. R. Mottelson	Nordita, Blegdamsvej, 17, DK-2100 København Ø, Denmark
L. Radicati	Scuola Normale Superiore, Classe di Scienze, Piazza dei Cavalieri, Pisa, Italy
D. H. Wilkinson	University of Oxford, Nuclear Physics Laboratory, Oxford, England
L. Rosenfeld	Nordita, Blegdamsvej, 17, DK-2100 København, Ø, Denmark
A. Sandage	Mount Wilson and Palomar Observatories, 813, Santa Barbara Street, Pasadena, California 91106, U.S.A.

Contents

M. Heinemann Epse Libert Université Libre de Bruxelles

Ch. Quesne Université Libre de Bruxelles

M. Rayet Université Libre de Bruxelles

M. Reidemeister Université Libre de Bruxelles

M. Rouserez Epse Fuld Université Libre de Bruxelles

Ch. Willain Epse Leclercq Université Libre de Bruxelles

Invited Foreign Listeners

R. Arvieu Faculté des Sciences de Paris et d'Orsay, Institut de Physique Nucléaire, B.P. n^0 1, F-91 Orsay, France

G. Bertsch Princeton University, Department of Physics: Joseph Henry Laboratories, Post Office Box 708, Princeton, New Jersey 08540, U.S.A.

J. Blomqvist Royal Institute of Technology, Division of Theoretical Physics, Stockholm 70, Sweden

P. Camiz Universita degli Studi Roma, Istituto de Fisica "Guglielmo Marconi", Piazzale delle Scienze, 5, Roma, Italy

J. Flores Universidad Nacional Autonoma de Mexico, APDO Post. 20-364, Mexico 20, D.F., Mexico

J. Mehra The University of Texas at Austin, Center for Statistical Mechanics and Thermodynamics, Austin, Texas 78712, U.S.A.

G. Ripka Centre d'Etudes Nucléaires de Saclay, B.P. n^0 2, F-91 Gif-sur-Yvette, France

M. Veneroni Institut de Physique Nucléaire, Faculté des Sciences de Paris et d'Orsay, B.P. n^0 1, F-91 Orsay, France

Invited Belgian Listeners

M. Bouten Vrije Universiteit Brussel

R. Ceuleneer Université de Mons

A. Gribaumont Université Libre de Bruxelles

J. Humblet Université de Liège

P. Macq Université de Louvain

J. Pelseneer Université Libre de Bruxelles

P. Van Leuven Rijksuniversitair Centrum, Antwerpen

J. L. Verhaeghe Rijksuniversiteit Gent

E. Tamm	Lebedev Physical Institute of the Academy of Sciences, Leninsky Prospekt, 7/9, Moscow, B-137, U.S.S.R.
S. Tomonaga	The Tokyo University of Education, 24, Otsuka-Kubomachi, Bunkyo-ku, Tokyo, Japan.
E. P. Wigner	Princeton University, P.O.B. 708, Princeton, New Jersey 08540, U.S.A.
J. Géhéniau (*Secretary*)	Université Libre de Bruxelles, 50, avenue F. D. Roosevelt, 1050 Bruxelles, Belgium

Invited Members

L. C. Biedenharn	Duke University, Department of Physics, Durham, North Carolina 27706, U.S.A.
C. Bloch	C.E.A., Centre d'Etudes Nucléaires de Saclay, Direction de la Physique, B.P. n° 2, F-91 Gif-sur-Yvette, France
G. E. Brown	Nordita, Blegdamsvej, 17, DK-2100 København Ø, Denmark
H. Frauenfelder	University of Illinois at Urbana-Champaign, Department of Physics, Urbana, Illinois 61801, U.S.A.
J. B. French	The University of Rochester, River Campus Station, Rochester, New York 14627, U.S.A.
D. R. Inglis	University of Massachusetts, Department of Physics and Astronomy Hasbrouck Laboratory, Amherst 01002, Massachusetts, U.S.A.
M. Jean	Institut de Physique Nucléaire, Division de Physique Théorique, B.P. n° 1, F-91 Orsay, France
B. R. Judd	The Johns Hopkins University, Physics Department, Baltimore, Maryland 21218, U.S.A.
H. J. Lipkin	The Weizmann Institute of Science, Department of Nuclear Physics, Rehovot, Israel
I. Talmi	The Weizmann Institute of Science, Department of Nuclear Physics, Rehovot, Israel
C. van der Leun	Fysisch Laboratorium, Rijksuniversiteit Utrecht, Bijlhouwerstraat, 6, Utrecht, Nederland

Secretary Members

M. Demeur	Université Libre de Bruxelles
J. Deenen	Université Libre de Bruxelles
M.-C. Denys Epse Bouten	Centre Nucléaire de Mol
Do Tan-Si	Université Libre de Bruxelles

Symmetry of Cluster Structures of Nuclei

D. M. BRINK
Department of Theoretical Physics
Oxford University

1 INTRODUCTION

The nucleus ^{8}Be is not quite stable and decays into two α-particles with a Q-value of 94 keV. The neighbouring nucleus ^{9}Be is stable in its ground state; but its energy of dissociation into $\alpha + \alpha + n$ is only 1.5 MeV. These experimental facts suggest that ^{8}Be might consist of two α-particles interacting by a force which is just too weak to bind them; and that ^{9}Be consists of two α-particles bound together by an extra neutron.

Certain excited states of ^{16}O have large reduced widths for decaying into ^{12}C + α. These states might consist of an α-particle bound to a ^{12}C-nucleus by some mutual interaction. In ^{12}C the twelve nucleons might be grouped together into three α-clusters, and the nucleus ^{6}Li could consist of a deuteron bound to an α-particle. These are examples of some cluster structures which could exist in light nuclei.

How can such cluster structures be described theoretically? Is a cluster description of nuclear states complementary with a shell-model description or is it distinct? How do symmetries of a cluster structure affect properties of nuclear states? These are several of the questions which will be discussed in this lecture.

The simplest theories treat clusters as elementary objects without internal structure. For example the low states of ^{12}C would consist of three elementary α-particles, obeying Bose-Einstein statistics, bound together by some $\alpha - \alpha$ interaction. This model was introduced by Wefelmeier[1] in 1937 and developed by Dennison[2] and others. Such a theory can account for the positions of the low excited $T = 0$ states of ^{12}C and for electromagnetic form factors and transition rates[3] in terms of a few parameters. This model is limited because there is no way of relating the parameters of the model to properties of individual nucleons and the interactions between them.

Microscopic nucleon cluster models aim to construct many-nucleon wave-functions which satisfy the requirements of the Pauli exclusion principle and which contain clustering effects. The resonating group method suggested by Wheeler[4] and used extensively by Wildermuth[5] and Neudatchin[6] is one way of constructing microscopic cluster models. The generator co-ordinate method of Hill and Wheeler[7,8] is another. The mathematical formulation of the two methods is different, but in fact the two theories are very closely related[9].

The discovery of rotational bands in ^{16}O has demonstrated the coexistence[10] of deformed states with spherical ones. The deformed states are described theoretically as $4p - 4h$ (4-particle, 4-hole) excitations and are strongly excited in α-transfer and α-scattering reactions. This fact suggests[11] a link between shell model states with 4-particles outside a shell and α-clustering effects in these nuclei.

2 CLUSTERING AND PERMUTATION SYMMETRY

It seems that nuclei in the first half of the p-shell are close to LS-coupling, so that permutation symmetry specified by an orbital Young's partition $[f]$, the total orbital angular momentum L, and the total spin s are good quantum numbers in addition to J and T, irrespective of whether the nucleus has a shell structure or not.

An orbital wave function with a given $[f]$ is constructed in the following way. The nucleon numbers are arranged in a definite order in a Young tableau. The wave function is then symmetrized with respect to nucleon numbers in the same row and finally antisymmetrized for nucleons whose numbers are in the same column. In a nucleon cluster model, a nucleus is divided into α, t, ^3He and d clusters each of which corresponds to a row of the Young tableau $[f]$. If there is a strong Majorana force this results in repulsion between clusters and the energy required to dissociate a nucleus into clusters corresponding to the Young tableau rows is rather small. The ground state of ^9Be has orbital symmetry $[f] = [441]$ and the dissociation energy into the corresponding clusters $\alpha + \alpha + n$ is only 1.5 MeV. Similarly the permutation symmetry of the lowest states of ^6Li is $[f] = [42]$ corresponding to a cluster structure $\alpha + d$. These important connections between cluster structure and permutation symmetry have been studied especially by Neudatchin and Smirnov[6].

3 CLUSTERING IN SHELL-MODEL WAVEFUNCTIONS

In the remaining sections of this lecture we discuss the relation between phenomenological and microscopic descriptions of cluster structure in light nuclei, and show how the symmetry of cluster configurations is a common feature of both descriptions.

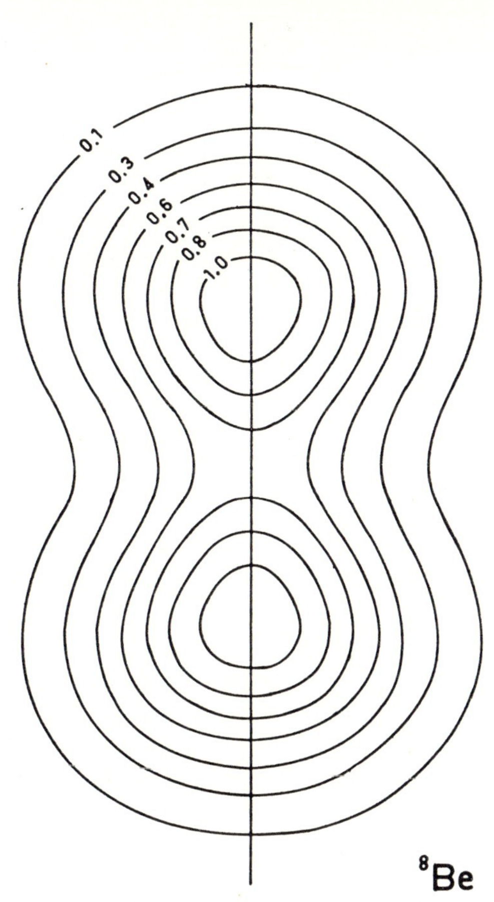

FIGURE 1 Contour plot of the nucleon density in ^{8}Be from Hartree-Fock calculations of Ripka[12]. The figure shows the nucleon density in a plane containing the symmetry axis of the intrinsic state

3.1 The Hartree-Fock model

The independent particle model or shell model is the simplest microscopic model of nuclear structure. Do clustering effects occur in shell model wavefunctions? At first sight it seems that clustering effects cannot be present in shell model wavefunctions, because an independent particle model does not contain many particle correlations. A closer investigation shows that this argument is not correct. We discuss the question in terms of Hartree-Fock (HF) theory. The HF self-consistent potential is a single-particle potential of the kind postulated in the shell-model, and the HF wavefunction is an independent particle wavefunction for the nucleus.

In HF theory one assumes a many-particle Hamiltonian $H = T + V$ for

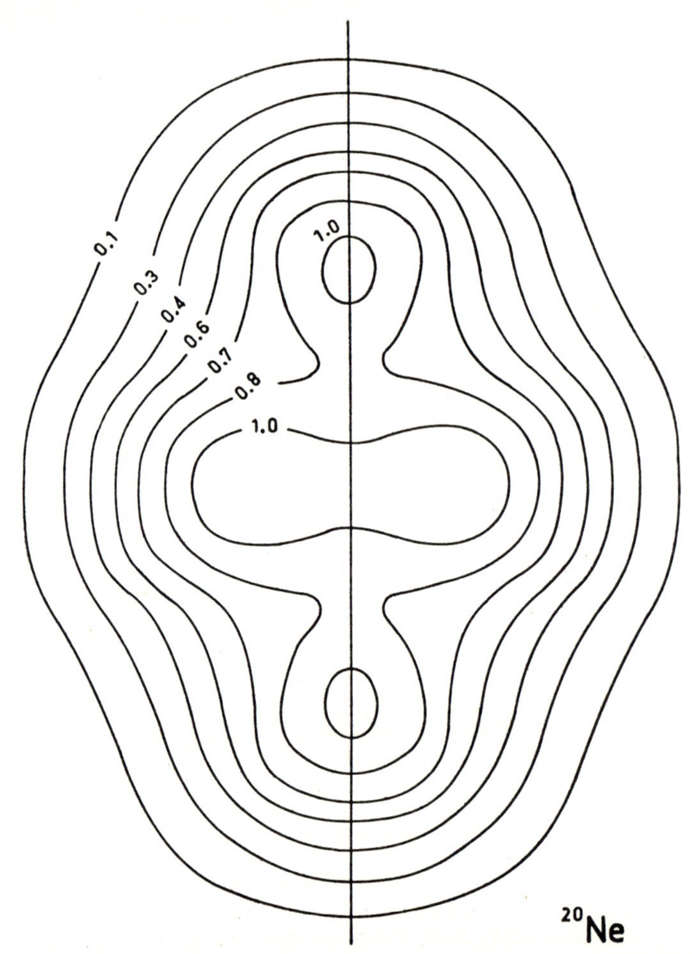

FIGURE 2 Contour plot of the nucleon density in ^{20}Ne from ref. 12. The figure shows the nucleon density in a plane containing the symmetry axis of the intrinsic state

the nucleus. It consists of the kinetic energy T of the nucleons and the potential energy V of interaction between them. The HF wave-function is a Slater-determinant (SD) which satisfies the variation principle

$$\delta E(\Phi) = 0 \tag{3.1}$$

where

$$E(\Phi) = \langle \Phi | H | \Phi \rangle / \langle \Phi | \Phi \rangle \tag{3.2}$$

is the energy expectation value in the state.

Hartree-Fock calculations have been made in light nuclei by many authors[12]. Figures 1 and 2 show contour plots of the nucleon density in ^{8}Be and ^{20}Ne taken from a review by Ripka[12]. There is a tendency for nucleons to bunch into regions of high density in both these nuclei. If one interprets these regions of high density as clusters, then nucleons tend to form clusters even in the HF model which is an independent particle model. Ripka's HF calculations allow only axially symmetric solutions. A recent calculation by Faessler and Eichler[13] relaxes this restriction. These authors find that the HF wave-function in ^{12}C corresponds to an oblate nucleon density distribution. The contour plot of the density distribution shown in Fig. 3 shows that the nucleons tend to cluster at the corners of an equilateral triangle. The amount of clustering depends on the nucleon-nucleon interaction used in the HF calculation.

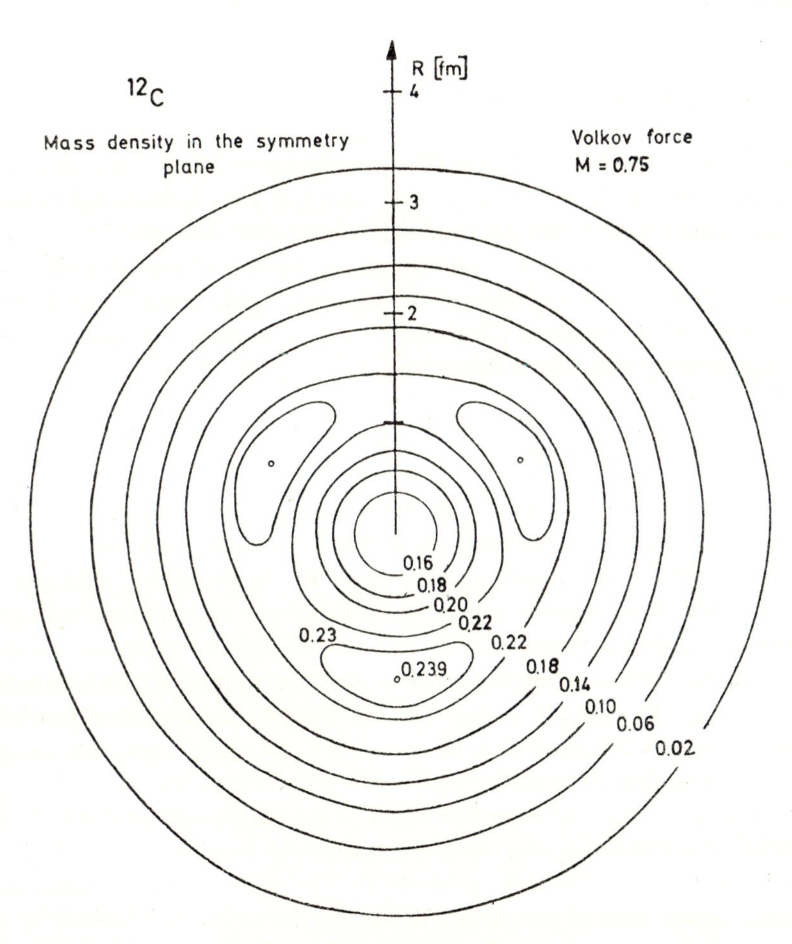

FIGURE 3 Contour plot of the nucleon density in ^{12}C from Hartree-Fock calculations of Eichler and Faessler[13]

The many-particle Hamiltonian $H = T + V$ describing the nuclear dynamics has many symmetries. It is invariant with respect to translations, rotations and reflections, and is approximatly charge independent. Exact eigenstates of the nuclear Hamiltonian share these symmetries. However the HF wave-function does not normally have all the symmetries of the many particle Hamiltonian. For example the HF wave-functions of ^{8}Be and ^{20}Ne, whose density distributions are shown in Figs. 1 and 2 are not eigenstates of angular momentum. What then is the relation between an HF wave-function of a nucleus and the nuclear energy levels? It turns out that the HF wave-function can be interpreted as an intrinsic state of the nucleus and that many nuclear energy levels correspond to rotations and vibrations of this intrinsic state.

3.2 α-Cluster Wave-functions

In section 3.1 we showed that HF wave-functions show some clustering effects. Now we show that a class of wave-functions constructed to show α-clustering effects explicitly, have simple oscillator shell-model wave-functions as limiting cases. We may construct wave-functions containing α-clustering effects by using the method suggested by Margenau[14]. To describe a system with $4n$ nucleons ($2n$ neutrons and $2n$ protons) we take a set of n single-particle orbital wave-functions.

$$\phi_i(\mathbf{r}) = A_i \exp\left(-\frac{1}{2b_i^2}(\mathbf{r} - \mathbf{R}_i)^2 \right), \; i = 1, \ldots, n; \; A_i = [b_i^3 \pi^{3/2}]^{-1/2}$$

depending on n vectors $\mathbf{R}_1, \ldots, \mathbf{R}_n$ and n numbers $b_1, \ldots, b_n$ as parameters. The wave-function ϕ_i describes the motion of a single nucleon in a 1s harmonic oscillator orbit centred at the point $\mathbf{R}_i$. A $4n$-nucleon state can be constructed from these n orbital wave functions by requiring that each orbital state should be occupied by two protons and two neutrons, and then forming the corresponding $4n$-particle Slater-determinant (SD) wave-function $\Phi(x_1 \ldots x_{4n};\ \mathbf{R}_i \ldots \mathbf{R}_n,\ b_1 \ldots b_n)$. The resulting wave function depends on the $4n$ nucleon coordinates x_i (x_i stands for the space, spin, and isospin coordinates of the i-th nucleon); and on $4n$ parameters: the vectors $\mathbf{R}_1 \ldots \mathbf{R}_n$ and the $b_1, \ldots, b_n$. The wave function Φ is completely antisymmetric in the particle coordinates x_i and therefore satisfies the requirements of the Pauli exclusion principle. There is a tendency for two protons and two neutrons to cluster about each of the points $\mathbf{R}_i$ but the clustering is inhibited by the Pauli principle if the points $\mathbf{R}_1 \ldots \mathbf{R}_n$ are close together. The size of the i-th cluster is determined by the parameter b_i. If there is a tendency to form

α-clusters in light nuclei then the class of wave-functions described here contains the possibility of describing such clustering. The basic set of orbital functions ϕ_i is not orthogonal, but this does not create any difficulties for the construction of the SD wave-function Φ or for evaluating matrix elements.

As an example we consider a system of two α-clusters (^{8}Be). We take two orbital wave-functions

$$\phi_1 = A \exp\left[-\frac{1}{2b^2}(\mathbf{r} - \mathbf{d})^2 \right]$$

$$\phi_2 = A \exp\left[-\frac{1}{2b^2}(\mathbf{r} + \mathbf{d})^2 \right]$$

The 8-nucleon SD wave-function constructed from these two orbital wave-functions represents two α-clusters separated by a distance $2d$. We choose the vector $\mathbf{d}$ parallel to the z-axis and investigate the behaviour of the SD wave-function Φ as $d \to 0$. In this limit the two orbital functions $\phi_1(\mathbf{r})$ and $\phi_2(\mathbf{r})$ become identical and the unnormalized SD wave-function vanishes. If, however, Φ is normalized to unity then it tends to a definite non-zero limit as $d \to 0$. To see what the limit is it is convenient to introduce orthogonal linear combinations of ϕ_1 and ϕ_2:

$$\chi_1 = \phi_1 + \phi_2, \quad \chi_2 = \frac{1}{d}(\phi_1 - \phi_2)$$

as $d \to 0$

$$\chi_1 \to 2A \exp\left(-\frac{r^2}{2b^2} \right), \quad \chi_2 \to \frac{2Az}{b^2} \exp\left(-\frac{r^2}{2b^2} \right)$$

or χ_1 tends to a $1s$ spherical state and χ_2 to a $1p$ state. The limiting form of the cluster wave-function Φ is just a harmonic oscillator shell model wave-function with 4 nucleons in the $1s$-shell and 4 nucleons in the $1p$-shell in the orbital state $(n_x, n_y, n_z) = (0, 0, 1)$. This is deformed intrinsic state of the type considered by Elliott[23] in his SU_3 model. If we project out angular momentum eigenstates with $L = 0, 2, 4$ we obtain the lowest states in the shell model description of ^{8}Be in (L, S) coupling.

A similar situation occurs in other nuclei. If we take a system of 3α-clusters at the corners of an equilateral triangle of side d as a model for ^{12}C, then in the limit as $d \to 0$ we get a shell model state with 4 nucleons in the $1s$ shell and 8 nucleons in the $1p$ shell. Projecting states with definite angular momentum gives the lowest (L, S) coupling shell model wave-functions for ^{12}C. A configuration of 3α-clusters in a line gives a $4p - 4h$ excited state of ^{12}C in the limit of overlapping clusters. A tetrahedral configuration of 4α-clusters gives the double closed shell state of ^{16}O in the limit of overlapping clusters.

3.3 Variational calculations with α-cluster wave-functions

We suppose that the dynamics of a nucleus may be described by a many particle Hamiltonian $H = T + V$, where $T = \Sigma t_i$ is the total kinetic energy of the nucleons and V is some effective nucleon-nucleon interaction which is suitable for use in Hartree-Fock (HF) calculations. The expectation values $\langle H \rangle$ of H with respect to the cluster wave-function $\Phi(x_i; \mathbf{R}_j, b_j)$ is a function of the cluster positions $\mathbf{R}_1, \ldots, \mathbf{R}_n$ and the cluster radii $b_1, \ldots, b_n$. We may use the cluster wave-functions as trial wave-function in a Ritz variational principle and minimize $\langle H \rangle$ with respect to the parameters $\mathbf{R}_i$ and b_i. In this way we find the equilibrium sizes and geometrical configuration of the α-clusters. In some cases the equilibrium configuration might correspond to overlapping clusters. Then a shell model description of the nucleus would be more appropriate than a cluster description.

The α-cluster wave-functions defined in Section 2.1 form a restricted set of SD wave-functions. In HF theory one minimizes the expectation value of the Hamiltonian H in the set of all SD wave-functions. The kind of variational calculation described in this section is therefore a restricted HF theory. It would always be less good than a complete HF calculation. Recently two groups[15,16] have made extensive variational calculations with α-cluster wave-functions. In ref.[15] the α-cluster calculations have been compared with HF calculations by Faessler *et al.*[17] using the same interaction. The equilibrium energies are almost the same for ^{12}C, ^{16}O, ^{20}Ne and ^{24}Mg showing that a cluster variational calculation is almost as good as a complete HF calculation in these nuclei.

4 ROTATIONS AND ANGULAR MOMENTUM PROJECTION

In the present section we study the rotations of nuclei with a cluster structure in the framework of a phenomenological theory and a microscopic theory and try to establish a relation between the two kinds of theory. We discuss the α-cluster model of ^{20}Ne as a special example.

4.1 Phenomenological Theory

In this theory we suppose that ^{20}Ne is made up from five elementary α-particles which obey Bose-Einstein statistics. Interparticle forces bind the

α-particles to form a trigonal-bipyramid structure with D_{3h} point symmetry (Fig. 4). Thus ^{20}Ne is like a molecule built up from five α-particles.

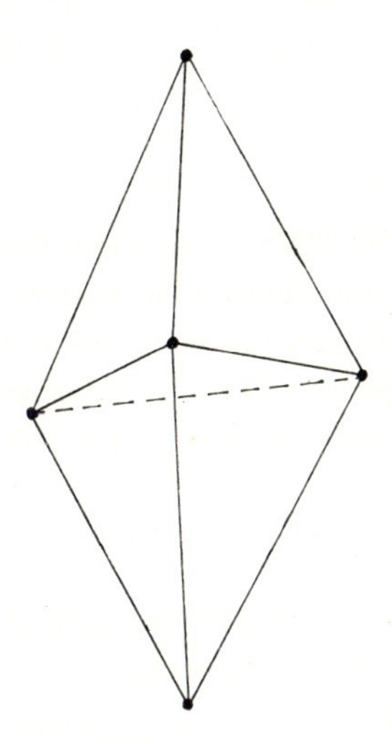

FIGURE 4 Possible equilibrium α-structure for ^{20}Ne with D_{3h} point symmetry

Such a nucleus would be expected to have excited states corresponding to rotations of the molecular-like structure. For the trigonal-bipyramid structure of ^{20}Ne the excitation energies are those of a symmetric-top and are given by the formula

$$E_{IK} = \frac{\hbar^2}{2J}(I(I+1) - K^2) + \frac{\hbar^2}{2J_3}K^2 \tag{4.1}$$

where J_3 is the moment-of-inertia of the nucleus about the 3-axis (Fig. 4) and J is the moment-of-inertia about an axis perpendicular to the 3-axis. The quantum number I is the total angular momentum of a rotational state and K is the projection of this angular momentum on the 3-axis. The allowed values of I, K and parity are limited by the Bose-Einstein statistics of the α-particle and the D_{3h} symmetry of the intrinsic state[18]. The permitted values can be found by the methods of molecular physics and are

$$\begin{aligned} |K| &= 0, \quad I^\pi = 0^+, 2^+, 4^+, 6^+, \dots \\ |K| &= 3, \quad I^\pi = 3^-, 4^-, 5^-, \dots \\ |K| &= 6, \quad I^\pi = 6^+, 7^+, \dots \end{aligned} \tag{4.2}$$

In the language of group theory the number of times a state with total angular momentum I and parity π iccurs in the rotational spectrum of ^{20}Ne is equal to the number of times the identity representation of the groups D_{3h} is contained in the representation $D^{I\pi}$ of the full rotation-reflection group.

4.2 Microscopic Theory

A cluster wave-function for ^{20}Ne of the kind discussed in section 3.2 is not an eigenstate of angular momentum or parity, but it does have a lower symmetry. In ^{20}Ne the clusters are centered at the vertices of a trigonal-bipyramid with D_{3h} symmetry, and the equilibrium cluster wave-function Φ_0 is invariant for rotations and reflections of the D_{3h} group. Thus

$$R(s)\,\Phi_0 = \Phi_0 \quad \text{for } s \text{ in } D_{3h} \tag{4.3}$$

If Φ_0 is an α-cluster wave-function which minimizes the energy expectation value $E(\Phi)$, and if R is an arbitrary rotation or reflection operator then the transformed wave-function

$$\Phi' = R\Phi_0 \tag{4.4}$$

has the same energy expectation value as Φ_0. Thus Φ' also minimizes the energy expectation value $E(\Phi)$. This degeneracy of solutions of the cluster variation problem is associated with the symmetry breaking character of the approximation. The same kind of degeneracy exists with deformed HF wave-functions.

Wave-functions which are eigenstates of angular momentum with quantum numbers (I, M) can be generated from Φ_0 by angular momentum projection[19]. The states

$$\Phi_{KIM} = \int (D^I_{MK}(s))^*\, R(s)\, \Phi_0 \, \mathrm{d}s \tag{4.5}$$

are all eigenstates of angular momentum and the cluster state Φ_0 is a linear combination of them. The class of projected wave-functions 4.5 contains $2I + 1$ states corresponding to different values of K but in general not all these states are linearly independent. The number of linearly independent states with quantum numbers (I, M) which can be projected from a single intrinsic state is related to the symmetry of the intrinsic state. It turns out[20] that the number of linearly independent states with angular momentum (I, M) which can be projected from Φ_0 is not greater than the number of times the identity representation of the rotation symmetry group of the intrinsic state is contained in the representation D^I of the full rotation group. These are exactly the states (4.2) allowed by the phenomenological theory.

5 VIBRATIONS

5.1 Phenomenological Theories

We may generalize the phenomenological model of ^{20}Ne discussed in section 4 by assuming the nucleus is a semi-rigid structure of five α-particles bound to an equilibrium configuration by harmonic forces. The nucleus can have excited states corresponding to vibrations and rotations of the α-particle structure. This model of ^{20}Ne was studied by Bouten[18]. He assumes that the equilibrium configuration of the α-particles is a trigonal-bipyramid with D_{3h} symmetry. In this model ^{20}Ne is like a symmetric top molecule with Hamiltonian

$$H = \frac{\hbar^2}{2J}(I_1^2 + I_2^2) + \frac{\hbar^2}{2J_3} I_3^2 + \sum_{k=1}^{9} (P_k^2 + \lambda_k Q_k^2) \tag{5.1}$$

I_1, I_2, I_3 are the projections of the total angular momentum on the principle axis of inertia. The Q_k are normal coordinates and the P_k are conjugate momenta. Because the α-particles are bosons and the equilibrium state has D_{3h} symmetry for ^{20}Ne it follows that at most six of the λ_k are different, and that the normal coordinates Q_k transform according to irreducible representations of the symmetry group D_{3h}. Table 1 of *ref* 18 gives the symmetry types of the normal modes.

Each vibrational mode has a rotational band associated with it.

The allowed values of the total angular momentum and parity are determined by the symmetry of the normal mode. They are

1) no vibration at one quantum A_1'

a) $K = 0, \quad I^\pi = 0^+, 2^+, 4^+, 6^+, \ldots$

b) $K = 3, \quad I^\pi = 3^-, 4^-, 5^-, \ldots$

2) one quantum E'

a) $K = 1, \quad I^\pi = 1^-, 2^-, 3^-, 4^-, 5^-, \ldots$

b) $K = 2, \quad I^\pi = 2^+, 3^+, 4^+, \ldots$

3) one quantum A_2'' (5.2)

a) $K = 0, \quad I^\pi = 1^-, 3^-, 5^-, \ldots$

b) $K = 3, \quad I^\pi = 3^+, 4^+, 5^+, \ldots$

4) one quantum E''

a) $K = 1, \quad I^\pi = 1^+, 2^+, 3^+, \ldots$

b) $K = 2, \quad I^\pi = 2^-, 3^-, 4^-, \ldots$

The (b) bands are expected to be rather higher than the (a) bands because the moment of inertia $J_3 \ll J$ and because of "tunneling". Bouten[18] tried to fit experimental levels of ^{20}Ne with the α-particle λ_k as parameters. Several fits are possible because of the large number of parameters in the theory.

5.2 Microscopic Theory of Vibrations

States with good angular momentum projected from a given intrinsic state can be interpreted as arising from rotations of the intrinsic state. There are other states which are associated with vibrations of the intrinsic state. These vibrational states can be constructed by the generator coordinate method of Hill and Wheeler.[7,8,21]

We assume that the nucleus can be described by a many-particle Hamiltonian H, and that $\Phi(x, \alpha)$ is a set of A-particle wave-functions depending on the nuclear coordinates $x = (\mathbf{r}_1, \ldots, \mathbf{r}_A)$ and on parameters (generator coordinates) $\alpha = (\alpha_1, \ldots, \alpha_m)$. An approximate eigenstate of the Hamiltonian H is constructed as a superposition of the wave-functions $\Phi(x, \alpha)$ with a weight function $f(\alpha)$

$$\Psi(x) = \int \Phi(x, \alpha) f(\alpha) \, \mathrm{d}\alpha \tag{5.3}$$

Where the right hand side of e.g. (5.3) is an m-dimensional integral over the space of parameters $\alpha_1, \ldots, \alpha_m$. Hill and Wheeler find the generator wave-function $f(\alpha)$ from the variation principle

$$\delta_f\langle E\rangle = \delta_f[\langle\Psi|H|\Psi\rangle/\langle\Psi|\Psi\rangle] \tag{5.4}$$

Equation (5.3) leads to an integral equation for $f(\alpha)$

$$\int [H(\alpha, \alpha') - EN(\alpha, \alpha')] f(\alpha') \, \mathrm{d}\alpha' = 0 \tag{5.5}$$

where

$$H(\alpha, \alpha') = \langle\Phi(\alpha)\,|H|\,\Phi(\alpha')\rangle \tag{5.6}$$

$$N(\alpha, \alpha') = \langle\Phi(\alpha)|\Phi(\alpha')\rangle \tag{5.7}$$

The lowest eigenvalue E_0 of Eq. (5.5) gives an approximation to the ground state energy of the system and height eigenvalues to excited states. The accuracy of the approximation depends on the choice of generator wave-functions $\Phi(x, \alpha)$.

It is convenient to assume that the parameters (α_i) are chosen so that

$$E(\alpha) = H(\alpha, \alpha)/N(\alpha, \alpha) \tag{5.8}$$

has a minimum at the point $\alpha_i = 0$ in the space of generator coordinates. Thus the state $\Phi(x, 0)$ is the best approximation to the exact ground state of the nucleus amongst the class of wave-functions $\Phi(x, \alpha)$ in the sense of the energy variation principle. The point $\alpha_i = 0$ is an equilibrium point in the space of generator coordinates and the wave-function (5.3) describes vibrations of the system about this equilibrium point.

The angular momentum projection formula (4.5) can be derived from a special application of the generator coordinate method. One defines the generator wave-functions $\Phi(\Omega)$ by rotating the HF intrinsic state Φ_0 through Euler angles $\Omega = (\alpha, \beta, \gamma)$. The Euler angles are the generator coordinates. Solutions of the HW integral equation (5.5) can be found analytically and turn out to be a linear combination of rotation matrices.

$$f(\Omega) = \sum_K [D^I_{MK}(\Omega)]^* A_K \tag{5.9}$$

and the HW wave-functions are eigenstates of angular momentum.

To discuss vibrations using the HW method we take the generating wave-functions to be α-cluster wave-functions of the kind discussed in section 3. Then the generator coordinates α_i are the displacements

$$\delta\mathbf{R}_j = \mathbf{R}_j - \mathbf{R}_j^{(0)}$$

of the α-clusters from their equilibrium positions $\mathbf{R}_j^{(0)}$. Thus for ^{20}Ne there are 15 generator coordinates. Three combinations of these coordinates describe translations of the center-of-mass ofthe nucleus and three the rotations of the equilibrium state. The remaining 9 combinations describe deformations of the intrinsic state.

The HW integral equation (5.5) can not usually be solved analytically. Even a numerical solution is out of the question if there are 15 generator coordinates because then the HW integral equation contains 15 dimensional integral. There is an analytic solution[22] to equation (5.5) if the overlap function $N(\alpha, \alpha')$ can be approximated by a Gaussian function in the parameters α_i and α'_i, and if the ratio $k(\alpha, \alpha') = H(\alpha, \alpha')/N(\alpha, \alpha')$ can be approximated by a polynomial in α_i and α'_i containing at most quadratic terms.

$$N(\alpha, \alpha') = \exp\left\{-\tfrac{1}{2}\sum_{ij} n_{ij}(\alpha_i - \alpha'_i)(\alpha_j - \alpha'_j)\right\} \tag{5.10}$$

$$k(\alpha, \alpha') = E(0) + \tfrac{1}{2}\sum(b_{ij}\alpha_i\alpha_j + 2a_{ij}\alpha_i\alpha'_j + b^*_{ij}\alpha'_i\alpha'_j) \tag{5.11}$$

We call the approximations (5.10) and (5.11) the Gaussian overlap approximation (GOA). The second part (5.11) of the GOA is analogous to the small vibration approximation in classical mechanics.

In the GOA the HW integral equation can be shown[21] to be equivalent to an eigenvalue problem

$$hg = Eg \tag{5.12}$$

where the equivalent Hamiltonian is

$$h = E_0 + \tfrac{1}{2} \sum_{i,j=1}^{15} (T_{ij} \pi_i \pi_j + V_{ij} q_i q_j) \tag{5.13}$$

with $\pi_j = -i \dfrac{\partial}{\partial q_j}$ are momenta conjugate to the coordinates q_j. The coordinates q_j in turn correspond to the generator coordinates α_j. The generator coordinates α_j in the generating wave-function $\Phi(x, \alpha)$ are parameters, while the q_j in (5.13) are dynamical variables.

The equivalent Hamiltonian h (e.g. 5.13) has the form of a small-vibration Hamiltonian. It can be put into normal form by introducing normal coordinates Q_j. Then it corresponds exactly to the phenomenological Hamiltonian (5.1) of the classical α-particle model.

The equilibrium α-cluster state $\Phi(x, 0)$ for ^{20}Ne is left invariant by the operations s of the symmetry group D_{3h}. If $\Phi(x, \alpha)$ is a cluster state near the equilibrium state (ie. α_i small), then it is transformed into another cluster state $\Phi(x, \beta)$ near the equilibrium state by an operation s of D_{3h}. Thus the transformations of the wave-functions induce a transformation of cluster parameters. This transformation is linear for the cluster parameters

$$\beta_i = \sum_j \bar{s}_{ij} \alpha_j$$

The transformation matrices transforming the cluster parameters form a representation of the symmetry group D_{3h} of the equilibrium state. These transformation properties of the cluster wave-functions imply certain invariance properties for the overlap functions $N(\alpha, \alpha')$ and $H(\alpha, \alpha')$

$$N(\alpha, \alpha') = N(\bar{s}\alpha, \bar{s}\alpha'), \quad H(\alpha, \alpha') = H(\bar{s}\alpha, \bar{s}\alpha')$$

for each transformation s in the symmetry group of the intrinsic state. If we assume that the GOA holds this invariance has the consequences that

$$\bar{s}^+ n \bar{s} = n, \quad \bar{s}^+ a \bar{s} = a, \quad \bar{s}^+ b \bar{s} = b,$$

for each matrix $\bar{s}$ representing a symmetry operation of the intrinsic state.

The above results are consequences of the symmetry of the intrinsic state (D_{3h} symmetry for ^{20}Ne) and the invariance of the many-particle Hamiltonian H for rotations and reflections. They result in the equivalent Hamiltonian h (e.g. 5.13) having the same symmetries. These symmetries are the same as those of the phenomenological theory of section (5.1). Thus the predictions of the microscopic theory (Hill Wheeler method plus Gaussian overlap approximation) are qualitatively the same as those of the phenomenological theory.

In the phenomenological theory the quantities J_j and λ_k in the Hamiltonian (5.1) are parameters which must be fitted to experimental data. They can not be calculated from theory. In the microscopic theory the properties of vibrational states can be calculated from the kinetic energy and potential matrices T and V in the equivalent Hamiltonian (5.13). These in turn can be computed from the matrices n, a and b of the GOA which are found by fitting the overlap functions N and H. The overlap functions $N(\alpha, \alpha')$ and $H(\alpha, \alpha')$ are determined by the generating wave-functions and the many particle Hamiltonian. Thus it is possible to calculate directly the energies of vibrational states of the α-cluster structure from an assumed nucleon-nucleon interaction.

Results of such a calculation for ^{20}Ne have been given by Brink and Weiguny[21].

REFERENCES

1. W. Wefelmeier, *Naturwiss.* **25**, 525 (1937)
2. D. Dennison, *Phys. Rev.* **57**, 454 (1940); *Phys. Rev.* **96**, 378 (1954)
   S. L. Kamery, *Phys. Rev.* **103**, 358 (1956)
3. L. J. McDonald, *Nucl. Phys.* (1970)
4. J. A. Wheeler, *Phys. Rev.* **52**, 1083 (1937)
5. K. V. Wildermuth and Th. Kanellopoulos, *Nucl. Phys.* **9**, 449 (1958)
6. V. G. Neudatchin and Yu. F. Smirnov, *Atomic Energy Review*, Vol **3**. No. 3 p 157 (1965)
   Prog. Nucl. Phys. **10**, 273 (1969)
7. D. L. Hill and J. A. Wheeler, *Phys. Rev.* **89**, 1102 (1953)
8. J. J. Griffin and J. A. Wheeler, *Phys. Rev.* **108**, 311 (1957)
9. H. Horiuchi, *Prog. Theo. Phys.* **43**, 373 (1970)
10. G. E. Brown and A. Green, *Nucl. Phys.* **85**, 87 (1966)
11. V. Gillet, *Proc. of Int. Conf. on Properties of Nucl. States*, Montreal, Canada 1969 p. 483
12. G. Ripka "Hartree-Fock theory of deformed light nuclei", *Advances in Nuclear Physics* (1968)
13. J. Eichler and A. Faessler, preprint (1970)
14. H. Margenau, *Phys. Rev.* **59**, 37 (1941)
15. Y. Abgrall, E. Caurier, and G. Monsonego, *Phys. lett.* **24B**, 609 (1967)
16. D. M. Brink, H. Friedricks, A. Weiguny, and C. W. Wong, preprint (1970)
17. P. U. Sauer, A. Faessler, H. H. Walter, and M. M. Stingl, *Nucl. Phys.* **A125**, 257 (1969)
18. M. Bouten, *Nuovo Cim* **26**, 63 (1962)
19. R. E. Peierls and J. Yoccoz, *Proc. Phys. Soc.* **A70**, 381 (1957)
20. D. M. Brink, International School of Physics 'Enrico Fermi' course XXXVI (1965)
21. D. M. Brink and A. Weiguny, *Nucl. Phys.* **A120**, 59 (1968)
22. B. Jancovici and D. H. Schiff, *Nucl. Phys.* **58**, 678 (1964)

DISCUSSION

G. Bertsch I would like to comment on the discussion of vibrations in the alpha model. The model you described by Bouten and the model of Dennison assume small vibrations about the symmetric configuration. I have made calculations with the classical α-model with empirical interactions and found that the vibrations were not small but completely distort the structure of the configuration. This would invalidate the classification of excited states by the same point group as the ground state.

D. M. Brink Could I make a short comment about that? I think that the microscopic theory has a limitation, one can carry through calculations really only in cases where one has the small vibration approximation, but it shows that there is a strong connection between the microscopic theory and the classical α-particle model. So when one is looking at anharmonic effects like the one you mention may be it is good to try to calculate them in a classical α-particle model.

A. Bohr The cluster model may be viewed in terms of a deformation in the one-particle density with a very high degree of structure. One can envisage experiments directly testing the various components in this deformation. We can be rather confident that a nucleus, such as ^{12}C and ^{20}Ne involves a large quadrupole deformation, since this is directly indicated by the $K = 0$ rotational band. The picture of the density distribution obtained from H.F. calculations also indicates the tendency towards a deformation of Y_{33} symmetry, as involved in the α-model. However, it is a question whether this deformation exceeds the zero point fluctuations in the system; the question amounts to whether the 3^-excitation is of rotational character ($K = 3$) or is better described as a vibrational mode of excitation. Various experiments can answer this question.

C. Bloch I would like to ask Dr. Brink whether he could comment on the dependence on the force, of the tendency to cluster. I have the impression that the formation of clusters is most sensitive to the force. If the force is slightly varied one way or the other, one immediately obtains either a collapse of the α-particle structure, or the nucleus breaks into α-particles. On the other hand, the gain in energy produced by the clustering is always very small. Is it not then a little difficult to understand from a theoretical point of view how the α-particle structure can play an important role?

D. M. Brink Yes, I think it is certainly true that the clustering is very sensitive to the force. If one changes the force in an arbitrary way, that can affect the clustering a lot. A very important thing in choosing the forces is to try to have the relative binding energies of the separated clusters and the composite system correct. If, for example, you have a force which gives the binding energy of O^{16} many times more than the binding energy of four α-clusters, then, there is no tendency to cluster. If you have a force which is enough to bind α-clusters but not enough to bind O^{16} and if you make α-clusters calculations like this, all the clusters fly apart. Perhaps the criterion for choosing the forces is that they should give the more or less correct relative binding energy of the separate clusters in the composite system.

Another remark is that even if one takes the cluster wave functions in the complete shell model limit where they correspond to oscillator wave functions then in certain cases when one looks at the density distribution there is still clustering effect. For example, if one takes the $SU(3)$ intrinsic state in the Ne^{20} and plots the density distribution one does find still a tendency to cluster. So that, perhaps even in the shell model limit the clustering is not completely gone.

C. Bloch Having made a critical comment on the α-particle model, I would like now to make one in favor of it by reporting briefly some experimental results recently obtained at the Saclay Tandem Van de Graaf[1]. The idea, suggested by V. Gillet, was to look for the nuclear quartet states corresponding to a close clustering of four nucleons by "injecting" an α-particle into a nucleus. This can be done by means of the (^{16}O, ^{12}C) reaction. An ^{16}O beam of about 50 MeV was then used for bombarding target nuclei in the Ni region, and the energy of the outgoing ^{12}C was observed (with a 250 keV resolution) in order to identify the states most strongly populated in the residual nucleus.

The observed reactions were Fe(^{16}O, ^{12}C)Ni and Ni(^{16}O, ^{12}C)Zn, for the isotopes 54, 56 of Fe, and 58, 60, 62, 64 of Ni.

On all cases, strong peaks are observed corresponding to the selective excitation of a few levels in the Ni or Zn residual nucleus between 4 and 10 MeV of excitation energy. At these excitation energies, the level density is much higher than shown by the peaks. The width of the peaks is the same as that corresponding to the low lying well-isolated states, suggesting the possibility that single levels are being excited. Above 10 MeV the cross-section decreases due to Coulomb barrier effects. Examples of the obtained spectra are shown on Figs. 1 and 2.

Angular distributions as well as the dependence on the bombarding energy have been studied. Without entering into any detail, the situation may be

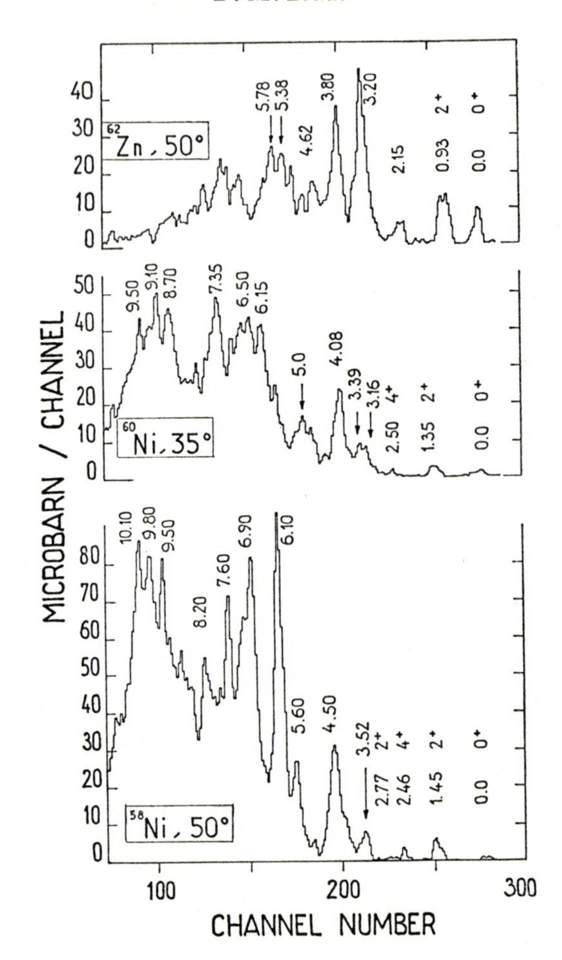

FIGURE 1 Carbon spectra of $^{54,56}Fe$ (^{16}O, ^{12}C) $^{58,60}Ni$ and ^{58}Ni (^{16}O, ^{12}C) ^{62}Zn obtained at 48 MeV ^{16}O incident energy.

summarized by saying that the results are consistent with the direct transfer of an α-particle. It seems also that the possible excitation of the emitted ^{12}C does not play any appreciable role. It does not seem possible to identify the spin of the residual nuclear states from the observed cross sections.

The most interesting feature of these experiments is the systematic behaviour of the excitation spectra which appears when the various target nuclei are compared, and which may be interpreted in terms of quartet states. This is shown schematically on Fig. 3 obtained by evaluating the peak areas for all measured angles, after substraction of a smooth continuum back-

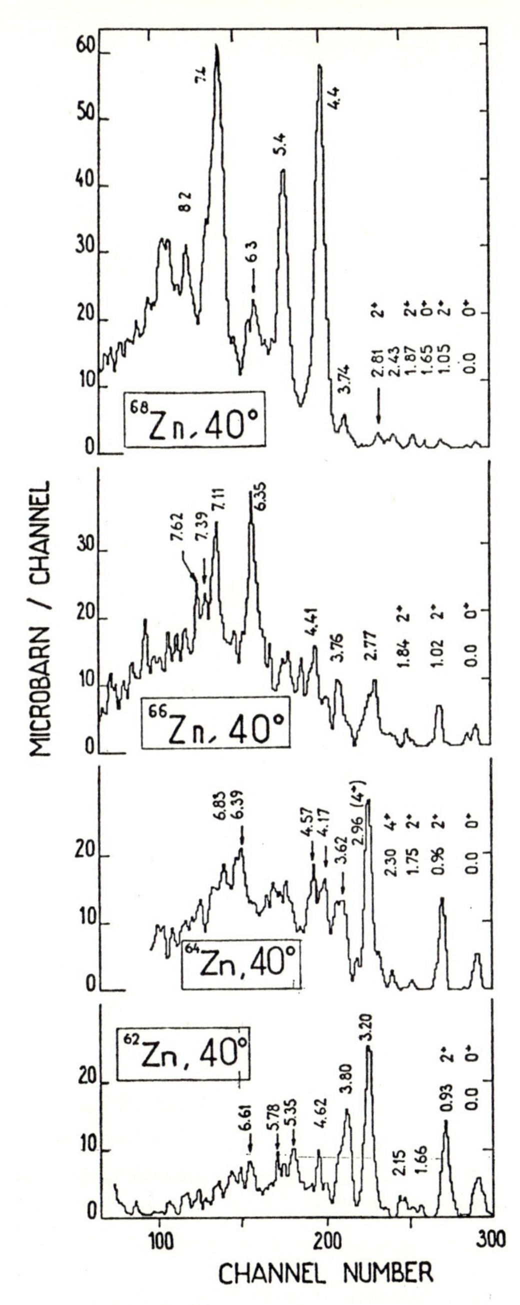

FIGURE 2 Carbon spectra obtained for the various Ni isotopes 58, 60, 62, 64 with 48 MeV ^{16}O incident energy.

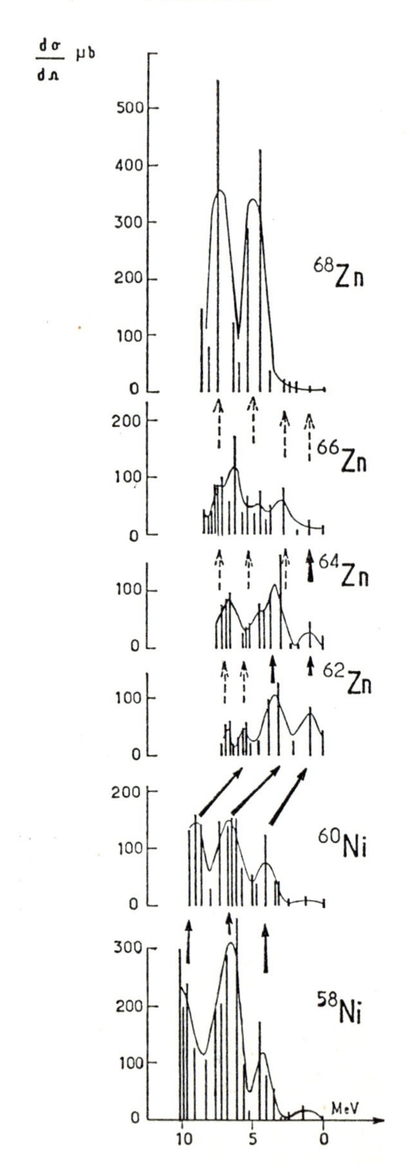

FIGURE 3 Schematic representation of the systematic behaviour of the excitation spectra in the various residual nuclei.

ground. A striking continuity appears when going from one nucleus to a neighbouring one:

a) Overall mean cross sections decrease slowly from ^{58}Ni to ^{66}Zn and rise very strongly for ^{68}Zn.

b) For the same number of neutrons, and between ^{60}Ni and ^{62}Zn, the strongly excited groups are shifted by about 3 MeV towards the ground state. This may be explained by the fact that the $f7/2$ proton shell is just closed in Ni, whereas Zn has two protons in addition of the closed shell. These two protons may directly form the quartet state.

c) For fixed proton number, there is a gradual evolution from ^{62}Zn to ^{68}Zn. There is a gradual decrease of the excitation in the low energy part of the spectra, below $\sim$ 4 MeV, and at the same time a relative increase in the high energy part, above $\sim$ 6 MeV. Thus, the spectrum of ^{66}Zn is quite smooth, whereas ^{62}Zn and ^{68}Zn differ very much: the ground state 0^+ and first 2^+ levels are strongly excited in ^{62}Zn and ^{64}Zn, little in ^{66}Zn and hardly at all in ^{68}Zn. On the contrary, the higher levels are little excited in ^{62}Zn, but very much in ^{68}Zn. This may be understood as a gradual pushing up of the quartet produced by the neutrons gradually filling the $2p3/2$, $1f5/2$ shell when going from ^{62}Zn to ^{68}Zn. In each nucleus the injected neutrons of the α-particle seem to fall preferably into the lowest available state, and therefore the introduction of neutrons into the target nucleus raises the energy of the strongest peaks.

It seems to me that experiments of this type are of particular importance in as much as they show the persistence of states having "simple structures" even in the presence of a background of many other excitations. The levels with an α-particle structure are probably very much favoured by the symmetry and tight structure of the α-particle. Nevertheless, the existence of such types of simple excitations at relatively high energies seem to me to raise one of the most interesting problems in the nuclear structure theory.

M. Bouten I would like to ask a question about the contour plots which you showed from Ripka's thesis, indicating that Hartree-Fock calculations suggest clustering of nuclei into α-particles. If I am not mistaken, Ripka's calculations for ^{28}Si considered only the 12 valence particles, or is that wrong?

G. Ripka The density countour plots shown by Prof. Brink come from a Hartree-Fock calculation in which all the nucleons are in deformed orbits, in other words, in which the closed ^{16}O shell is allowed to have a quadrupole polarization. In fact the α-structure suggested by these countour plots is accentuated by the quadrupole polarization of the closed shells.

M. Bouten I would like to report some results of PHF calculations for ^{8}Be which answer somewhat the question about cut-off's of rotational bands which has been discussed several times this week. In our calculations, we calculate the expectation value of a Hamiltonian between functions of the

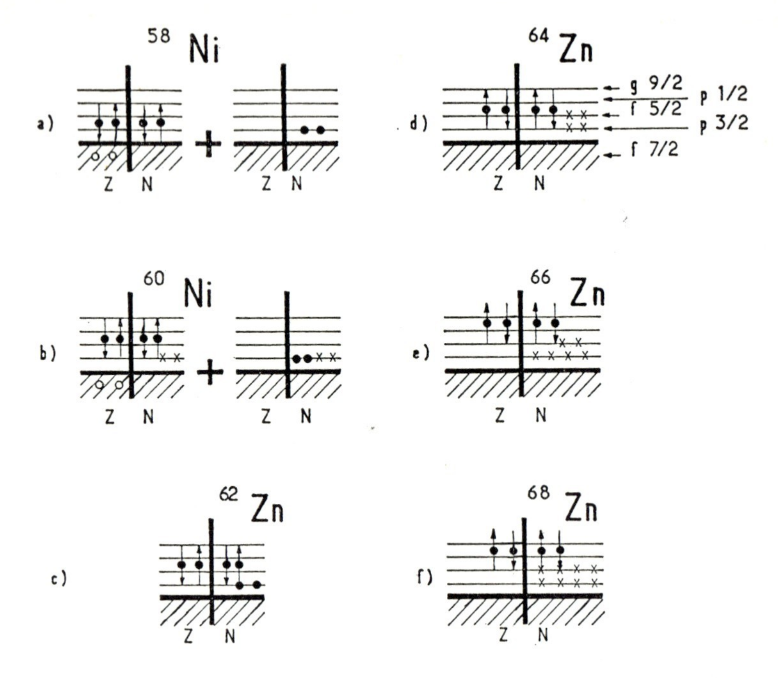

FIGURE 4 Naive shell model description of the quartet states.

form

$$P_L \det|[\chi_{000}(b_\perp b_\parallel)]^4 [\chi_{001}(b'_\perp b'_\parallel)]^4|$$

where χ_{000} and χ_{001} are deformed oscillator functions respectively with no quanta, or one quantum in the Z-direction. As interaction we used the Yale-Shakin matrix elements. We calculated the energies for L up to $L = 8$. The energy spectrum comes out to be very similar to a rotational model. We also calculated all quadrupole moments and $B(E2)$'s, which indicate that the $L = 6$ and $L = 8$ level are connected to the lower L-values as in a rotational nucleus. This shows that the rotational band does not cut off after $L = 4$ but continues to higher L-values.

J. B. French It seems to me that the conclusion must depend on the single-particle spacings. If the orbits higher than p are too far above, then the wave function used which gives the band may not be reasonable.

M. Bouten In our calculations, single particle energies do not enter as we only calculate expectation values of the total Hamiltonian and minimize them. The results will of course depend on the interaction which is used, but for all reasonable forces I believe that a similar result will be found.

G. Ripka I wish to point out that the single particle energy gap, which in ^{16}O is the energy separation of the *p*-shell and of the *d*-shell orbits, does not determine the excitation energy of the 4*p*–4*h* state of ^{16}O. This energy gap gives some information on the stability of the closed shell against very small deformations. The results of calculations using various forces show that forces which give a good estimate of the excitation energy of the 4*p*–4*h* state are those which also estimate correctly the binding energy differences between ^{20}Ne and ^{16}O and between ^{16}O and ^{12}C. There are examples of two forces of which one gives a larger energy gap and at the same time a lower excitation energy of the 4*p*–4*h* state than the other.

L. Radicati I would like to know what is the contribution to the current and therefore to the magnetic moments and transition probabilities that come from these α-particle clusters. In fact obviously if there are α-particles they will only contribute to the isoscalar part of the current but one can possibly think of isovector contributions arising from four nucleon configurations. Is this contribution increasing for heavy nuclei not necessarily α-particle nuclei? The single particle contributions which I discussed yesterday are certainly not the only ones; there are the lower mass contributions mentioned by Biedenharn and I presume there are these heavy mass contributions too. How important are they?

D. M. Brink Well, the α-cluster wave function which I wrote down correspond to (444) permutation symmetry in $SU(4)$, or scalar in $SU(4)$.

Therefore, there is not any spin contribution to the magnetic moment for the cluster states which I wrote down. This is only an orbital contribution. For H.F. wave functions in some nuclei, for example nuclei at the beginning of the $s - d$ shell, the overlap with a $SU(4)$ scalar wave function is quite large.

What happens when one gets up to nuclei at the end of the $s - d$ shell or beyond? I do not know.

L. Radicati Would these 4 particle contributions be the most important contributions to the oscillation and vibration mentioned by Bohr and Mottelson?

These are certainly collective motions but there are many other collective motions of course beyond these 4 particles oscillations. I wonder whether this would be a fairly important contribution or not?

D. M. Brink For many of the excitations, the α-particle model is very like to rotational model, for example for the first rotational band of Ne^{20} the wave function of the α-cluster model is very similar to a wave function in the rotational model. The α-cluster model has some modes of excitation which are not explicitly included in the rotational model. That particular mode of oscillation which I draw on the board is not something you will get easily in the rotational model without introducing explicitly some new coordinates.

L. Radicati But do you not restrict yourself to light nuclei?

D. M. Brink Everything I have said is restricted to light nuclei.

D. H. Wilkinson Have any attempts been made to understand the $T = 1$ states of your "alpha-particle" nuclei in terms of the $T = 1$ excitations of the constituent "alpha particles"? One finds, for example, in $A = 12$, strong beta-decay from ^{12}B and ^{12}N to states of ^{12}C that are well described by the alpha-particle model but that are not well described by the simple shell model. Should it not be possible to relate at least some of the $T = 1$ states of the alpha-particle nuclei to $T = 1$ states of the alpha-particle (although of course the beta-decay will not go via the alpha-particle components).

D. R. Inglis In the density distribution in the principal plane of ^{12}C, as calculated by Faessler *et al.* and shown by Dr. Brink, the very interesting three-fold azimuthal symmetry was really very weak. In going around the circle of mean-maximum density, one crosses only one contour line indicating a density variation from perhaps 225 to 215 or 5%. If a 3^- state were to arise from this slight deviation from azimuthal symmetry, it would be expected to be at very high energy, in keeping with the remark of Prof. Bohr, and the lack of low 3^- state remains evidence against a well-developed α-cluster structure.

Dr. Brink justified the appearance of a 0^+ state at 6 MeV in ^{16}O as a $4p$–$4h$ state by the gain in energy of the four particles in forming an α-cluster to compensate some of their excitation energy. One should also adduce the gain in energy from increased correlation of the unexcited nucleons, corresponding to the four holes, mainly that associated with an oblate spheroidal deformation if the azimuthal 3-fold structure is weak as in ^{12}C. Thus the 0^+ state at 6 MeV is the one that may resemble a tetrahedron while the ground state is seen as a shell-model state.

The most intriguing evidence for an α-cluster model has been the success Dennison had in explaining the relative energies of rotational energies in ^{16}O—the part of his work that depends only on having a structure of tetrahedral symmetry and not on questionable parameters. The ratio of 4^+ to 3^- excitation is $4(4 + 1)/3(3 + 1) = 5/3$ which, with 3^- near 6 MeV, puts the first 4^+ state near 10 MeV and, remarkably enough, within about 1% of the observed level. This, however, presumes the ground state to have the same tetrahedral structure as have the rotational states and the agreement must be seen as spurious now that, with Dr. Brink, we see the lowest state as a shell model state from which the 0^+ state at 6 MeV similar to a tetrahedron is generated by $4p$–$4h$ excitation.

C. Bloch I would like to comment on what Professor Inglis said about the amazing accuracy of the Dennison model for ^{16}O. If one looks at it from a critical point of view, it is very difficult to understand, because the model is based on the linear approximation for the vibration of the tetrahedron. This assumes small amplitudes. Now if one looks at the parameters obtained from fitting the levels, the zero point amplitude of the vibration turns out to be of the same order as the size of the tetrahedron. How can one obtain such a good agreement with experiment under such circumstances?

D. R. Inglis In my comment, I tried to pick up the one feature of Dennison's results, which is not dependent on these parameters, namely a relation between two rotational states, which depended only on the distance between the particles and the rotational symmetry.

C. Bloch Vibrational states also fit very well.

D. R. Inglis Yes, I agree, that's remarkable. But there one has less reason to believe it than in the rotational part, which requires only the tetrahedral symmetry.

P. Kramer Apparently there seems to be different approaches to clustering that lead to similar qualitative and quantitative conclusions on the ordering of levels. In ^{12}C for example one may arrive at the level sequence from the idea of α-clusters interacting through harmonic oscillator forces. When comparing these approaches, one should be rather careful as most of the states are obtained from projection procedures, and it is clear that the same state may be projected in many different ways.

L. Radicati I come back to the question asked by Dr. Wilkinson about the $T = 1$ states. It seems to me that when you want to compare the β decay from Li^8, it is probably very hard to imagine that the α-particle contribution to the isovector current which is involved in this decay could be of any great importance. The main contribution there almost certainly comes from the single nucleon model though there might be other contributions too. I imagine these would be very small indeed, since they would correspond to the contributions from excited α-particles, excited in isotopic spin too. So, I would imagine that there the model will have very little to say, that is my impression.

A. Bohr The α-clustering is a special deformation effect of very high structure, partly in the one-particle density, but also in the parentage channel, where it leads to collective transitions for 4-particle transfer. In the latter respect, it is an extreme version of pairing effects. Also in heavy nuclei, we see the tendency to α-clusters in the hedge enhancement of the α-decay. However, the enhancement seems to be well accounted to by the pair correlations of the superfluid type.

P. Kramer I would like to point out that in one respect some recent calculations of the Tübingen group differ from the ones discussed by Prof. Brink and also from the resonating group method, and this is the introduction of the mutual cluster distortion when the clusters approach each other. The calculations of ^{12}C for example show that the internal states of the α-clusters become equal to those of the free clusters for large distances between the clusters while in the region of strong overlap they are distorted. This may be relevant with respect to clustering in heavier nuclei as apparently we can expect a simple cluster structure only at the nuclear surface. For the interior region I would prefer to speak of clustering as a correlation of low nucleons with respect to permutational symmetry.

REFERENCES

1. H. Faraggi *et al.*, *Annals of Physics* **66**, 905 (1971)

Vibrational Motion in Nuclei

B. R. MOTTELSON
Nordita,
Copenhagen, Denmark

The problem of finding a proper balance and of understanding the deeper connections between collective and independent particle degrees of freedom is, perhaps, the central issue in all many particle systems. In the study of nuclear structure, because of the possibility of detailed investigation in sharp quantum states, these questions are encountered in an especially concrete form and the fundamental issues involved have been a recurring theme throughout the history of the subject*. In this development the questions of symmetry and of symmetry breaking have continued to reveal interesting new aspects. I shall begin by briefly reminding you of some of the major turning points in the development of our current ideas concerning the role of independent particle and collective motions in the nucleus.

The earliest discussion of nuclei built out of neutrons and protons was based on an independent particle picture similar to that which had been so successfully developed in the description of atomic structure (see, for example, the contributions of Heisenberg and Gamow at the Solvay meeting of 1933); this picture brought into focus the problem of saturation of the nuclear interactions and in addition attempts were made to interpret the early data on nuclear decay schemes in terms of the orbits of individual particles.

A sharp change in the development was produced by the investigation of neutron induced reactions (Fermi 1934); the capture cross sections were much larger than could be accounted for in terms of the motion of the neutron in an average nuclear potential and the spacings of the resonances were orders of magnitude smaller than could be accounted for in an independent

---

* A more complete discussion of the issues considered in the present report, will appear in Vol. II of *Nuclear Structure* by A. Bohr and B. Mottelson. Within the scope of the present report it is hardly possible to make detailed references to the many different contributers to this development; such references may also be found in this forthcoming volume.

particle description. These many-particle features of the neutron reactions implied a significant coupling between the motion of the incident neutron and the many degrees of freedom of the nucleons of the target nucleus (Bohr 1936, Breit and Wigner 1936). An extreme model of a closely coupled system is provided by the dynamics of a liquid drop and on the basis of this comparison it was suggested that the fundamental modes of excitation of the nucleus might be related to the quanta of surface and density waves of such a droplet (Bohr and Kalckar 1937).

The possibility of such collective motion in the nucleus was strikingly confirmed by the discovery of the fission process (Hahn and Strassmann 1938). The low value of the threshold energy for the fission process in heavy nuclei could be immediately understood in terms of the near balance between the electrostatic repulsion and the surface tension (Frisch and Meitner 1939, Frenkel 1939, Bohr and Wheeler 1939). Further evidence for collective motion was provided by the discovery of the giant dipole resonance in the nuclear photo effect (Baldwin and Klaiber 1948), which could be interpreted as the fundamental mode of motion of the neutrons with respect to protons (Goldhaber and Teller 1948; Jensen *et al.* 1950).

The discussion of all these phenomena was put in a new light by the unambiguous demonstration of the nuclear shell structure and its interpretation in terms of independent particle motion (Haxel, Jensen, and Suess 1949, Mayer 1949). One was then faced with the problem of reconciling the occurence of both particle and collective degrees of freedom and of exploring the effects of the collective modes on the motion of the individual nucleons and the significance of the shell structure for the collective dynamics.

For example, it was recognized at an early stage that the single particle excitations of the independent particle model exhaust the dipole oscillator sum rule, but the collective neutron-proton oscillations also exhaust this sum rule. Therefore it is not at all justified to simply combine these two modes of excitation; rather, one is forced to recognize that the existence of the collective dipole mode implies major modifications in the motion of a single particle (especially in the dipole moment carried by such a particle) and at the same time that the collective dipole mode itself can be expressed in terms of appropriate linear combinations of one particle excitations.

PARTICLE-VIBRATION COUPLING THROUGH DEFORMED FIELD

The understanding of these issues has come gradually with the appreciation of the central role played by the one particle field generated by a collective distortion in the average nuclear density*. This concept is analogous to the displacement potentials employed, for example, in the discussion of the electron-phonon coupling in metals, the electron plasma interaction in an electron gas and the particle-phonon coupling in liquid ^{4}He or in ^{4}He–^{4}He mixtures. To leading order in the vibrational amplitude, α, the increment in the one particle potential can be expressed in the form

$$\delta V = \varkappa \alpha F(x) \tag{1}$$

The one particle field, F, depends on the nucleonic variables, x, (space, spin, and isospin). The problem of determining the structure of the field F in terms of these variables is the same as that of determining the structure of the collective mode and is based ultimately on the conditions of self-consistency. However, it is often possible to obtain valuable guidance from symmetry arguments based on a macroscopic picture of the expected collective mode. Thus, for example, a shape oscillation of multipole order $\lambda\mu$ may be associated with a field

$$F_{\lambda\mu} = \sum_{i=1}^{A} f(r_i)\, Y_{\lambda\mu}(i) \tag{2}$$

where the radial form factor f is nodeless and peaked at the nuclear surface†. In (2) we have assumed that the collective field will act equally on neutrons and protons (i.e. an isoscalar field, $\tau = 0$) as suggested by the classical picture of a shape oscillation in which the neutron-proton ratio remains unaffected at each point. In a nucleus with a large neutron excess, the one particle excitations produced by the field (2) will be different for neutrons and protons (as a result of the neutron excess) and thus the iso-symmetry of the collective mode must be completely violated at the microscopic level. However, when considering the long wave length, average, effects of the collective mode, as, for example, the average one particle potential (1) generated by the collective motion, it may still be possible to assign a macroscopic isobaric symmetry quantum number as assumed in the field (2) (see, the experimental evidence on this point, referred to below in connection with the examples of different modes).

---

* The role of this field in providing the link between single particle and collective motion in the nucleus was recognized from somewhat different points of view by Rainwater (1950), A. Bohr (1952) and Hill and Wheeler (1953); the appreciation of the great scope and variety of these couplings has been the result of a long and continuing development.

† For a nucleus with spherical equilibrium shape, the $(2\lambda + 1)$ components of the field (2) are each associated with a corresponding component of the deformation amplitude $\alpha_{\lambda\mu}$ so that the resulting field coupling is rotationally invariant.

$$\partial V_\lambda = \varkappa_\lambda \sum_{\mu=-\lambda}^{\lambda} F^*_{\lambda\mu}\, \alpha_{\lambda\mu}$$

The coupling constant, $\varkappa$, in Eq. (1) determines the strength of the coupling between density and field. For the simplest modes (which indeed includes all those that have been at all well studied so far) the coupling constant can be estimated from the observed static fields by assuming that the ratio of density to potential remains unaltered under the collective oscillation, and the available evidence seems to be in rather good agreement with such estimates. It should be emphasized, however, that it would be possible to make these estimates with considerably greater certainty on the basis of a deeper understanding of the velocity dependence and density dependence of the average nuclear potentials.

The basic matrix element of the coupling (1) involves a single particle transition accompanied by the creation of a single quantum of the collective oscillation

(a)

This matrix element can be directly measured in an inelastic scattering experiment, in which case the initial and final one particle states correspond to scattering states in the continuum; the first order matrix element also gives directly the amplitude for decay of the collective mode into a particle-hole pair

(b)

The field coupling can also be rather directly observed in the renormalization of the one particle matrix element of the field *F*, as described by the diagrams

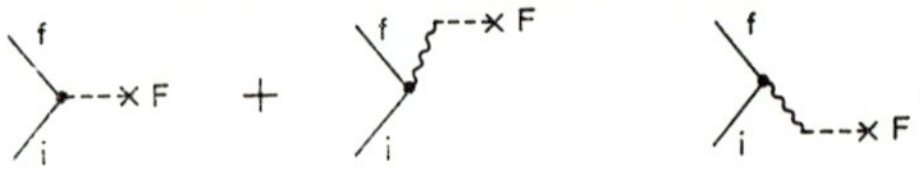

(c)

The contribution to the "effective charge" implied by the last two terms in (c) is closely related to the renormalization in the effective interaction between two particles

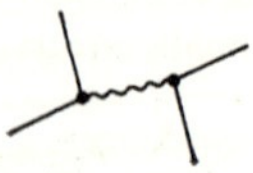

(d)

It is also possible to obtain estimates of the various anharmonic effects from these basic couplings; for example, the leading order terms in the effective particle-phonon interactions are obtained from

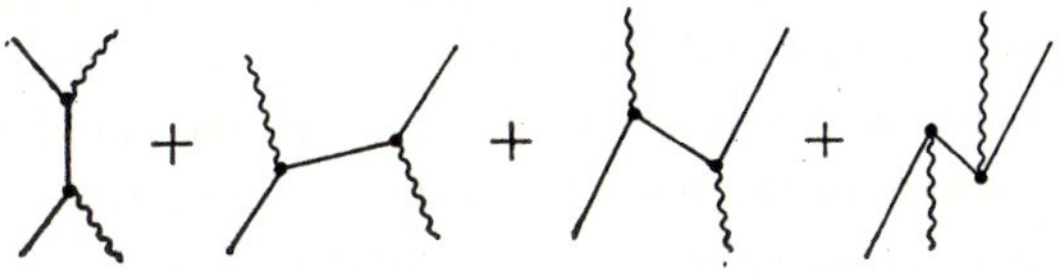

(e)

and the cubic terms in the effective phonon Hamiltonian from

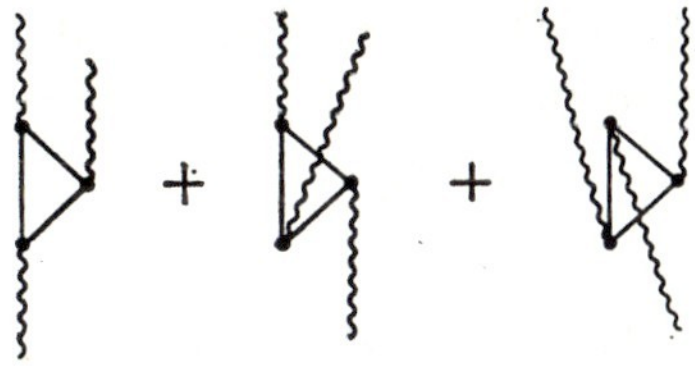

(f)

The field coupling not only describes the effect of the collective motions on that of the particles, and *vice versa*, as illustrated in the diagrams (a), (b), (c), and (e), (f), respectively, but is at the same time the basic organizing force that generates the collective oscillation itself out of the excitations of the individual nucleons as illustrated by the diagrams of the random phase approximation.

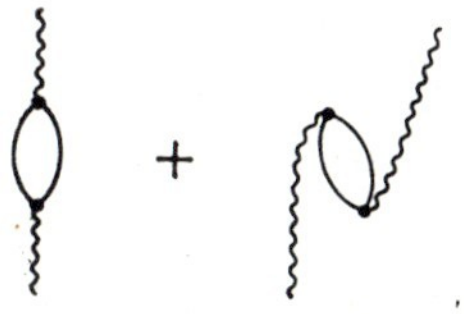

(g)

The field (1) is, of course, a consequence of the two body interactions. The connection is similar to that between these interactions and the static potentials studied in the bound state spectrum and elastic scattering processes; however, so far, this connection has been somewhat elusive since it depends in a subtle way on many correlation effects. In exploring the properties of the nuclear vibrations we encounter these fields directly and may therefore attempt to establish connections between different phenomena that are to a considerable extent independent of the detailed features of the underlying interaction.

The properties of these fields are so central to our interpretation of the nuclear collective modes, and the variety and the possible combinations are so rich a field, that we must attempt to gain insight from all possible lines of attack: partly from comparison with the static fields, partly from an analysis of the two boby forces, and partly from the hints provided by the available data on the various phenomena such as indicated by the diagrams sketched above.

In writing the interaction (1) and in the Feynman diagrams considered, the collective variables (in α) and the particle variables (in F) are considered on quite the same footing. A consistent treatment of the coupling (1) ensures that the anti-symmetry of the particles, including those that are involved in the collective motion, and the orthogonality of the different excitation quanta are correctly taken into account. In such a formulation the collective modes appear just as "elementary" as do the particle degrees of freedom from which they are built. The anharmonic terms, such as (e) and (f), contain the natural limitation to the treatment in terms of *independent* elementary modes of excitation.

SURVEY OF VIBRATIONAL MODES

In the nuclear dynamics, the simplest collective modes are the shape oscillations and the neutron-proton oscillations that have been referred to above. The classical macroscopic picture of these modes has provided a valuable starting point for the analysis (in particular has supplied the basic symmetry quantum numbers to be expected), but as we shall see, the description in terms of quantized one-particle excitations gives an important, non-classical structure to these excitations. In addition to the modes with classical analogues, there may occur deformations and corresponding oscillations in the abstract spaces describing a nucleon's more quantal aspects; thus we expect to find modes involving charge exchange or spin flip of nucleons, as well as oscillations in the pairing field involving the creation or annihilation of two nucleons.

The rich possibilities for vibrational modes in nuclei and the problems associated with the interaction of the excitation quanta among themselves and with other elementary modes of excitation presents a field that is potentially of great scope, but has so far only been explored to a very limited extent. In the remainder of this talk I shall be able to only briefly mention a few examples from this extensive field of investigation.

SHAPE OSCILLATIONS

From the earliest discussions of nuclear collective oscillations it was recognized that the low compressibility of nuclei and strong neutron-proton exchange forces implied that the lowest mode of collective motion should correspond to shape oscillations of quadrupole type. The symmetries of such a picture imply a field of the type (2). If we ask for the independent particle excitation spectrum produced by such a field, F, we obtain a picture that depends somewhat on the particular nucleus we consider, but is qualitatively like that shown in Fig. 1

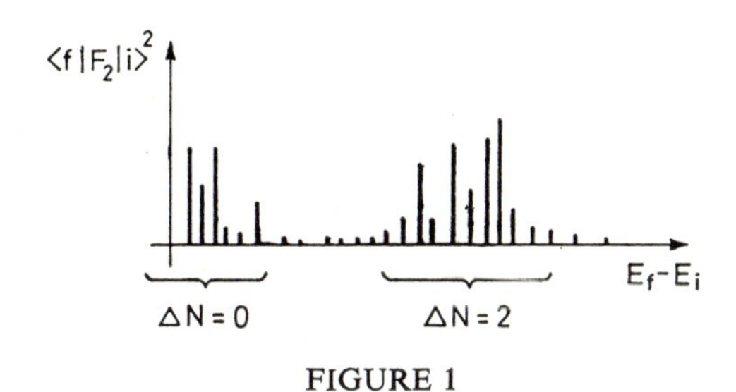

FIGURE 1

The low energy excitations correspond to transitions between orbits within the unfilled shells (transitions in which the total number of nodal surfaces does not change, $\Delta N = 0$) while the high frequency strength corresponds to transitions which increase the number of nodal surfaces, by one, ($\Delta N = 2$). In this situation we are led to expect two collective modes; one built out of the low frequency and one out of the high frequency excitations. It is important to emphasize that both of these modes have nodeless form factors; it is the non-local character of the density matrix in a quantal system that makes possible the occurrence of two orthogonal modes both of which have nodeless form factors when expressed in terms of the diagonal matrix elements of the density. I am not sure that we have understood at a sufficiently deep level the significance of this non-classical element in the description of nuclear shape oscillations; we lack a useful symmetry classification or geometrical picture of the distinction of these two modes. In any case, it is quite clear that the low frequency quadrupole mode, about which we know so much, is mainly built out of the low frequency excitations. The high frequency mode is expected to carry almost 90% of the quadrupole oscillator strength, should occur at an excitation energy of roughly 60 $A^{-1/3}$ MeV, and has so far never been convincingly identified in any experiment. The location and characterization of this main component in the nuclear quadrupole shape oscillation is clearly a task of considerable importance in testing our understanding of the structure of shape oscillations in the nucleus.

The low frequency quadrupole shape oscillations have been recognized as a systematic feature in the spectra of almost all nuclei, since the data were systematically collected and interpreted by Scharff-Goldhaber and Weneser (1955). The experimental data provides significant evidence on the symmetries of these excitations. From a comparison of the cross section for excitation by $(\alpha\alpha')$ (which acts symmetrically on neutrons and protons), by ($E2$) (which acts only on protons) and (pp') and (nn') (which act somewhat differently on neutrons and protons and with an opposite relative difference) we can conclude that the low energy excitations involve a field that is approximately the same for neutrons and protons as in the classical picture of a shape oscillation and as assumed in the field (2). Thus, the observed mode has, within the accuracy of the available data, a good macroscopic isobaric quantum number, $\tau = 0$, although, as discussed above, this symmetry is completely violated at the microscopic level in a heavy nucleus. It should be mentioned that a rather sensitive test for small isovector components in the macroscopic field of the shape oscillations may be detected by studying the (pn) process (or the analogous (^{3}He, t) reaction) leading to the isobaric analogue ($M_T = T - 1$) of the quadrupole vibration; and such experiments have recently been initiated. The field (2) has been assumed independent of the spins of the nucleons, but not much direct evidence is available on this point. The study of polarization following (pp') reactions has provided some evidence for the expected deformation of the spin orbit potential associated with the collective quadrupole oscillations (Sherif and Blair 1970). The success of the interpretation of the particle-vibration coupling effects involving the octupole mode (see below) provides evidence that in this case the collective field is predominantly spin independent as in (2).

In nuclei not too far from closed shells, the low frequency quadrupole oscillations correspond to deformations of the spherical equilibrium shape. However, in most of these nuclei the spherical shape is only weakly stable against the tendency to establish a large static equilibrium deformation, and the spectra exhibit rather large anharmonicities. There is at present a considerable experimental and theoretical effort concerned with characterizing these anharmonicities and developing appropriate methods for relating them to the nuclear shell structure. This program extended to still larger values of the deformation makes contact with the exciting developments that have followed the discovery of the fission isomers (Polikanov *et al.* 1962) and their interpretation in terms of a fission barrier involving an intermediate minimum. One is here concerned with the general problem of studying the effect of the quantized orbits of the individual nucleons on the collective potential and kinetic energy functions in the many dimensional deformation space, (Kumar and Baranger, Strutinski, Nilsson, Swiatecki, Nix and others).

The lowest frequency component of the octupole shape oscillation has been systematically identified in almost all nuclei (the first evidence for this mode was obtained by Cohen 1957, and interpreted by Lane and Pendlebury

1960). For this mode the anharmonicities are expected to be appreciably less than for the low frequency quadrupole oscillations because there are more particles involved and the spherical shape is more stable against this higher multipole deformation; in the last few years, there has been impressive success in interpreting the available data in terms of the perturbation diagrams of the particle-vibration coupling (Hamamoto 1968–70); in such an analysis, it is possible to take all of the properties of the collective mode from the experimental data on even-even nuclei, and then to test in considerable detail the consequences of the presence of an extra particle as revealed in the spectra of the neighbouring odd-A nuclei.

ISOVECTOR MODES

The dipole resonance observed in the nuclear photo effect is the main example, which has been studied so far, of a collective isovector oscillation in the nucleus. The classical picture of this mode implies a collective field of the form

$$F = \sum_{i=1}^{A} \tau_3(i) f(r_i)\, Y_{1\mu}(i) \tag{3}$$

where the radial form factor $f(r)$ must be approximately linear in r over most of the nucleus, since the observed dipole mode exhausts the dipole oscillator sum rule to within the available experimental accuracy. The isovector symmetry follows directly from the structure of the dipole operator but again it must be emphasized that this is a macroscopic symmetry of the average field that is generated, but, because of the neutron excess, the symmetry is completely violated if we consider the individual configurations involved in the excitation. In this connection it would be of interest to have direct evidence on the isoscalar field that may be associated with the dipole mode; such evidence could be obtained, for example, from a study of the $(\alpha\alpha')$ scattering leading to the dipole resonance state.

The strength of the dipole field can be estimated by assuming that the ratio of density to potential for the dipole oscillation is the same as that for the average isovector potential observed in the bound states and scattering states of a single nucleon. This estimate of the dipole field coupling leads to a prediction of the dipole frequency that is in good agreement with the available data in heavy nuclei. An additional test of this coupling is provided by the effective charge for $E1$ transitions. Until recently there existed almost no quantitative evidence on the $E1$ effective charge since the nuclear shell struc-

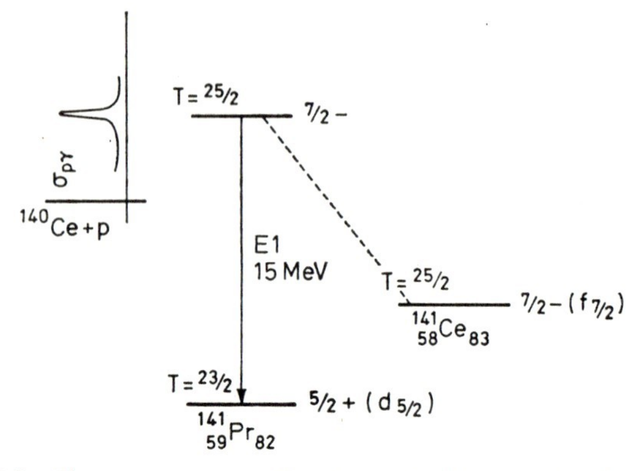

FIGURE 2 The states connected by the dotted line are related by rotation in isobaric spin space (analogue states)

ture does not provide any strong $E1$ transitions between nearlying one particle orbits. This difficulty has been circumvented by considering transitions between one quasi-particle states with different total isobaric spin as in the experiment of Ejiri *et al.* (1968) indicated in Fig. 2. In this experiment the measured value of Γ_γ was about ten times smaller than calculated for an $E1$ single particle transition between configurations with the indicated quantum numbers. This reduction agrees quantitatively with that estimated from the effective charges estimated from the diagrams (c) above. Because of its fundamental role in the understanding of the dipole mode it would be valuable to have additional more direct evidence on the strength and structure of the dipole field such as could be obtained from direct inelastic scattering, more extensive measurements of the effective charge phenomena, or other of the coupling effects referred to above.

The isospin, $\tau = 1$, of the dipole excitation can be coupled to the isospin, T_0, of the ground state configuration to produce states with $T = T_0 - 1$, T_0, and $T_0 + 1$. The expected pattern of these states is shown in Fig. 3.

The levels connected by dotted lines are rigorously related to each other by rotations in isobaric space (microscopic symmetry), provided we can ignore the Coulomb field, but the states with different total T are expected to differ in major respects in heavy nuclei as a result of the isobaric spin T_0 which provides a symmetry breaking boundary condition on the vacuum with respect to which the collective mode is defined (see Fig. 4).

The quantitative experimental investigation of these patterns would be a valuable test of our understanding of these symmetries and symmetry breaking effects. (Fallinos *et al.* 1967; Petersen and Veje 1967).

The problem of the width and line shape of the dipole resonance is not well understood. We encounter here a variety of coupling effects between the

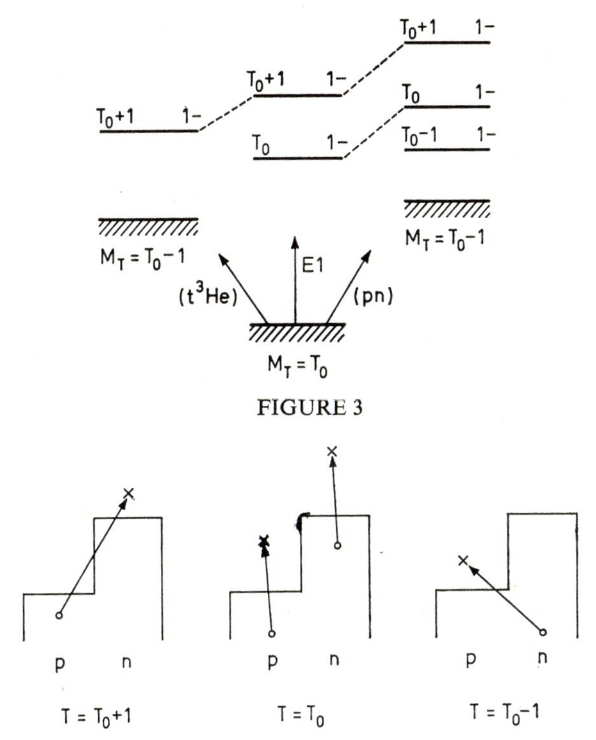

FIGURE 3

FIGURE 4 The transition shown for $T = T_0 - 1$ is not available for $T = T_0$ or $T_0 + 1$, and similarly those shown for $T = T_0$ are not available for $T = T_0 + 1$

dipole motion and other collective oscillations such as the quadrupole deformations and possible spin dependent collective modes, as well as couplings to the more random degrees of freedom which leads eventually to the formation of the compound nucleus.

At the present time the dipole mode is the only isovector mode that has been clearly established but considerable interest is attached to the search for the other expected isovector modes. The monopole isovector mode plays a central role in the discussion of the breaking of isospin symmetry by the Coulomb field. The quadrupole isovector mode is important in the discussion of quadrupole effective charges.

SPIN-DEPENDENT MODES

Because of the strong spin-orbit coupling acting on the one particle motion, the total spin quantum number does not even approximately exist as a constant of the motion in heavy nuclei. However, one can consider collective fields labeled by a macroscopic spin quantum number, σ, as in the above discussion of the iso-spin of the collective oscillations; calling the orbital angular momentum $\varkappa$, and the total angular momentum λ we have for each $\lambda\mu$ two possibilities

$$\begin{aligned} \lambda\pi &= (-1)^{\lambda} & \varkappa &= \lambda \quad \sigma = 0 \\ & & \varkappa &= \lambda \quad \sigma = 1 \end{aligned} \qquad \text{(I)}$$

$$\begin{aligned} \lambda\pi &= (-1)^{\lambda+1} & \varkappa &= \lambda + 1 \quad \sigma = 1 \\ & & \varkappa &= \lambda - 1 \quad \sigma = 1 \end{aligned} \qquad \text{(II)}$$

The two fields (I) are coupled by interactions, like the spin-orbit force, which mix singlet and triplet states, while the fields (II) are coupled by the tensor forces.

Evidence for rather strong spin-dependent fields in the nucleus is provided by the observed renormalization of the spin g-factors in the effective $M1$ operators and a similar renormalization of the effective Gamow-Teller coupling for a single nucleon. Thus, considerable interest attaches to the search for the collective spin dependent oscillations that are expected to result from these spin fields. There exists some evidence for the expected $I^{\pi} = 1^{+}\ \tau = 1$ spin flip mode ($\sigma = 1\ \varkappa = 0\ \lambda = 1$) in light nuclei but otherwise there is almost no data on the rich variety of spin dependent collective oscillations that may be expected in the nuclear spectra.

The breaking of the spin symmetry by the spin-orbit force is in some ways much more through than the violation of iso-symmetry by the Coulomb interaction since the total spin is in no sense a good quantum number for the ground states of heavy nuclei while the total iso-spin appears to be rather accurately preserved in even the heaviest nuclei. This difference is exhibited in an interesting manner if we consider the U_4 symmetry (supermultiplet symmetry; Wigner 1937). This symmetry requires that the fields

$$\begin{aligned} F_{\tau\mu_\tau} &= \Sigma\tau_{\mu_\tau} \\ F_{\sigma} &= \Sigma\sigma \\ F_{\sigma\tau\mu_\tau} &= \Sigma\sigma\tau_{\mu_\tau} \end{aligned}$$

should all have the same coupling strength $\varkappa$. (The available evidence on this point is inconclusive due to the fact that the collective modes due to F_σ and $F_{\sigma\tau}$ have not yet been experimentally identified; tentative evidence is provided by the renormalization effects mentioned above and at least the sign of these couplings all seem to be the same). However, even if the field couplings were exactly the same, the spin-orbit force implies violation of U_4 symmetry in the ground state that leads to collective vibrational excitations that are expected to deviate in a major way from the pattern corresponding to a microscopic U_4 symmetry. For example, in ^{208}Pb the expected collective modes are

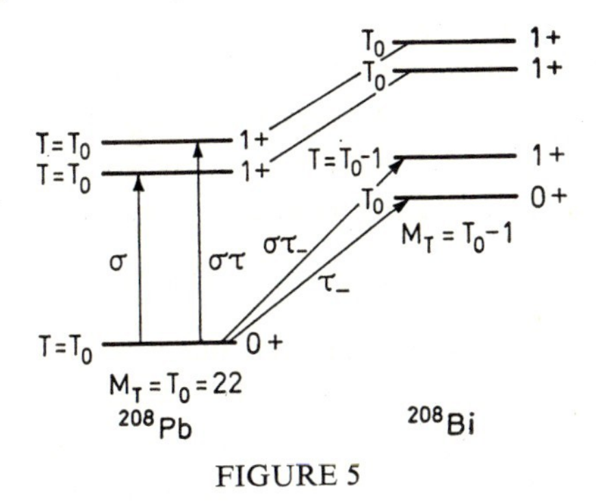

FIGURE 5

where the two 1^+ states in ^{208}Pb result from the spin flip excitation of the ground state configuration of ^{208}Pb, $(i_{13/2}^{-1}\, i_{11/2})$ (neutrons) and $(h_{11/2}^{-1}\, h_{9/2})$ (protons). This pattern may be compared with that obtained in the limit of microscopic U_4 symmetry.

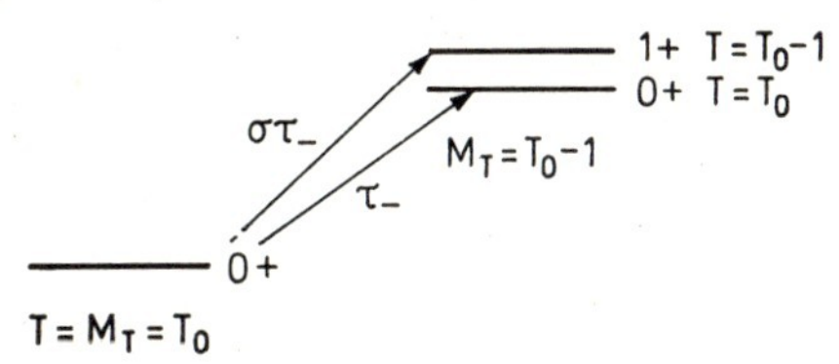

FIGURE 6 Low lying states in the supermultiplet $[f_1 f_2 f_3 f_4] = [^N/_2 {}^N/_2 {}^Z/_2 {}^Z/_2]$: There is only one state with $T = T_0$ and one with $T = T_0 - 1$ in this supermultiplet

Thus the expected collective excitations with $\mu_\tau = 0$ $(\lambda^\pi = 1^+)$ are the characteristic indication of the violation of U_4 symmetry in the ground state wave function; with increasing strength of the U_4 symmetric field couplings the excitation energy of these states increases and the transition strength decreases (conservation of oscillator strength) and so the excitations gradually fade away as U_4 symmetry is imposed. These collective 1^+ states have not yet been experimentally identified*; although I think that there is little doubt that they will be found, it is clear that their identification and quantitative characterization is crucial in an evaluation of the position of the symmetry quantum numbers σ and U_4 in the nucleus.

* Note added after the meeting.

Recent studies of the (γn) process by Bowman, Baglan, Berman and Phillips (submitted to Phys. Rev. Lett.) have identified the expected $\sigma = 1$ $\tau = 1$ $\lambda^\pi = 1^+$ resonance at 7.9 MeV excitation in ^{208}Pb. The M1 strength is distributed over a number of fine structure lines spanning an energy interval of somewhat less than 1 MeV.

PAIRING MODES

The quanta that we have considered so far have all preserved the total number of nucleons, (the corresponding density fields are labeled by the baryon quantum number $\alpha = 0$) but we also encounter in the nuclear spectra collective excitations that involve addition or removal of correlated pairs of particles ($\alpha = \pm 2$). The density field associated with these modes can be written

$$\varrho_{\alpha=2}(r) = a^{+}(\mathbf{r}, m_s = +\tfrac{1}{2})\, a^{+}(\mathbf{r}, m_s = -\tfrac{1}{2}) \tag{5}$$

where, for simplicity, we have confined our attention to pairs of identical particles; the local pair density involves the amplitude for spatial coincidence of the two particles and thus is restricted to the singlet state. The density (5) can be expanded in multipoles, of which the monopole moment is that which has been most extensively studied

$$\begin{aligned} M_{\alpha=2,\lambda=0} &= \int \varrho_{\alpha=2}(\mathbf{r})\, d\tau \\ &= \sum_{\nu} a^{+}(\nu)\, a^{+}(\bar{\nu}) \end{aligned} \tag{6}$$

where the states ν form a complete set of one particle states and the pair $(\nu\bar{\nu})$ are related by time reversal invariance. The field coupling[1] for the pairing vibrations involves the moments for the non-hermetian density function (5); for example, for the monopole mode, we can write

$$\delta V = -G\alpha \sum_{\nu} a^{+}(\nu)\, a^{+}(\bar{\nu}) \tag{7}$$

This coupling (the pairing interaction) generates a correlated collective state out of the available two particle configurations $(\nu\bar{\nu})$ and the extent of this correlation can be measured by the two particle transfer reaction. In some cases the observed cross-sections are enhanced by factors as great as 30 times those that would correspond to uncorrelated states.

In the neighbourhood of closed shell configurations, one has the opportunity to study the superposition of several correlated pairs (labeled with either the same or different quantum numbers) and thus to explore the anharmonicities in the pair vibrational mode. With the addition of many pairs beyond closed shells, the interactions between the pair quanta become large, but it is still possible to obtain a simple description of this mode by considering the potential energy as a function of the magnitude of the pair field. This function is found to exhibit a rather narrow minimum (for sufficiently many particles outside of closed shells) and thus in first approximation one may consider a static pair field as in the description of a superconductor (Bardeen, Cooper, and Schrieffer 1957). Such a field violates the gauge symmetry and implies collective excitations of rotational character which link the ground (and excited) states of successive nuclei differing by $\Delta A = 2$ (see the discussion by A. Bohr at this meeting). In this situation, one can construct collective excitations that are strongly enhanced both with respect to $\alpha = 0$ operators (inelastic scattering and $E\lambda$ transitions) and at the same time with respect to $\alpha = \pm 2$ operators (two particle transfers); recent evidence on the β-vibration in deformed nuclei indicates that these low lying $K^{\pi} = 0^{+}$ excitations that are enhanced in $E2$ excitations, are also very strongly enhanced in two particle transfer (Schiffer 1970). It remains to be seen whether our present understanding of the field couplings will provide an adequate basis for a quantitative description of these modes that involve a mixing of $\alpha = 0$ and $\alpha = 2$ symmetries.

REFERENCES

Baldwin and Klaiber (1948) *Phys. Rev.* **73,** 1156
Bardeen, Cooper, and Schrieffer (1957) *Phys. Rev.* **108,** 1175
Bohr (1936) *Nature* **137,** 344
Bohr and Kalckar (1937) *Mat. Fys. Medd. Dan. Vid. Selsk.* **14,** No. 10
Bohr and Wheeler (1939) *Phys. Rev.* **56,** 426
A. Bohr (1952) *Mat. Fys. Medd. Dan. Vid. Selsk.* **26,** No. 14
Breit and Wigner (1936) *Phys. Rev.* **49,** 519, 642
Cohen (1957) *Phys. Rev.* **105,** 1549
Ejiri *et al.* (1969) *Nucl. Phys.* **A128,** 388
Fallieros *et al.* 1967 *Can. Jou. Phys.* **45,** 3221
Fermi (1934) see, for example, The Nobel Lectures publ. by Elsevier Publishing Co. (1965)
Frenkel (1939) *J. of Phys. Acad. Sci. USSR,* **1** No. 2
Frisch and Meitner (1939) *Nature* **143,** 239
Goldhaber and Teller (1948) *Phys. Rev.* **74,** 1046
Hahn and Strassmann (1939) *Naturwiss.* **27,** 11
Hamamoto (1968–70) *Nucl. Phys.* **A126,** 545, **A135,** 576, **A141,** 1
Haxel, Jensen, Suess (1949) *Phys. Rev.* **75,** 1766
Hill and Wheeler (1953) *Phys. Rev.* **89,** 1102
Jensen *et al.* (1950) *Z. Naturforsch.* **5a,** 343, 413
Lane and Pendlebury (1960) *Nucl. Phys.* **15,** 39
Mayer (1949) *Phys. Rev.* **75,** 1969
Petersen and Veje 1967 *Phys. Lett.* **24B** 449
Polikanov *et al.* (1962) *Zh. Eksp. Teo. Fiz.* **42,** 1464
Rainwater (1950) *Phys. Rev.* **79,** 432
Scharff-Goldhaber and Weneser (1955) *Phys. Rev.* **98,** 212
Schiffer (1970) private communication
Sherif and Blair (1970) *Nucl. Phys.* **A140,** 33
Wigner (1937) *Phys. Rev.* **51,** 106

DISCUSSION

E. P. Wigner It would be good to have a more detailed explanation of the statement that if the U_4 symmetry were accurate for the normal state of Pb^{208}, the 1^+ states would "fade away".

B. R. Mottelson In the pure $j - j$ coupling limit there are strong transitions to 1^+ states in the neighbourhood of 6 MeV produced by operators of the type $\Sigma\sigma_i$ and $\Sigma\sigma_i\tau_z(i)$. The occurence of these transitions represents a flagrant violation of the selection rules that would be expected in a U_4 symmetric system. In the model that I briefly considered, the introduction of the field couplings described by the coupling constants $\varkappa_\tau = \varkappa_\sigma = \varkappa_{\sigma\tau}$, implies that the strongly excited 1^+ states increase in energy and decrease in transition strength with increasing coupling constant. Ultimately with very strong field coupling the U_4 symmetry becomes rather accurate and the 1^+ states carry very little strength. The available evidence on these couplings suggest that there are significant modifications with respect to the pure $j - j$ coupling limit, but the U_4 symmetry is still far from an accurate symmetry of the nuclear ground state in a heavy nucleus.

E. P. Wigner Could we have a more down-to-earth definition of what you mean by the "existence" of the fields $V = \varkappa\alpha F(x)$?

B. R. Mottelson The field coupling $\partial V = \varkappa\alpha F(x)$ can be directly measured in an inelastic scattering process leading to an excited vibrational state in exactly the same sense that the average static potential (field) of the shell model can be measured in the elastic scattering processes or the binding energies and density distributions of the one particle states. These couplings also appear in the determination of the collective frequence and amplitude itself in terms of the spectrum of one particle excitations. By comparing these many different nuclear phenomena we can test the simple form of the coupling that has been assumed.

R. Brout Let me try to bridge the communication gap between Professors Mottelson and Wigner.

What is supposed in Mottelson's scheme is that there exist terms in the nucleus's hamiltonian which are *simulated* by density-density interactions such as

$$\int v(\varkappa - \varkappa')\mathscr{S}(\varkappa)\,\mathscr{S}(\varkappa') \quad \text{or} \quad \int v(\varkappa - \varkappa')\,\varrho(\varkappa)\,\varrho(\varkappa')$$

where $\mathscr{S}$ is spin (or isospin) density, ϱ number density etc. These interactions play an important role in the study of the response of the nucleus to external disturbances which are in turn coupled to these various densities. These same interactions often give rise to collective excitations. One may refer to these special densities as interesting (collective) *co-ordinates* or *operators* as one wishes. It is by careful analysis of experimental data on spectra and response functions that one learns which ones of these simulating terms are important, even in ignorance of the true hamiltonian.

D. H. Wilkinson I should like to make two comments on the utility of muons and pions as probes that may directly or indirectly bring us news of some of the less-well explored collective oscillations to which Dr. Mottelson has alluded. Consider the familiar electric dipole giant resonance that may picturesquely be described as a swinging of all the neutrons of the nucleus against all the protons. The nucleon spin is not involved in this process that may therefore be labelled $L = 1$, $S = 0$. Now this same oscillation may be excited by negative muon capture through the Fermi interaction so that the absolute probability for muon capture through this mechanism is given directly by the giant dipole cross section. But muon capture can also go through the Gamow-Teller interaction, in which the spins are involved, and so excite collective $L = 1$, $S = 1$ oscillations if such indeed exist. They will form a triplet: $J^{\pi} = 0^{-}, 1^{-}, 2^{-}$ as against the familiar single electric dipole resonance. These $L = 1$, $S = 1$ resonances are degenerate with the $L = 1$, $S = 0$ resonance under $SU(4)$ and, as Foldy and Walecka first indicated, constitute a major component of the total muon absorption probability. It is fair to say that we cannot understand muon absorption without involving these new collective oscillations and that we must furthermore assume their approximate degeneracy with the $L = 1$, $S = 0$ resonance. Now although we cannot go further and determine the location of these resonances directly by measuring the spectrum of the emitted neutrinos we may exploit the dose parallel between muon capture and radiative pion capture, first pointed out by Ericson, and determine the absorption spectrum by examining the spectrum of the (π^{-}, γ) process. This is now possible.

My second comment concerns the highly-abusive and highly-important collective monopole oscillation or "breathing mode" that is essentially determined by the nuclear compressibility about which we are astonishingly ignorant. A very indirect approach to this may come from a determination of the nuclear polarization shift to the 1 s-level of muonic atoms. This shift may itself be found from a sufficiently-refined analysis of muonic X-ray energies combined with electron scattering data. From present estimates it appears that the two most important factors in the shift are the E_1 and the E_0 collective states: the former is well-known empirically and so we have the chance of learning, at last, something about the latter.

B. R. Mottelson I completely agree with Pr. Wilkinson on the very fundamental importance of the monopole mode and the unfortunate fact that at the present time, we know almost nothing about it. Perhaps it is worth mentioning that there are of course two monopole modes one with the isoscalar and one with isovector structure. The isoscalar can also in principle be studied and the effective charge resulting from that is involved in the question of isotope shift in nuclei, the change in radius as we add an additional particle. The isovector mode also contributes in such isotopic shift effects but most explicitly in the analysis of the validity of the isobaric quantum number because, after all, the Coulomb field in the nucleus which breaks the isobaric symmetry has spherical symmetry and therefore acts directly on the isovector monopole mode. Therefore all the experiments which test the validity of the isobaric symmetry, are also studying the coupling to these modes and there have recently been very interesting developments for example in the quantitative study of the Coulomb energy differences between nuclei. It has been identified that if you calculate assuming simple one particle wave functions you simply don't get the right answer because the extra particle, if it is a neutron, acts differently on the neutrons and the protons of the rest of the system and changes then the proton distribution of the rest through its coupling to this monopole mode. The rich variety of phenomena and the crucial nature of these modes through their coupling to the compressibility and the validity of isospin, makes them important topics for further investigation.

H. J. Lipkin The two types of quadrupole oscillations are simply understood in a harmonic oscillator model. There are two possible single particle operators which have quadrupole transformation properties under rotations, one depending on co-ordinates, one on momenta. The particular linear combination used by Elliott gives the $\Delta N = 0$ excitations. The two are most simply described by oscillator creation and destruction operators. The product a^+a gives $\Delta N = 0$, the product a^+a^+ gives $\Delta N = 2$.

I do not know exactly how to generalize this beyond the harmonic oscillator. But in any model with shell structure, where shells have alternating parity, the matrix elements of the operators p and x would be expected to be dominated by $\Delta N = \pm 1$ transitions. The operators p^2, x^2 and px therefore would be dominated by the transitions $\Delta N = 0$ or ± 2.

L. C. Biedenharn I would like to remark in connection with Prof. Wilkinson's comment that we have rather more pedestrian ways to investigate the $0^-1^-2^-$ operator multiplet—namely in inelastic electron scattering. For *light* nuclei the existence of a giant magnetic quadrupole excitation has been unequivocally demonstrated experimentally. The 0^- excitation goes via dispersion-like terms, but should be easier to demonstrate for (e, e') excitation than for μ-capture.

Next a question: why does Professor Mottelson feel that isospin is violated microscopically? In his lecture he mentioned the neutron excess, but surely this feature compensates for the Coulomb isospin breaking so as to *restore* isospin symmetry. The situation is rather like an isospin deformation, but with isospin symmetry preserved.

B. R. Mottelson It was not my intention to imply that the *total isospin* quantum number was badly violated; rather it attempted to focus attention on the fact that, even though this quantum number appears to be quite accurate, the large neutron excess and resulting isospin in the ground state implies a very non-isotropic siuation in iso-space when we come to consider the *excitation* of the system. Thus the excitations can only be described approximately (macroscopically) in terms of a symmetry between neutron and proton contribution even though they take place between nuclear states each of which has a sharp total isobaric spin.

R. Brout In recent work on classical fluids, Drs. Feder, T. Schneider, H. Thomas and myself* have come up with a prescription on how to refine RPA type calculations (i.e. those in which the polarization function in Mottelson's talk are calculated by -ww○ww-) so that the long wavelength, low frequency oscillations are correctly described.The usual RPA calculation is recoverable from the equation of motion of the single particle density matrix by approximating the two particle density $\varrho^{(2)}(x, x', t)$ as $f^{(1)}(x, t)$ $f^{(1)}(x', t)$ where $f^{(1)}$ is single particle density. When one tries to improve this statistically, one's first guess is to replace $\varrho^{(2)}$ by $f^{(1)}f^{(1)}\, g(x - x')$ where g is the static correlation function. The result is that one recovers the usual RPA expressions with the "bare interaction v" replaced by a v_{eff} given

$$\nabla v_{\text{eff}} = g \nabla v$$

i.e. the force is appropriately weighted by the correlation, not the potential. One finds in the classical fluid at long wavelength the dispersion law $\omega^2 = c^2 g^2$ where c^2 is *almost* what it should be, the macroscopic compressibility

---

* *On the Dynamics of the Liquid Solid Transition* IBM preprint July 1970.

modulus $\left[c^2 = \frac{1}{M}\left(\frac{\partial P}{\partial n}\right)\right]$. Here P is pressure, n number density, M mass per particle. There is one thing that goes wrong; namely a term in $\partial g^{(z)}/\partial n$ is missing. On further analysis it turns that the ansatz which yields correct results in the classical fluid at small wave vector is

$$\nabla v_{\text{eff}} = g\nabla v + \frac{1}{2} n \frac{\partial g}{\partial n} \nabla v$$

The second term is vital.

Dr. Schneider has informed me that when these ideas are applied to the many electron correlation problem, the improvement over existent theories is marked; and he now hopes to have reliable estimates over a range of metaller densities.

I have presented these ideas here because it is quite possible that they will be relevant in ab initio calculations of collective motion in nuclei of the sort discussed by Professor Mottelson.

W. Heisenberg In your lecture you distinguished between the motion of single particles on the one hand and collective modes on the other hand in the nucleus. How is this term "single particle" defined? Do you mean a nucleon as it is outside of the nucleus, or do you take into account that the nucleon inside the nucleus is surrounded by a "cloud" of (collective!) deformations? (I do not refer to the meson cloud). The nucleon inside of the nucleus of course has a slightly different mass from that outside. Or do you think that such differences are negligible for your purpose?

B. R. Mottelson The considerations that I have presented are mainly concerned with the calculation of the "cloud" of collective excitations surrounding a nucleon and the evaluation of the effect of this "cloud" on the energies, transition probabilities etc. in the coupled system. Thus the "single particle" from which we start is a quasi particle in the sense that it is fully "dressed" with respect to all degrees of freedom expect that collective mode that is being explicitly considered. Thus, for example, in a system with a static pairing field, the one particle matrix elements that describe the coupling to the collective field are modified by the factors U and V that describe the effect of the pair-correlation. The magnitude of the one-particle self energy terms associated with the couplings to the strongest low energy collective modes is typically of the order 0.5 MeV and thus quite important in the description of the low energy spectra.

G. E. Brown First, I would like to make a remark. Considerable work has gone on at a more microscopic level to describe the vibrations, beginning

from two-body forces. Such work involves the marrying of Brueckner theory and many-body techniques. The problem of the hard core is taken account of by the Brueckner theory. Just this hard core makes the approach proposed by Prof. Brout difficult, since his approximation involves the static two-body density matrix $\varrho(x)$, whereas fluctuations in the time-dependent functions bring the particles into the region of the hard core interaction in this approximation. These matters are handled properly in the Brueckner theory.

I believe that if Mottelson had had time to develop also the description beginning from two-body forces, Prof. Wigner would not have had trouble with the language.

While I have the floor, I would like to add a comment about Prof. Heisenberg's question. Without the collective excitations, the effective mass of the nucleon—as would follow from the Brueckner theory—would be

$$m^* \simeq 0.7\,\mathrm{m}.$$

Bertsch (who is here) and Kuo have shown that the coupling with the collective excitations raises this to $m^* \cong m$. It is observed empirically that $m^* \cong m$. Otherwise the Nilsson model, etc., just wouldn't work.

A. Bohr Only a few comments. I was feeling that the discussion that was initiated by Pr. Wigner was left at a point that didn't fully clarify the basis for the description. The starting point is the discovery of a variety of collective modes. One attempts to understand the occurrence of these modes and their coupling to the motion of individual particles in terms of the time dependent average potentials generated by the collective motion. In this manner a self consistent description in terms of elementary excitations, including collective and particle degrees of freedom, is obtained. As emphasized by Professor Mottelson, although the quanta of the collective motion are built out of the particle degrees of freedom, the requirements of orthogonality between the different excitations are systematically taken care of in terms of interactions of exchange character. It should be emphasized that the description is fully microscopic and leads to nuclear wave functions that can be expressed in terms of co-ordinates of the individual particles.

Lie Groups in Atomic Spectroscopy

B. R. JUDD
Physics Department, Johns Hopkins University
Baltimore, Maryland U.S.A.

1 MOTIVATION

The aim of this survey is to discuss, in a general way, the role of Lie groups in atomic spectroscopy. It is planned to describe not only the actual usefulness of Lie groups in treating problems of direct physical significance, but also those features whose main attraction seems to lie exclusively in the mathematics. Over the years, these two aspects of the analysis have functioned in a mutually stimulating way. For example, a knowledge of group theory often suggests a particularly elegant and satisfying way of approaching a problem of physical interest. As the calculations proceed, one very often finds that the results are simpler than one would have anticipated. Unexpected selection rules frequently occur, and multiplicity problems are often less troublesome than one has any right to expect. This clearly points to the existence of a group-theoretical substratum to the mathematics. Its elucidation deepens one's understanding of atomic theory and enlarges the mathematical techniques that can be drawn on when another problem of physical interest arises.

Racah's celebrated 1949 article "Theory of Complex Spectra IV" is a good example of how the solution to one problem unearths a host of new puzzles.[1] That article concerns itself with the calculations of the matrices of the inter-electronic Coulomb interaction for the *f* shell. Until that time, only a few *f*-electron configurations had been treated, though the analysis for the *d* shell was complete. However, many suggestive features were already apparent. For example, the relative energies of the terms of maximum multiplicity were known to be expressible as multiples of just one linear combination of Slater integrals. Again, remarkable degeneracies occurred if special ratios of the Slater integrals were selected.[2] These and other features received direct and transparent explanations when Racah set up the analysis in terms of Lie groups. But the internal structure of many of the group-theoretical calculations turned out to be unexpectedly simple.

2 RACAH'S GROUPS

To describe a few of these surprising results, something of Racah's actual analysis must be outlined. The basic group structure is given by the sequence of inclusions

$$U_7 \supset R_7 \supset G_2 \supset R_3.$$

The seven-dimensional space of U_7 and R_7 is spanned by the seven orbital functions of an f electron. The group R_3 is the rotation group in ordinary three-dimensional space. The group G_2 is an exceptional group of Cartan that, by an astonishing piece of good fortune, can be inserted between R_7 and R_3. Basis states are defined by the irreducible representations of these groups. The Coulomb interaction, although a scalar in R_3, does not correspond to a single irreducible representation of any of the other groups. Racah effected a decomposition into suitable components by writing

$$\sum_{i>j} e^2/r_{ij} = e_0 E^0 + e_1 E^1 + e_2 E^2 + e_3 E^3,$$

where the operators e_i are described by unique pairs of irreducible representations W and U of R_7 and G_2, and where the parameters E^i are linear combinations of Slater integrals. The operators e_0 and e_1 are total scalars [i.e., $WU \equiv (000)(00)$]; e_2 corresponds to $WU \equiv (400)(40)$; and e_3 to $WU = (220)(22)$. To evaluate the matrix elements of e_3, Racah found that it is extremely useful to introduce another operator, Ω, that, like e_3, belongs to $(220)(22)$, and whose matrix elements are given simply by

$$\tfrac{1}{2}L(L+1) - 12G(G_2),$$

where $G(G_2)$ in Casimir's operator for G_2. For it turns out—and here is the remarkable result—that, for a given U and U', all matrix elements

$$\langle \alpha UL | e_3 + \Omega | \alpha' U'L \rangle$$

exhibit the same dependence on L, no matter what the additional quantum numbers α and α' are. From the Wigner-Eckart theorem, this would only be expected if U occurs once in the reduction of the Kronecker product $(22) \times U'$; and this is often not the case.

3 TENSOR OPERATORS

The simplifications of the kind just decribed that one continually comes across in Racah's article have proved to be a major stimulus for theoretical work in atomic shell theory. However, it took a long time for their significance to be appreciated. The immediate concern of atomic spectroscopists was to extend Racah's methods to other operators of physical interest. The simplest are those that are sums of single-electron operators—such as the spin-orbit interaction. Work in this direction was completed in 1963 with the publication of Nielson and Koster's tables of matrix elements of tensor operators for the p^N, d^N, and f^N configurations.[3] But, although designed for a purely practical role in atomic-structure calculations, these tables, like Racah's article, are full of surprises. The casual reader cannot fail to be struck by the number of null matrix elements—and this in spite of the fact that Nielson and Koster automatically rejected all zeros arising from the violation of the triangular conditions of ordinary angular-momentum theory. Of course, many of the tabulated zeros can be given a straightforward explanation; but, until quite recently, the residue of inexplicable zeros was uncomfortably large.

4 ALTERNATIVE SCHEMES

The search for new groups is facilitated by the use of annihilation and creation operators. These connect configurations differing in the number N of electrons and thus considerably widen the scope for possible group structure. The 2^{4l+2} states of the l shell can be regarded as basis states for the vector representation [1] of the unitary group $U_{2^{4l+2}}$. So, if we restrict ourselves to the f shell, the search is limited to subgroups of U_{16384}. At the same time, we do not want to choose subgroups that are too remote from the original scheme of Racah, for, if we do, we shall obviously come no closer to a resolution of the unexpected simplifications that occur when that scheme is used. In listing the various sequences of groups and subgroups, it is convenient to restrict attention to f electrons, although all groups (with the exception of G_2) can be generalized to arbitrary l. To date, the following inclusions have proved useful:

(1) Racah's scheme, extended:

$$U_{16384} \supset R_{29} \supset R_{28} \supset U_{14} \supset \begin{Bmatrix} R_3^S \times U_7 \\ Sp_{14} \end{Bmatrix} \supset R_3^S \times R_7$$

$$\supset R_3^S \times G_2 \supset R_3^S \times R_3^L \supset R_3^J \supset R_2^J.$$

(2) Intervention of quasi-spin:

$$R_{28} \supset R_3^Q \times Sp_{14} \supset R_3^Q \times R_3^S \times R_7 \supset R_3^S \times R_7.$$

(3) Separation of spin-up and spin-down spaces:

$$R_{29} \supset R_{15}^+ \times R_{15}^- \supset R_{14}^+ \times R_{14}^- \supset U_7^+ \times U_7^-$$
$$\supset R_7^+ \times R_7^- \supset G_2^+ \times G_2^- \supset R_3^+ \times R_3^- \supset R_2^S \times R_3^L \supset R_2^J.$$

(4) Quasi-particle factorization:

$$R_{14}^+ \times R_{14}^- \supset R_7^\lambda \times R_7^\mu \times R_7^\nu \times R_7^\xi$$
$$\supset R_3^\lambda \times R_3^\mu \times R_3^\nu \times R_3^\xi \supset R_2^S \times R_3^L \supset R_2^J.$$

It is now proposed to take each of these reduction schemes in turn and discuss some of their more interesting features.

5 SPIN AND QUASI-SPIN

The scheme (1) differs from that given earlier only in making the presence of the spin space apparent through the group R_3^S and in extending the sequence of inclusions. The coupling of the total spin **S** and the total orbital angular momentum **L** to a resultant **J** leads to R_3^J and its subgroup R_2^J. The unitary group U_{14} acts in the 14-dimensional space spanned by all products of single-electron spin and orbital functions; its subgroup Sp_{14} was known to Racah, but he did not develop its properties. The possibility of introducing the groups R_{28} and R_{29} was recognized only within the last few years.[4] It turns out that the 16384 states of the f shell can be regarded as basis functions for the simplest spin representation $(\frac{1}{2}\,\frac{1}{2}\,\ldots\,\frac{1}{2})$ of R_{29}; and this decomposes into the two spin representations $(\frac{1}{2}\,\frac{1}{2}\,\ldots\,\frac{1}{2}\pm\frac{1}{2})$ of R_{28}, corresponding to basis states with even numbers and odd numbers of electrons respectively. At first sight, this description may seem rather uninteresting, having little apparent use and contributing to the traditional scepticism of spectroscopists to group theory. However, an operator O, sandwiched between a bra and a ket described by spin representations, can only belong to irreducible representations of the type

$$(0\ldots 0),\ (11\ldots 10\ldots 0),\ \text{or}\ (11\ldots 1\pm 1),$$

if the matrix element so formed is not to vanish. This follows from the Wigner-Eckart theorem and the application of the notion of stretched weights to the representations involved.

The usefulness of this result becomes apparent when we consider the scheme (2), in which a three-dimensional rotation group R_3^Q in quasi-spin space is introduced. The generators of R_3^Q are scalars in the spin and orbital spaces, and they create or annihilate pairs of electrons or none.[5] This use of quasi-spin is a natural extension of the work carried out for a nuclear shell by Kerman,[6] Helmers,[7] and by Lawson and Macfarlane.[8] It turns out that the decomposition of the representations of R_{28} comprising ones and zeros into representations of $R_3^Q \times Sp_{14}$ is often very simple; moreover, a given irreducible representation of Sp_{14} often occurs with a single irreducible representation of R_3^Q. This means that an operator that transforms according to a specified irreducible representation of Sp_{14} often corresponds to a unique quasi-spin rank.[9] Racah's combination $e_3 + \Omega$ provides a nice example. A study of its matrix elements in f^2 shows that it belongs to (1111000) of Sp_{14}, and this in turn fixes its quasi-spin rank at 2. By applying the Wigner-Eckart theorem in quasi-spin space, it is possible to give elegant explanations of many of the peculiar properties of $e_3 + \Omega$ (but not, it seems, the multiplicity simplification mentioned earlier).

The scheme (3) above involves the abandonment of S as a good quantum number and the separation of those electrons for which the spin-projection quantum number $m_s = \frac{1}{2}$ from those for which $m_s = -\frac{1}{2}$. The two spaces are distinguished by plus and minus signs. This separation is by no means a new idea. As long ago as 1937, Shudeman[10] used it to count the LS terms of the configurations g^N, h^N, and i^N. In the inclusion scheme set out above, the quantum number L is recovered by coupling the angular momenta (L^+ and L^-) that define representations of $R_3^+ \times R_3^-$. However, we could equally well recover the representations of G_2 by using the inclusion $G_2^+ \times G_2^- \supset G_2$: all that is required is to add corresponding generators of G_2^+ to those of G_2^-.[11]

6 AN EXAMPLE

It is probably worthwhile to describe an example in some detail. Last year, Armstrong and Taylor published the matrix elements in f^4 of the magnetic spin-spin interaction between electrons.[12] A number of unexpected selection rules appeared; in particular, they found

$$(f^4(211)\,(21)^3\,L \| z_3(220)\,(21)^5\,D \| f^4(111)\,(10)^5\,F) = 0$$

for all L. The operator z_3 is a part of the spin-spin interaction for which $WU \equiv (220)(21)$. The designation ^5D is a reminder that the spin and orbital ranks are both 2. Now, there is no obvious reason why the above matrix element should be zero. Let us, however, select the state $M_S = 0$ for the bra and $M_S = 2$ for the ket, and consider the G_2 coupling. For the ket,

$$|f^4(111)\,(10)^5\,F, M_S = 2\rangle = |[(10)_4 \times (00)_0]\,(10)\,F\rangle,$$

the subscripts to the representations U^+ and U^- showing the numbers of electrons in the spin-up and spin-down spaces. For the bra,

$$\langle f^4(211)\,(21)^3\,L,\,M_S = 0| \\ = \langle[(10)_2 \times (11)_2]\,(21)\,L| \pm \langle[(11)_2 \times (10)_2]\,(21)\,L|,$$

the representations (10) and (11) being the only allowed possibilities. The choice of sign is not important to us (the opposite combination corresponds to $\langle f^4(220)(21)^1L,\, M_S = 0|$, for which the selection rule under investigation is equally valid). As for the operator, we note that it must be capable of connecting a ket of f^2 for which $M_S = 1$ to a bra of f^2 for which $M_S = -1$; and being a two-electron operator, this is sufficient to determine its structure as

$$z_3\{[(10) \times (11)]\,(21) \pm [(11) \times (10)]\,(21)\},$$

where both representations (10) and (11) derive from (110) of R_7. It can be shown that (10) does not occur in the reduction of $(11) \times (11)$, so the matrix element reduces to

$$\langle[(10)_2 \times (11)_2]\,(21)|[(10) \times (11)]\,(21)|[(10)_4 \times (00)_0]\,(10)\rangle$$

which, by analogy with ordinary angular-momentum theory, can be written as

$$\begin{Bmatrix} (10) & (11) & (21) \\ (10) & (11) & (21) \\ (10) & (00) & (10) \end{Bmatrix} ((10)_2\|(10)\|(10)_4)\,((11)_2\|(11)\|(00)_0),$$

provided we ignore the irrelevant multiplicity difficulties in defining properly the 9-U symbol.

The first reduced matrix element in this expression can be written, in greater detail, as

$$((110)\,(10)\|(110)\,(10)\|(111)\,(10)).$$

In this form it is easy to see that it must be zero. For, if we take products of the representations associated with bra and operator, we find (111) occurs in the symmetric part of $(110) \times (110)$, whereas (10) occurs in the asymmetric part of $(10) \times (10)$. Hence the reduced matrix element is zero and the selection rule follows. This final step is basically the same as de Swart's method for proving certain isoscalar factors for SU_3 are zero.[13]

The separation of a Kronecker square into symmetric and antisymmetric parts is a particularly simple example of a plethysm, the usefulness of which has been stressed by Wybourne[14] in his recent book "Symmetry Principles and Atomic Spectroscopy." Plethysms were first introduced into shell theory by Elliott for studying the nuclear $s + d$ shell.[15] It is worth mentioning here that Wybourne's book contains extensive tables of branching rules and decompositions of Kronecker products. These tables are of great value in performing the kind of manipulations just described.

7 PHYSICAL APPLICATIONS

In f^6 there are three ${}^5D$ terms. Bases in Racah's scheme are

$$|(210)\,(20)\,{}^5D\rangle,\quad |(210)\,(21)\,{}^5D\rangle,\quad |(111)\,(20)\,{}^5D\rangle$$

whereas, if the group $R_3^+ \times R_3^-$ is used, the three D states for which $M_S = 2$ are

$$|(P_5 \times F_1)\,D\rangle,\quad |(F_5 \times F_1)\,D\rangle,\quad |(H_5 \times F_1)\,D\rangle,$$

there being five electrons in the spin-up space and one in the spin-down space. It is a remarkable fact, for which no really satisfactory explanation has been found, that the three ${}^5D$ states existing in nature, (for example, in Eu^{3+} $4f^6$) are almost pure states of the second type.[16] The peculiar spin-orbit splittings that these terms exhibit can be related to this property, the constant λ in the equivalence

$$\zeta \sum_i \mathbf{s}_i \cdot \mathbf{l}_i \equiv \lambda \mathbf{S} \cdot \mathbf{L}$$

being determined by the equation

$$\lambda = \zeta\,\frac{L^+(L^+ + 1) - L^-(L^- + 1)}{2M_S L(L + 1)}.$$

This connection between λ and ζ also explains very readily why $\lambda(f^6\,{}^5P) = 0$, since the only permitted coupling turns out to correspond to $L^+ = L^- = 3$. Thus the scheme (3) is of considerable value from a purely practical point of view.

8 QUASI-PARTICLES

The last scheme, listed as (4) above, is a comparatively recent discovery.[17] The spin-up space can be factored into two equal parts by introducing quasi-particle creation operators of the type

$$\lambda_m^\dagger = \sqrt{\tfrac{1}{2}}[a_{\frac{1}{2}m}^\dagger + (-1)^{l-m}a_{\frac{1}{2}-m}],\quad \mu_m^\dagger = \sqrt{\tfrac{1}{2}}[a_{\frac{1}{2}m}^\dagger - (-1)^{l-m}a_{\frac{1}{2}-m}],$$

where $a_{\frac{1}{2}m}^\dagger$ is a creation operator of an electron for which $m_s = \frac{1}{2}$, $m_l = m$. Every component of $\mu^\dagger$ anticommutes with every component of $\lambda^\dagger$, and the two sets of operators can be used to from the generators for $R^{\lambda}_{2l+1} \times R^{\mu}_{2l+1}$. The spin-down space can be similarly factored. The principal advantage of this scheme is the enormously rich classification scheme that is obtained when states are written as

$$|(L^{\lambda}L^{\mu})\, L^{+},\, (L^{\nu}L^{\xi})\, L^{-},\, LM_{L}\rangle.$$

For the l shell, the representations of R^{θ}_{2l+1} ($\theta \equiv \lambda, \mu, \nu$, or ξ) that intervene are the spin representations $(\frac{1}{2}\,\frac{1}{2}\,\ldots\,\frac{1}{2})$; and these decompose into representations of R_3 without any duplication for $l < 9$. In other words, we have a unique way of defining states of such extraordinarily extensive configurations as k^N. If such configurations were observed in nature, the quasi-particle scheme would represent a major computational breakthrough; for no coefficients of fractional parentage are needed in the calculations. As it is, it leads to new ways of analyzing the structure of an atomic shell, and familiar properties appear in a fresh light. For example, the selection rules

$$\langle d^N(20)\, L |\sum_i \mathbf{s}_i \cdot \mathbf{l}_i \,|\, d^N WL'\rangle = 0,$$

where W stands for the representations (00), (20) or (22) of R_5, can be interpreted in terms of the group S_4 obtained by permuting among themselves the four spaces λ, μ, ν, and ξ.[18] Extensions of the quasi-particle factorization to nuclear shells have been made by Elliott and Evans,[19] and by Hecht and Szpikowski.[20]

9 EFFECTIVE OPERATORS

It would be a mistake to give the impression that the use of groups in atomic shell theory merely serves to broaden one's point of view or to provide pretty explanations for null matrix elements. The firm grip we now have on rare-earth and actinide spectra is directly due to Racah's pioneering efforts. The magnetic tapes that are at present in use and that carry the complete Coulomb and spin-orbit matrices for the configurations f^N utilize the basis provided by Racah's group-theoretical scheme. A more recent example of a practical problem that would be difficult to cope with without using group theory is Feneuille's analysis of three-electron operators in the d shell.[21] Configuration interaction associated with single-electron excitations can be represented within a configuration d^N by the addition of effective operators. Some of these are two-electron operators and can be easily handled; in addition, there are the four three-body scalars

$$V(kk'k'') = \sum_{q,q',q''} \sum_{h \neq i \neq j} \begin{pmatrix} k & k' & k'' \\ q & q' & q'' \end{pmatrix} (v^{(k)}_q)_h\, (v^{(k')}_{q'})_i\, (v^{(k'')}_{q''})_j$$

for which

$$(kk'k'') \equiv (222), (224), (244), (444).$$

(The operators $v_p^{(t)}$ are the components of the single-particle tensors $\mathbf{v}^{(t)}$). If the operators $V(kk'k'')$ are to be introduced into the Hamiltonian, it would seem that four separate parameters would be required. However, $\mathbf{v}^{(2)}$ and $\mathbf{v}^{(4)}$ together transform like the 14-dimensional representation (20) of R_5; and the four R_3 scalars in the Kronecker cube $(20)^3$ belong to (00), (22), (42), and (60). A straightforward application of the Wigner-Eckart theorem reveals that the particular linear combination of the operators $V(kk'k'')$ corresponding to (60) always possesses null matrix elements, while that corresponding to (00) can be absorbed by terms already present in the Hamiltonian. It follows that the three-body terms can be accounted for by (at most) two parameters; and this number is in fact necessary as well as sufficient. Shadmi, Caspi and Oreg have shown that the fit with the observed energy levels is markedly improved when these two additional terms are included in the analysis.[22]

At present, considerable attention is being paid to configuration interaction and its representation by effective operators. Feneuille's analysis concerned itself solely with the coupling of configurations by the Coulomb interaction; but this is not the only term in the Hamiltonian. The so-called electrostatically correlated spin-orbit interaction gives rise to effective operators of the same tensorial character as the Breit interaction, and the group-theoretical decomposition of the combined operator has been studied for f electrons.[24] Actual matrix elements have been made available for all configurations up to f^5,[25] and many of the remaining ones are now known, particularly those involving states of high S.

The effects of configuration interaction on hyperfine structure are also under active investigation. The magnetic field at the nucleus produced by an electron with a non-zero azimuthal quantum number l has two sources: the first derives from the purely orbital motion of the electron, and the second is associated with the electron's spin. In the normal way, both these terms would involve the single electronic radial integral $\langle r^{-3}\rangle$. It is now common practice, however, when fitting experimental data, to use separate radial integrals, $\langle r_l^{-3}\rangle$ and $\langle r_{sC}^{-3}\rangle$, for each of these terms.[26] The fractional changes defined by the equations

$$\langle r_l^{-3}\rangle = (1 + \Delta_l)\langle r^{-3}\rangle,$$

$$\langle r_{sC}^{-3}\rangle = (1 + \Delta_{sC})\langle r^{-3}\rangle,$$

can be calculated from second-order perturbation theory. This has recently been done by Bauche-Arnoult[27] for terms of maximum multiplicity of d^N

and f^N. Although the inter-configurational radial integrals are unknown, she has succeeded in obtaining some striking dependences on N. Some of her results, particularly those concerning the interaction of f^N with $f^{N-1}f'$, can be understood by introducing R_7 and G_2 groups whose generators are the sums of the generators of the corresponding groups acting in the spaces of the f and f' electrons separately. Other results are more intractable. For example, it turns out that, for all single-electron excitations except $d \rightarrow g$, the difference $\Delta_l - \Delta_{sC}$ is algebraically identical for all the lowest terms of both members of any one of the pairs (d, d^2), (d^3, d^4), (d^6, d^7) and (d^8, d^9). This unexpected simplification illustrates well the point made at the very start: group theory aids calculations, but, at the same time, the detailed results often point to the existence of a structure that requires further analysis for its elucidation.

ACKNOWLEDGEMENT

The analysis of Sec. 6 was carried out in collaboration with Dr. J. Bauche. Partial support for this work was received from the United States Atomic Energy Commission.

REFERENCES

1. G. Racah, *Phys. Rev.* **76**, 1352 (1949)
2. O. Laporte and J. R. Platt, *Phys. Rev.* **61**, 305 (1942)
3. C. W. Nielson and G. F. Koster, *Spectroscopic Coefficients for the p^n, d^n and f^n Configurations* (The MIT Press, Cambridge, Massachusetts 1963)
4. B. R. Judd, *Group Theory in Atomic Spectroscopy*. In *Group Theory and Its Applications* (Academic Press Inc., New York 1968)
5. B. R. Judd, *Second Quantization and Atomic Spectroscopy* (The Johns Hopkins Press, Baltimore 1967)
6. A. K. Kerman, *Ann. of Phys.* **12**, 300 (1961)
7. K. Helmers, *Nucl. Phys.* **23**, 594 (1961)
8. R. D. Lawson and M. H. Macfarlane, *Nucl. Phys.* **66**, 80 (1965)
9. B. R. Judd, *Phys. Rev.* **141**, 4 (1966)
10. C. L. B. Shudeman, *J. Franklin Inst.* **224**, 501 (1937)
11. B. R. Judd and L. Armstrong, Jr., *Proc. Roy. Soc.* **A309**, 185 (1969)
12. L. Armstrong, Jr. and L. H. Taylor, *J. Chem. Phys.* **51**, 3789 (1969)
13. J. J. de Swart, *Rev. Mod. Phys.* **35**, 916 (1963)
14. B. G. Wybourne, *Symmetry Principles and Atomic Spectroscopy* (Wiley-Interscience, New York, 1970)
15. J. P. Elliott, *Proc. Roy. Soc.* (London) **A245**, 128 (1958)
16. B. R. Judd, *Phys. Rev.* **162**, 28 (1967)
17. L. Armstrong, Jr. and B. R. Judd, *Proc. Roy. Soc.* **A315**, 27 and 39 (1970)

18. B. R. Judd, *A Quasi-Particle Approach to Shell Theory*. International Symposium on the Theory of Electronic Shells of Atoms and Molecules (Vilnius, 1969)
19. J. P. Elliott and J. A. Evans, *Phys. Letters* **31B**, 157 (1970)
20. K. T. Hecht and S. Szpikowski, *Nucl. Phys.* **A158**, 449 (1970)
21. S. Feneuille, *Compt. Rend. Ac. Sc. Paris* **B262**, 23 (1966)
22. Y. Shadmi, E. Caspi, and J. Oreg, *J. of Research*, Nat. Bur. of Stand. **73A**, 173 (1969)
23. K. Rajnak and B. G. Wybourne, *Phys. Rev.* **134**, A596 (1964)
24. B. R. Judd, H. M. Crosswhite, and H. Crosswhite, *Phys. Rev.* **169**, 130 (1968)
25. H. Crosswhite and B. R. Judd, *Atomic Data* **1**, 329 (1970)
26. W. J. Childs, *Phys. Rev.* **160**, 9 (1967)
27. Cl. Bauche-Arnoult, *Proc. Roy. Soc.* **A322**, 361 (1971)

DISCUSSION

M. Moshinsky I have several comments to make on your interesting contribution. The first concerns this $U(16384)$ group. Of course the general representations of $U(2^{4l+2})$ are of little interest, but conceptually the group is very important. For example it is useful for the discussion of the complementary group to any subgroup of $U(2^{4l+2})$. It permits also an analysis of the relative strengths of two, three, four, etc. body effective interactions in a given shell as was discussed by C. Quesne for the $f^7/_2$ nuclear shell (*Physics Letters* **31B** (1970) 7).

The second comment concerns chains of groups such as $U(2l + 1) \supset R(2l + 1) \supset R(3)$. As is wellknown the irreducible representations (IR) of the groups in the chain are not sufficient, in general, to characterize the states. It is necessary then to introduce other quantum numbers that are not related to the chain of groups, but that does seem to be associated with the plethysm procedure you mentioned in your talk. It is important to notice though that the states obtained by this procedure are in general non-orthogonal, as Racah suggested in his Istambul lectures. It is also relevant to note that the plethysm procedure and the states it gives rise to may provide an opening for a general discussion of the fractional parentage coefficients along lines similar to what has been available since many decades ago for Wigner and Racah coefficients.

Finally in relation with an observation of Prof. A. Bohr I will like to note that in the atomic problem, in which $O(4)$ is a symmetry group for the Coulomb potential, this symmetry is not maintained when we introduce Coulomb interactions as was illustrated recently in calculations in the atomic $2s$–$2p$ shell.

B. R. Judd In connection with the point on fractional parentage, I should emphasize that the quasi-particle factorization that I described eliminates the need for coefficients of fractional parentage.

Secondly, the extension of $O(4)$ to a many-electron system has been a subject of some controversy, mainly between Alper and Sinanoglu on the one hand, and Butler and Wybourne on the other. They disagree on whether the group $O(4)$ is useful or not.

H. J. Lipkin I am a bit confused about this formulation in which you use quasi-particle operators which change the number of particles. Have you used a Bogoliubov transformation with specific coefficients?

B. R. Judd It is important to realize that the general Bogoliubov transformation replaces a creation (or an annihilation) operator by some such linear combination as:

$$Ua_{\alpha}^{+} + Va_{\bar{\alpha}}^{-}$$

where α and $\bar{\alpha}$ are relatively time reversed. In the quasi-particle scheme that I described, time reversal only takes place in the orbital space. So the quasi-particles are not, strictly speaking, the same as the more familiar ones.

L. C. Biedenharn In connection with Prof. Moshinsky's remark, let me say that for compact groups it is always possible to embed in a large enough unitary group such that one can then apply the canonical Gelfand decomposition. For this chain there exists a *unique* (within phase) multiplicity splitting, defining thereby Wigner operators for the Clebsch-Gordan series. For the operators transforming as the adjoint representation, this has been proved recently by Louck and myself in an explicit construction for all unitary groups (SU_n, n arbitrary; to appear in *J. Math. Phys.*).

(It is worth noting, however, that this embedding construction is not directly applicable to unitary representations of *non-compact* groups. For example the relevant geometry if $SU(2, 1)$ versus $SL(3, R)$—both having the same compact Lie algebra of $SU3$—differs very greatly and seems to exclude any simple embedding technique.)

With regard to Prof. Judd's elegant introduction of separate spaces for spin-up electrons and spin-down electrons, it might be helpful to note that this bears a formal relationship to the "symplecton" formulation mentioned earlier. This is helpful in that it *seems* to explain the empirical vanishing of certain matrix elements as a consequence of a symplectic symmetry (found automatically in the symplecton construction) which Judd has explained in terms of the more difficult concept of a plethysm.

B. R. Judd It should be stressed that the introduction of spin-up and spin-down spaces dates from the work of C. L. B. Shudeman, which was published as long ago as 1937, though not from a group-theoretical point of view. However, the Journal of the Franklin Institute, in which the paper appeared, is not widely read by physicists, and as a consequence, Shudeman's work has been widely overlooked. He listed the terms of such complex configurations as g^N, h^N and i^N, which is quite remarkable.

A. Bohr Can one identify simple correlation effects of a systematic character in the many-electron wave-functions that are obtained after diagonalization of the Hamiltonian with inclusion of the Coulomb repulsion between the electrons.

B. R. Judd One can certainly determine spatial correlations from the many-electron wave-functions and connect such correlations to a general description of the energies. Since detailed calculations are performed for transition probabilities and other properties, these correlations are not often used in obtaining an overall picture of the atom.

Symmetry and Statistics*

J. B. FRENCH
Department of Physics and Astronomy
University of Rochester
Rochester, New York

STATISTICAL BEHAVIOR and symmetry are two of the concepts which have played major roles in many-particle spectroscopy. Statistical methods in spectroscopy were introduced by Wigner[1] in the middle 1950's, his major assumptions being that we may associate with a system an ensemble of Hamiltonians, that, for the corresponding matrices, there are independent probability distributions for the matrix elements, and that the results follow by means of an ensemble average. With these and certain auxiliary assumptions it has been possible, by the exercise of very considerable mathematical skill, to derive laws describing the statistical properties of energy levels and other quantities.

The earliest major application of group theory in spectroscopy is also due to Wigner,[2] and has led to the development of a very broad subfield of nuclear physics. The conventional method of making use of symmetries in spectroscopy has usually involved the assumption that one or a few representations are adequate to describe a state (we shall say in that case that the symmetry is "good" or "almost good"); in the space generated by these, one then proceeds with the construction and diagonalization of the Hamiltonian matrix, and so forth.

Each of these methods assumes that, for some purposes, complex systems may be regarded as exhibiting simple behavior, but they contrast strongly with each other concerning the nature and source of the simplicities, and concerning the kind of information with which they undertake to deal. The random matrix method ignores completely the microscopic structure of the states (the fact that they correspond to a certain number of particles distribut-

* An extension, by invitation of the editor, of informal remarks made at the Conference. The work reported here was supported in part by the U.S. Atomic Energy Commission.

ed over a certain number of single-particle states*, and does not therefore make any close contact with the more standard methods in which interactions are explicitly exhibited. The symmetry method, as we have defined it above, is much concerned about microscopic structures but is hardly able to deal with non-trivial cases of badly admixed symmetries (for then the spaces become too large for matrix construction and diagonalization), and moreover there are usually not available methods of justifying restrictions to spaces small enough to handle. As a result the goodness of several of the standard symmetries in various domains of the periodic table is quite uncertain.

The method which we describe briefly here makes a quite different use of symmetry and statistics.[3,4] It does not assume that symmetries are good or almost good, and it makes no simplifying assumptions about the nature of the two-body interaction or about statistical independence of many-particle matrix elements. Instead of dealing directly with the properties of single states, or transitions between pairs of them, it deals with distributions of the energy levels and transition strengths and with certain refinements of these quantities. It is not restricted to spaces small enough to admit matrix construction and diagonalization since it does not deal with matrices. While the most obvious domain in which these methods should apply is to a theory of level densities (in which shell-model structures and residual interactions are taken account of) it turns out, remarkably enough, that in very many cases they may be applied also in the ground-state energy domain to give information concerning the ground-state energies and the low-lying spectra.

The "distribution" method assumes that there is indeed a statistical "simplicity" in the behavior of many-particle systems, one strongly dependent on the microscopic structure, generated by the action of a central limit theorem which implies that the distributions of the levels and related quantities tend, as particle number increases, to characteristic (close to Gaussian) forms. These forms being defined by only a few parameters, we need calculate only a few parameters in order to learn much about the general properties of the system. Since the $U(N)$ group of transformations among the single-particle states really dominates the behavior of spectroscopic systems, symmetry is implicit from the beginning. More complicated symmetries enter naturally when we refine the distributions, dealing then not just with $U(N)$ but with related groups, usually $U(N)$ subgroups, or with chains or lattices of such groups. These may define either the exact symmetries of the problem or, for example, other symmetries whose goodness we wish to study. To evaluate

* It is in fact convenient to define "spectroscopy" as dealing with spaces of this kind ("spectroscopic spaces"). Then, on the one hand, random-matrix theories are outside spectroscopy, while, on the other, Wigner's SU(4) is applicable both inside and outside this domain.

parameters needed for the distributions, it turns out that we need knowledge of the invariants of the group structures involved, and indeed one sees no way whatever of dealing, in complicated many-particle systems, with distributions which are not connected with group structures. Thus symmetries enter in two ways, one because they are interesting, one because only with their introduction do we find ways of constructing distributions.

In order to see how the whole thing goes and what kinds of problems we may hope to solve, consider the following situation. We have already made a large shell-model calculation and have found the eigenfunctions Ψ_i ($i = 1, \ldots d =$ dimensionality) and eigenenergies E_i; this implies of course that we have a shell-model space and an effective Hamiltonian H which operates inside it. The calculation has been made in a *jj* representation and thus the eigenfunctions are given as expansions in terms of states belonging to definite *jj* configurations. We plot, versus the eigenenergy, the intensity of a given configuration as it is found in each eigenstate. We do this for each configuration and have thus a set of discrete distributions. We now represent each of these distributions by a continuous curve which makes, in some natural way, the "best" approximation to the discrete distribution. Let us ask now what we could learn from these continuous distributions.

First the distributions may be recognized as partial level densities (densities for single configurations), so that, within the accuracy with which we have made the discrete → continuous transition, the distributions give us these level densities. Of course, as we go up in excitation energy, we would have to include more and more configurations in our original shell-model calculation (and know correspondingly more about the effective Hamiltonian) if our interest is with the *total* level densities. On the other hand there is increasing interest nowadays in the detailed mechanism for the formation of compound nuclei (Griffin[5]) for which the configuration densities, perhaps refined to specify angular momentum and isospin, are in fact essential quantities.

Near the ground state the level density becomes a less useful concept; we are instead interested in the ground-state energy and low-lying spectra, discrete quantities not given immediately by the continuous distributions. The exact distributions of course give the eigenenergies as those energies at which the distribution functions $\left(F(E) = \int_{-\infty}^{E} f(E')\,\mathrm{d}E'\right.$, where $f(E')$ is a frequency or density function$\Big)$ have steps, the k'th level being that for which the total distribution function, the sum of the partial ones, takes on the value $(k - 1)$ or k according as we approach from below or above. A natural attempt then (Ratcliff[3]) to recover these energies from the smoothed-out distributions is

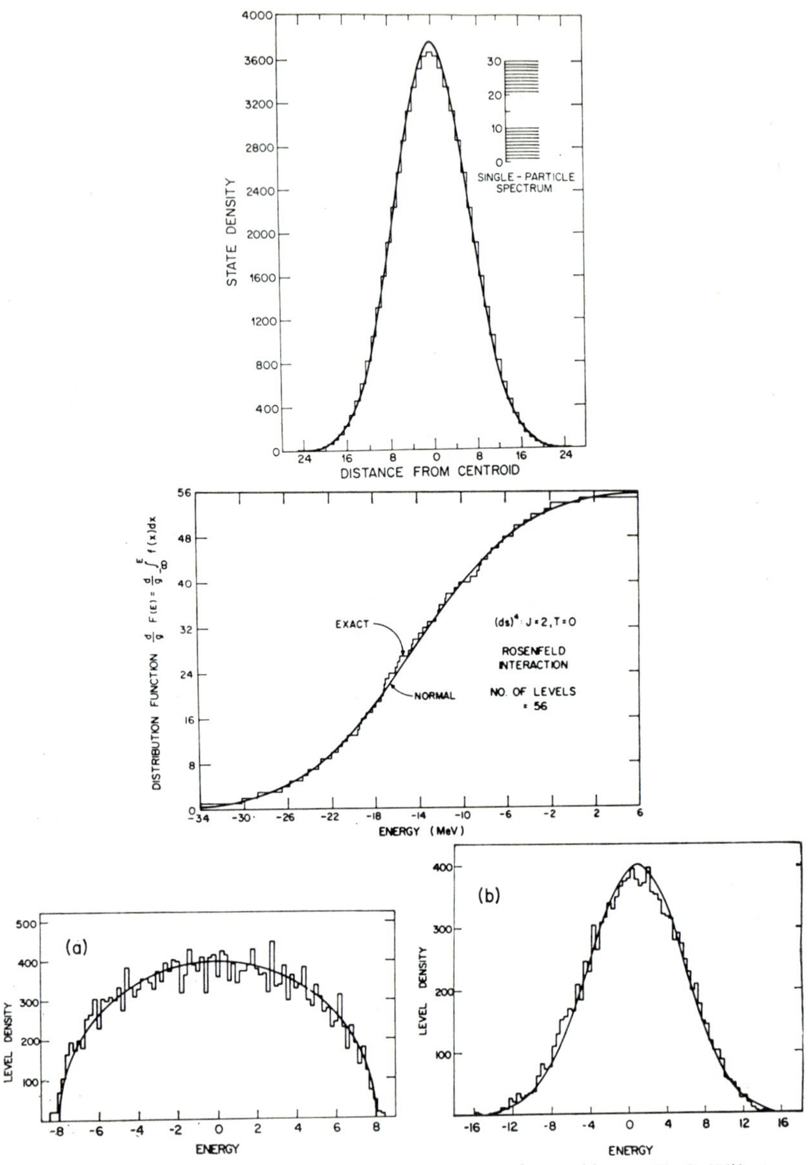

FIGURE 1 Evidence for normality. The upper figure (due to F. C. Williams, private communication) gives the exact and normal 5 particle-5 hole state density for 10 non-interacting particles, the single-particle spectrum being given in the inset. The middle figure gives the (cumulative) distribution function ($d =$ dimensionality, $g =$ degeneracy) for $(d, s)^4$ $J = 2$, $T = 0$ as calculated by detailed spectroscopy, compared with its normal approximation. The bottom pair give the cumulative spectra for two 100-member sets of 50 dimensional random matrices (S. S. M. Wong, private communication). (*a*) is for a conventional ensemble, which gives a semicircular spectrum, while (*b*) is for a random ensemble of two-body operators acting in f^7 with $T = J = 7/2$, which gives a normal spectrum

to take the energies at which the total distribution takes on the value* $(k - 1/2)$. It should be clear also that the contributions of the various configurations to the sum which defines the eigenenergy will, in principle, yield the relative intensities in that state of these configurations.

All this has been based on an analysis of a detailed shell-model calculation. But suppose now that we could find methods for directly calculating, from the specification of the space and the Hamiltonian, the parameters which define the continuous distributions. Then the low-lying energies and their high-energy extension, the level density, would emerge from that calculation, along with the configuration intensities for the low states and the configuration decomposition of the level densities. It should be immediately obvious that only if the distributions are defined by very few parameters will this be feasible, for these parameters must be moments of the distribution and the corresponding high powers of the Hamiltonian will be too complicated to handle (even if we could convince ourselves that spectroscopic models are adequate for their treatment).

The essential fact now is that distributions are close to being normal, a result which has become increasingly clear during the past few years. We need then only two non-trivial moments, defining the centroid and width (the zero'th moment gives simply a dimensionality). If we wish to verify normality in a given case we might consider calculating the third and fourth. Evidence for normality, derived from various sources, is presented in Fig. 1.

EVALUATION OF MOMENTS

We turn briefly to methods for evaluating the moments. Although one can recognize the nature of the formal structures encountered in the evaluation, there is as yet no general procedure but instead only a number of special ones. Badly needed in fact is a better general understanding of the representations associated with a lattice of groups.

If the distribution is over a set of states α then it is easy to see that the

---

* If there are degeneracies due to exact symmetries (for example $(2J + 1)$ for angular momentum) there is an obvious modification if the exact symmetry is specified in the distribution. If the distribution however sums over several exact-symmetry representations, as it does in examples discussed later, we can only proceed if we have, from elsewhere, knowledge of the exact symmetries of the low states of interest (as we might have for example with angular momentum for even-even nuclei). One could in principle "measure" the symmetry by studying the sensitivity of the low-lying distributions to the addition to H of a multiple of the 2-body Casimir operator, but in practice this would not work very well.

centroid energy $\mathscr{E}(\boldsymbol{\alpha})$ and the variance σ^2 (where then σ is approximately the 1/2-width) are given by

$$\mathscr{E}(\boldsymbol{\alpha}) = \langle H \rangle^{\boldsymbol{\alpha}},$$

$$\sigma^2(\boldsymbol{\alpha}) = \langle H^2 \rangle^{\boldsymbol{\alpha}} - \{\langle H \rangle^{\boldsymbol{\alpha}}\}^2$$

where $\langle 0 \rangle^{\boldsymbol{\alpha}}$ is the expectation value of 0 averaged over the set. Obviously now the trace of the density matrix for the set $\boldsymbol{\alpha}$ is going to be of consequence. Writing this, in second quantization, as

$$\varrho(\boldsymbol{\alpha}) = \sum_{\alpha \in \boldsymbol{\alpha}} \psi_\alpha \psi_\alpha^+,$$

we find easily that the many-particle *trace* $\langle\langle 0(k) \rangle\rangle^{\boldsymbol{\alpha}}$ of a k-body operator is

$$\langle\langle 0(k) \rangle\rangle^{\boldsymbol{\alpha}} = \langle\langle \tilde{\varrho}(\boldsymbol{\alpha}_c)\, 0(k) \rangle\rangle^{\mathrm{k}}$$

where $\tilde{p}(\boldsymbol{\alpha}_c)$ is a double involution of $\varrho(\boldsymbol{\alpha}_c)$, the $\boldsymbol{\alpha}_c$ indicating that we consider the $(N - m)$-particle density operator in which we use the hole $\rightleftarrows$ particle complementary states, and the symbol $(\sim)$ indicating that we use $\psi^+\psi$ in the density operator. Observe now that, provided we can construct the operator $\tilde{p}(\alpha_c)$, we have a simplification in that the m-particle results are expressed in terms of k-particle traces ($k \leqq 4$ for H^2). But beyond that, if the decomposition is into subsets (which we may write as $(\mathbf{m}, \boldsymbol{\alpha})$) with fixed particle number m, such that for all $m = 0 \ldots N$ the trace operator for one subset behaves as a multiple of unity in every other one, the result above decomposes further. We find then a remarkable "propagation" of information from the k-particle set throughout the entire lattice

$$\langle\langle 0(k) \rangle\rangle^{m,\boldsymbol{\alpha}} = \sum_{\boldsymbol{\alpha}'} \langle \dot{\tilde{\varrho}}(\boldsymbol{\alpha}_c) \rangle^{k,\boldsymbol{\alpha}'} \langle\langle 0(k) \rangle\rangle^{k,\boldsymbol{\alpha}'}$$

where the $\tilde{\varrho}$ averages, in which the number-dependence occurs, are propagation coefficients independent of the operator (except for its particle rank).

If we think of the sets $(\mathbf{m}, \boldsymbol{\alpha})$ as forming a lattice, then we have equations expressing values at any point in terms of values at a fixed set of "input" points defined by the value of k. The input values would of course have to be calculated from knowledge of the operator and the representations. There may be major advantages in changing from one input set to another, in which for example k-particle representations (and k could be as high as 4 for the usual H^2) are replaced by representations with $(t < k)$ holes. Moreover we may wish, as for example with H^2, to propagate traces without first making the decomposition according to particle rank k. Both of these problems are in the domain of a generalized combinatorial analysis in which one deals for example with the inversion of polynomials defined on lattices and the construction of lattice Green's functions (the propagation coefficients for definite and indefinite particle rank k are in fact analogous to the point-charge and surface-charge Green's functions of electrostatics). The major distributions in which the simple propagation described above is known to occur are those in which we average over all m-particle states, or over states belonging to a single configuration, with or without a specification of isospin, and distributions which involve a spin-isospin $SU(4)$ representation, or a specification of multi-shell identical-particle seniority. Most of the applications made so far have been for these cases with the distributions constructed by combinatorial methods.

For further progress we must however consider much more about the explicit group-theoretical nature of the things with which we are dealing. Symmetries have already entered implicitly, for it should be clear that the condition that operator traces in one set will be multiples of unity in the others will be satisfied only if the sets are irreducible representations of some group (in which case the trace operators are indeed invariant operators, being expressible for example as polynomials in T^2 and the number operator n, if we are dealing with isospin). Our interest is often not with a single group but with a chain or more complicated structure involving several groups. Ordinary SU_3 studies in light nuclei, involve, for example the structure (where $\Rightarrow$ denotes subgroup and $\times$ denotes direct product)

$$\begin{array}{ccccc} U(N) \Rightarrow & U\left(\dfrac{N}{4}\right) & \times & U(4) & \\ & \Downarrow & & \Downarrow & \\ & U(3) & & U(2) \times & U(2) \\ & \Downarrow & & (S) & (T) \\ & U(2) & & & \\ & (L) & & & \end{array}$$

and we would be interested further in the coupling of L and S to a resultant J. Our interest then is not with arbitrary SU_3 representations, or with all the equivalent ones combined together, but rather with representations in which various of the other symmetries indicated are specified. In such a case we may not be averaging over complete representations of the other groups, $U(N/4)$ for example in the case above and also $U(3)$ if we specify L, so that our density operators, while expressible in terms of the generators may not be so in terms of the invariants. Several different kinds of circumstances seem to arise, and a large number of questions concerning the construction of the invariants and the structural relationship between the representations. Such matters seem of considerable interest in themselves both for the light they would shed on symmetries in many-particle systems and because, with their answers, we would be in a better position to extend the distribution methods to a far larger domain than is presently accessible.

The important distribution for fixed particle number and angular momentum is of a rather different nature than the above example. A single level defines a representation, and so our interest is in a sum of distinct equivalent representations without any consideration of the larger groups which would, at least partially, distinguish between them. An inductive method is available for this, which, in principle can be applied to other symmetries, and which yields also a valuable partial-width decomposition of the width. The idea here is that a normal-form operator $\sim\psi_\alpha(k)\,\psi^+_\beta(k)$, acting on m-particle states, involves $(m - k)$-particle intermediate states. But we have also that $\langle\langle\psi_\alpha(k)\,\psi^+_\beta(k)\rangle\rangle^m = \langle\langle\psi^+_\beta(k)\,\psi_\alpha(k)\rangle\rangle^{m-k}$, this because the individual factors operate between the two sub-spaces. But now we introduce into the second trace an explicit commutation (or anticommutation depending on whether the individual factors are boson-like or fermion-like) and then have the m-particle trace expressed in terms of that for $(m - k)$ particles. The commutator term is a lower particle-rank operator than the original one and would be dealt with separately by the same procedure, which is then seen to be doubly inductive. The importance of this is that it works also for tensor products of tensor operators (rather than the simple product above), the isolation of a single intermediate representation being then achieved by a simple Racah transform and the coupled commutator similarly making use of the Racah algebra. This method has been used for fixed-J averaging, but, for other symmetries, sufficiently complete Racah coefficients are not yet available. The basic operation involved here can be generalized in terms of upward and downward shift operators which operate between appropriately ordered representations.

An important feature of this procedure is that it can yield a decomposition of the width into partial widths, partial in the sense that the intermediate states produced by the action of H acting on states of a given representation are themselves separated according to representation. Thus we learn to what extent the spreading of a representation is due to symmetry-preserving interactions and to what extent to symmetry-breaking ones, and, in the latter case, which representations contribute and how much to the width. It should be clear that this decomposition is of major importance when one is concerned with the goodness of symmetries.

There are various combinations of, and variants of, the techniques sketched above for evaluating many-particle moments. For identical-particle seniority we can for example propagate total widths by the combinatorial method applied to symplectic symmetry (in which the density operators are polynomials in the pairing operator), and then decompose into partial widths by using the quasispin R_3 Racah algebra. In the isospin case, and in principle for any direct-product subgroup of $U(N)$ including then the spin-isospin $SU(4)$, one can construct density operators which directly propagate the partial widths. The existence of these extended density operators is to be expected, for as soon as we introduce tensor couplings we are led naturally to non-scalar density operators

$$\varrho^{\nu}(\boldsymbol{\alpha}) = \sum_{\alpha \in \boldsymbol{\alpha}} (\psi_{\alpha}^{\boldsymbol{\alpha}} \times \psi_{\alpha}^{\boldsymbol{\alpha}\,+})^{\nu}$$

which are in fact the operators needed. Finally we mention that it may be advantageous to decompose H into operators of definite symmetry, $H = \Sigma H^{\nu}$, in which case $H^2 \equiv \sum_{\nu} (H^{\nu} \times H^{\nu})^0$, no cross terms coming in because only scalar operators can survive in a (scalar) trace. One finds, in the cases where this has been done, that the propagation is remarkably simpler, and, indeed, some of the unpleasant features of the combinatorial method (unpleasant in that complicated input traces are required) disappear. Moreover the decomposition of H is physically significant, the lower-symmetry parts for example being describable in terms of Hartree-Fock-like one-body operators induced into the space of one representation by the averaged interactions with another.

For an elementary example, consider the width of a configuration in the case of a number of spherical orbits, it being agreed for simplicity that all orbits are distinguishable by angular momenta, this ruling out for example orbital sets containing $1p_{3/2}$ and $2p_{3/2}$. We write the two-body matrix element as $W^{\Gamma}_{rstu} = \langle (j_r \times j_s)^{\Gamma} |H| (j_t \times j_u)^{\Gamma} \rangle$ where $\Gamma \equiv J$ or JT, the latter in isospin formalism, and where the orbits r, s, t, u, are ordered so that $r \leqq s$, $t \leqq u$; then for example W_{rrtt} with $r \neq t$ defines a two-orbit pairing term. The group of interest is the subgroup of $U(N)$ which describes separate transformations in the separate orbits, its irreducible representations being the configurations. It is obvious that interactions defined by matrix elements with different orbital structures are orthogonal to each other, contributing no cross terms in a configuration trace. The separate interactions moreover are all irreducible except for that defined by W^{Γ}_{rsrs}, which we render irreducible by eliminating a trace. Introducing

$$\mathscr{W}^{\Gamma}_{rsrs} = W^{\Gamma}_{rsrs} - (1 + \delta_{rs})\, N_r^{-1} (N_s - \delta_{rs})^{-1} \sum_{\Gamma} [\Gamma]\, W^{\Gamma}_{rsrs}$$

where $[\Gamma] = (2J + 1)$ or $(2J + 1)(2T + 1)$, and writing $\mathscr{W} = W$ for the other cases, we find for the variance

$$\sigma^2 = \sum_{\substack{r \leq s \\ t \leq u}} \frac{(N_r - m_r)(N_s - m_s - \delta_{rs})\, m_t\, (m_u - \delta_{tu})}{(N_r - \delta_{rt} - \delta_{ru})(N_s - \delta_{st} - \delta_{su} - \delta_{rs})\, N_t\, (N_u - \delta_{tu})} \sum_{\Gamma} \times [\Gamma] \{\mathscr{W}^{\Gamma}_{rstu}\}^2.$$

The separate terms have the characteristic form of a 2-particle variance multiplied by a propagation polynomial, the precise form of which follows by inspection. The single-particle energies (including the induced energies) do not come in here because the corresponding interaction is scalar. The general configuration case, allowing for radial "degeneracy" and for non-spherical orbits, is more complicated.

In the discussion we have focussed on the widths because, on the one hand, once the width problem is solved the simpler centroid problem becomes trivial, and, on the other hand, for most purposes the much more complicated higher moments are not needed. The purpose of this necessarily sketchy outline has been to give a feeling for the kinds of mathematical objects and structures encountered. It must be stressed that a general theory which will handle even most of the cases of interest is not available. It is badly needed for distributions, and would seem to be of considerable interest in its own right.

APPLICATIONS TO GROUND-STATE ENERGIES, SPECTRA, AND LEVEL DENSITIES

We turn now to some applications involving, first, configuration distributions with or without a specification of isospin. Table 1 gives, in MeV with respect to a Ni^{56} core, ground-state energies of (f, p) nuclei as calculated by configuration-isospin distributions* using a Brown-Kuo interaction in $f_{5/2}$, $p_{3/2}$, $p_{1/2}$ orbits. These are compared with coulomb-corrected empirical binding energies, and with shell-model calculations for the same orbits and interaction where these are available. Figure 2 gives low-lying spectra for Cu^{65} as deduced by configuration and configuration-isospin distributions, comparison with experiment being also shown. The agreements between the

* In most cases here, one does not calculate the ground-state energy directly but that of a low-lying excited state, afterwards subtracting the empirical excitation energy. This is of particular consequence for odd-odd nuclei where the spacings between the lowest levels (for different J values) fluctuate more strongly than could be handled by smooth distributions which involve various angular momenta. Observe also, as discussed above, that we must assume the J ordering in order to produce the spectra of Fig. 2, ahead. Fixed-J distributions would avoid this but whether they would solve the odd-odd problem is not known.

TABLE I Ground-state energies of (f, p) nuclei, as predicted by configuration—isospin distributions, by empirical binding energies, and by shell-model calculations.[4] Energies are in MeV with respect to Ni^{56}. Coulomb-correction has been applied to empirical energies. The interaction is B. K. supplemented by $\{.07\binom{n}{2} + .11T^2\}$. In cases marked (a) it has not been possible to use an excited state as a reference energy (see text) because of lack of data. In cases marked (b) the ground-state J value has been assumed to be 3/2.

Nucleus	Distribution	B.E.	S.M.	Nucleus	Distribution	B.E.
Ni^{60}	−42.8	−42.8	−42.9	Ni^{65}	−84.8	−83.9
Cu^{60}	−45.9	−45.4	—	Zn^{65}	−102.5	−101.9
Ni^{61}	−50.9	−50.7	−51.1	Ga^{65}	−119.1[b]	−107.9
Cu^{61}	−57.2	−57.0	−56.9	Ge^{65}	−113.4[b]	−111.0
Zn^{61}	−60.3[a]	−60.5	—	Ni^{66}	−92.1	−92.8
Ni^{62}	−61.8	−61.3	−62.1	Cu^{66}	−102.2[a]	−101.4
Cu^{62}	−66.3	−65.9	−66.6	Zn^{66}	−113.3	−112.9
Zn^{62}	−73.2	−73.2	—	Ga^{66}	−119.3[a]	−116.9
Ni^{63}	−68.8	−68.1	—	Ge^{66}	−126.5[a]	−123.3
Cu^{63}	−77.5	−76.7	−77.4	Cu^{67}	−109.6[a]	−110.4
Zn^{63}	−83.7[a]	−82.3	—	Zn^{67}	−119.5	−119.8
Ga^{63}	−86.9[b]	−86.1	—	Ga^{67}	−129.5[a]	−127.9
Ni^{64}	−78.6	−77.8	−78.6	Ge^{67}	−135.6[b]	−133.0
Cu^{64}	−86.0[a]	−84.5	—	Cu^{68}	−116.7	−116.6
Zn^{64}	−94.8	−94.1	—	Zn^{68}	−129.7	−129.9
Ga^{64}	−98.9[a]	−96.2	—	Ga^{68}	−138.0[a]	−136.0
Cu^{65}	−95.3	−94.4	−94.9	Ge^{68}	−146.9[a]	−144.9

calculations are altogether remarkable and quite unexpected. Equally surprising is the agreement between the orbit occupancies* in the target as deduced from shell-model calculations and from the partial distributions; we find for example that the Cu^{63} fractional occupancies, in percent, for the 5/2, 3/2, 1/2 orbits respectively, are 9, 58, 32 (distribution) compared with 9, 59, 32 (s.m.); for Ni^{64} the occupancies are 17, 50, 48 compared with 22, 48, 40, that being in fact about the worst agreement.

These results are probably better than typical, though we have so far nowhere encountered any significant failures. They tell us that in some cases at least there are statistical simplicities extending throughout the entire

* When the admixings are large, the configuration intensities, and presumably the representation intensities for other groups, show severe fluctuations from state to state, which cannot be reproduced by low moment distribution. One should therefore take averages for a few states, or, for one state, combine several representations. The fractional occupancies involve the latter kind of average; and of course they are directly comparable with experiment.

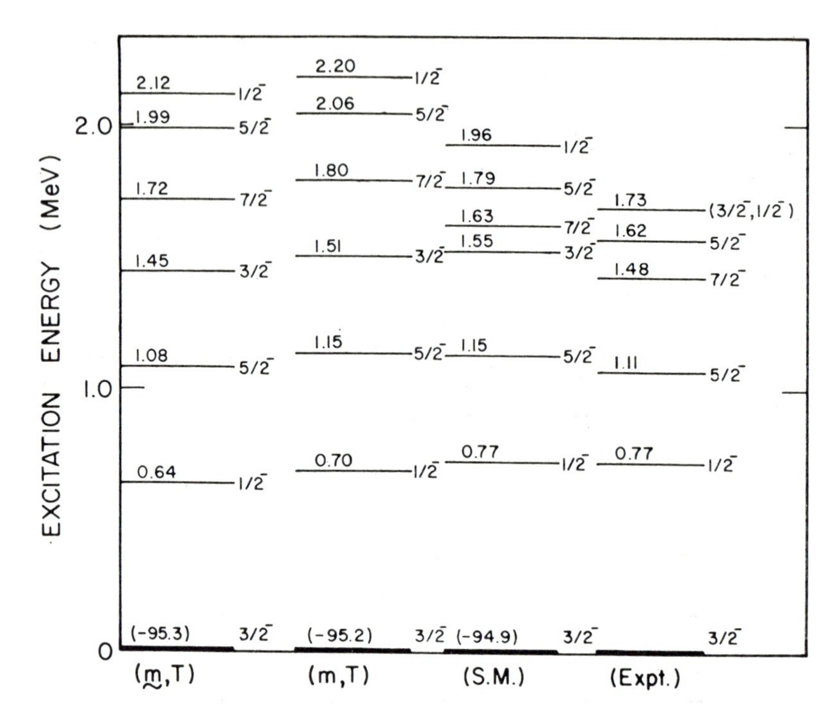

FIGURE 2 The low-lying configuration-isospin and isospin distribution spectra for Cu^{65} compared with shell-model results (S. S. M. Wong, private communication) and with experiment. The absolute energies of the ground states with respect to Ni^{56} are also given. [Taken from ref. (4)]

spectrum even to the ground-state domain. One simply does not know why things work out so well, and, not knowing, we should be cautious especially with distributions in much larger spaces. In Table 1 the largest matrix dimensionality is about 6000, while the distribution spaces (which involve various J values) go up to 35,000. Some recent very preliminary work suggests that it may be safe to work in spaces very much larger than this but that is not yet established. Perhaps it would be worth the considerable amount of trouble to improve the distributions by calculating third and fourth moments. We mention finally that the calculations described could be done by hand. The totality of the (f, p) energies and complete spectra takes about 20 seconds IBM-360-65 computing time.

The distributions which we have been discussing are partial level densities, and so we immediately have a theory of level densities which takes account of the residual interactions and which gives *a priori* evaluations of its parameters in terms of the parameters of the active orbits and the effective interaction. If we interpret the distributions in terms of discrete levels we have spectra, otherwise densities. In these ways, and others as well, we avoid the common "break" between theories used for low-lying excitations and those used for high. One might hope in fact eventually to learn, from level densities, some things about the effective interactions.

The conventional level-density theory relies on counting, by methods of combinatorial analysis or statistical mechanics, the states formed as products (or coupled products) of single-particle or quasiparticle states. The corresponding energies are then sums of single-fermion energies. This counting can in fact be more easily done, for finite single-particle spaces, by forming the appropriate distributions (very easy indeed since H is a one-body operator), and, just as easily, we can decompose the densities according to configurations, numbers of particle-hole pairs, and so on. An example is shown in Fig. 1.

Much more interesting is to ask whether, in a given case, there exists a single-particle basis which will give validity to the conventional theory: the question is whether there exists a decomposition of the single-particle space into orbits such that the effective single-orbit energies, induced by interactions between orbits (and including the self interaction) account for most of the Hamiltonian. This is in fact a group-theoretical question, for the separation of the effective single-particle H is part of the same unitary-group decomposition used above for configuration widths. For a given partition, things are easy but to find that partition which maximizes the "simple" part of H is an optimization problem which may be difficult. There are general reasons, based on the unitary decomposition of the norm of H, why a good basis might well exist in very large spaces with many active particles. On the other hand, if we insist on partitioning into spherical orbits, this is far less likely to be true, and we have in fact verified for the usual interactions that it is not at all so in the (d, s) shell. While the induced energies there are not small, especially for the more realistic interactions, they account for only a fraction of the total H and as a consequence the configuration widths are far larger than the configuration spacings.

The relationship between level densities, calculated with and without those residual interactions which cannot be transformed away, is interesting. However the distribution theory of level densities is self-contained and, as we have described it, suffers only in that it does not give level densities for fixed angular momentum. This happens because fixed-J distributions are technically difficult to construct, and it would in fact represent a serious deficiency if we could not circumvent it in some way.

The conventional procedure for dealing with J works directly with the component J_z which defines an *additive* quantum number M. Then we can apply the central limit theorem to derive*

$$\varrho(E, M) = (2\pi\sigma^2)^{-1/2}\,\varrho(E)\exp(-M^2/2\sigma^2),$$
$$\sigma^2 = m\langle J_z^2\rangle^1.$$

We cannot expect the variance to be properly given in this way because of Pauli-principle effects. If we average σ^2 over all the states formed with m particles in N single-particle states we find easily $\bar{\sigma}^2 = m(N-m)(N-1)^{-1}\cdot\langle J_z^2\rangle^1$ which shows the effect. Conventionally, then, one assumes the Gaussian form with a free value for σ. We shall do the same but will derive the σ value and verify the form as well by evaluation of the second and fourth moments. The latter leads to the "excess" given in terms of the moments by $\gamma_2 = \mu_2^{-2}\{\mu_4 - 3\mu_2^2\}$, whose value is zero for a normal distribution and small, say $|\gamma_2| < 0.4$, for "close to normal". To evaluate the moments as functions of the excitation energy we may use the configurations as an energy "indicator" by which we can convert the configuration moments into energy-dependent moments. If the configuration intensities are $I_{[m]}(E)$ we have

$$\langle J_z^k\rangle^m \simeq \frac{\sum_{[m]} I_{[m]}(E)\langle J_z^k\rangle^{[m]}}{\sum_{[m]} I_{[m]}(E)}.$$

The intensities come from the distributions and contain the interaction effects. The configuration averages come by the usual propagation as described above.

The excess turns out to be small. For a single shell of identical particles we find $\gamma_2 \xrightarrow{N\to\infty} -\dfrac{6}{5m}$ in line with the fact that we approach normality for large m. For $(d, s)^{12}$ we find $-0.20 \leqq \gamma_2 \leqq -0.17$ and similarly for other cases. Thus we can safely assume the conventional form but now we have also a good theory of the energy-dependent "spin cut-off factor" σ. Figure 3 shows level densities calculated in 1-MeV bands (except for the extreme ones which are 3 MeV) for the large shell-model space Cu^{65} considered in terms of $(f_{5/2}\cdot p_{3/2}p_{1/2})$ orbits. The absolute distribution calculations (no free parameters of any sort) are compared with the results of large shell-model calculations; agreement is excellent. The distribution densities are very easy to calculate and may be derived for much more complicated cases; and of course the decomposition by configurations is available also.

Fixed-isospin and fixed-(J, T) densities may be derived in the same way, the first also via fixed-isospin distributions. Level densities, partial with

---

* Do not confuse the two uses of the symbol ϱ.

respect to other symmetries (spin-isospin $SU(4)$ for example which should be useful in studying α-particle reactions) can also be calculated. We stress that

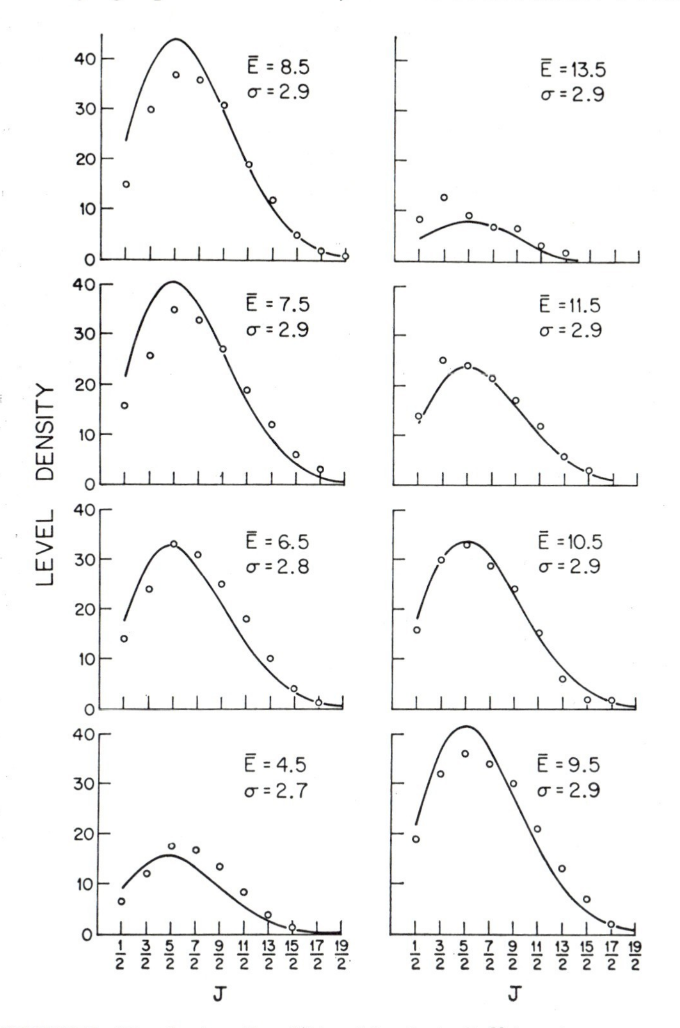

FIGURE 3 The absolute $T = 5/2$ level density in Cu^{63} (continuous curves) as derived from the configuration-isospin distribution with the calculated angular-momentum cut-off factor σ, compared with the exact shell-model results of Wong (private communication). Except at the lowest and highest energies, where 3-MeV bands are used, the exact counting is done in 1-MeV intervals. [Taken from ref. (4)]

our purpose here has been to give a theory of level densities which takes account of orbital structures and residual interactions, and which allows decompositions of the densities according to interesting symmetries.

GOODNESS OF SYMMETRIES

Consider next the goodness of symmetries, the extent to which Hamiltonian eigenstates belong to irreducible representations. If our interest is in a certain symmetry we will clearly form distributions based on its irreducible representations. If these representations are non-overlapping in energy, the symmetry clearly must be good. It may also be that they overlap but not in the neighborhood of the ground state, in which case the symmetry is good for the lowlying states. In a domain however where the distributions overlap we may need more information because we must distinguish between "interweaving" of representations (the states still belonging to irreducible representations), and admixing of them. If a symmetry is exactly good in a given representation it follows that the width of that representation comes exclusively from interactions between its own states; the representation space is then a Hamiltonian eigenspace, one in which H could be diagonalized, in which case we would recognize the width as arising from the spreading of the eigenvalues. Clearly then we need the separation of the widths into "internal" and "external" parts, and indeed the finer separation into partial widths which we have discussed above. If these were available we would have in fact, a perturbation theory of admixtures in which the average intensity of a representation admixed into a dominant one is essentially a partial variance divided by the square of a centroid difference.

We consider a few examples, to begin with, $(d, s)^{12}$ with the usual three spherical orbits, and we ask about the configuration symmetries. There are 45 configurations whose centroids more or less uniformly span ~ 50 MeV (the total spectrum span is ~ 100 MeV). The widths are ~ 8 MeV, so obviously the representations are strongly overlapping, and this is found to be true also near the ground state. We clearly do not expect the configuration symmetry to be at all good*, but to look closer we can make the partial-width decomposition. As described earlier all we have to do for this is to decompose H according to the orbital structure of its matrix elements, a particular term then connecting one representation only with a single other one. We can moreover use the method used above for fixed-J level densities to convert the set of partial widths for configurations into a plot against excitation energy, which displays the "effectiveness" of various parts of H (pairing, multipole, self-interactions, etc.) in different domains of excitation. Such a plot is given in Fig. 4. It is clear from this that configurations will be admixed everywhere, not a surprising result. These energy-dependent norms would appear to have many other uses, among them for example the com-

* Though not good it is very useful as demonstrated earlier.

parisons of Hamiltonians at given excitation energies, and the evaluation of interaction matrix elements usable in Griffin's cascade theory of the formation of the compound nucleus.

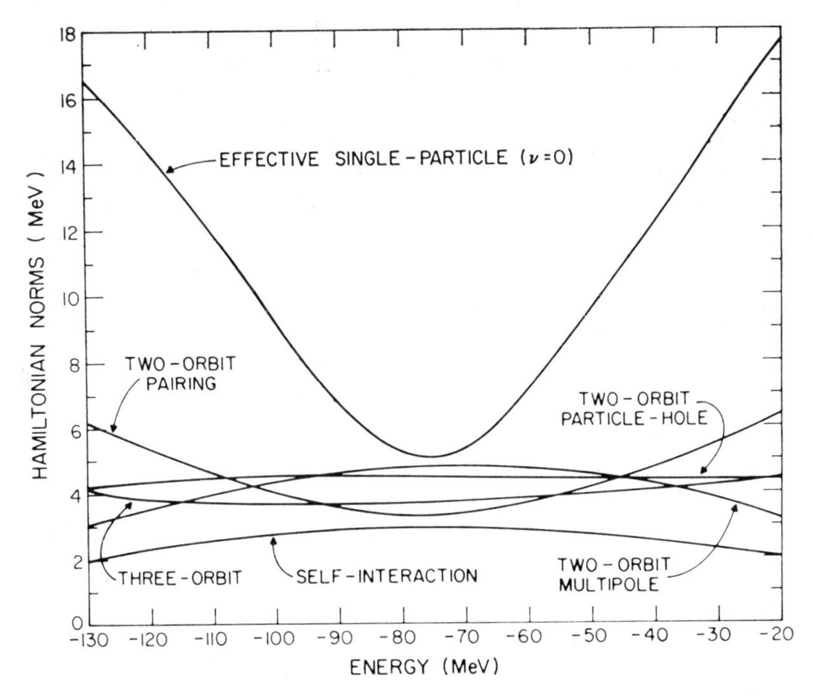

FIGURE 4 The variation with excitation energy of various orthogonal parts of a Rosenfeld interaction in $(d, s)^{12}$. The $(\nu = 0)$ label indicates scalar with respect to the configuration group. The interaction giving the overall centroid has been subtracted out. The ground state is at approximately —130 MeV. [Taken from ref. (4)]

We have mentioned earlier that methods available for evaluating partial widths rely heavily on Racah algebra for the groups involved. The goodness of identical-particle multi-shell seniority has been studied in this way for Ni and Sn isotopes. Rather than discuss this we turn briefly to the spin-isospin $SU(4)$ symmetry for which the Racah algebra is *not* available. However, as with any other direct-product unitary subgroup of $U(N)$, the (scalar) density operators can be constructed as polynomials in the number operator and the three $SU(4)$ invariant (Casimir) operators. The eigenvalues can be derived by standard methods and thus we can construct the many-particle $SU(4)$ distributions, make an independent estimate of the ground-state energies, and deduce the average irreducible $SU(4)$ contributions in the ground-state neighborhood. The (d, s)-shell ground-state energies compare fairly well with those found by configuration distributions, the 2–4 MeV differences being ascribable to the fact that in the $SU(4)$ case we are unable so far to specify the isospin, the spaces then being very large in the middle of the (d, s) shell.

The widths and centroids for a central (Rosenfeld) interaction, given earlier by Parikh,[6] are such as to make a single $SU(4)$ representation, the one with the maximum allowed space symmetry, dominant in the ground-state region. This would be in agreement with results of Akiyama *et al.*[7] for the nuclei up to Mg^{24}; however with more realistic interactions (in particular the $(K + 12FP)$ of Halbert *et al.*[8,4] for the lower half shell and the $BK(12.5)$ interaction[4] for the upper half) Parikh finds, except near the low end of the shell, a strong admixing of two representations which usually contribute, more or less equally, 80–90% of the total intensity.* For $(d, s)^{12}$ these are space symmetry ($SU(6)$) [444] and [4431], the total number of states of these symmetries being about 4,000 and 170,000 respectively. The difference between the central and non-central cases comes about because in the latter the two dominant representations are effectively much closer to each other than in the former. It happens that the low-lying representations have, rather closely, the same widths, that value being then the natural unit with which to measure the energies; in that unit the central interaction separates the two $(d, s)^{12}$ representations by $1.0\,\sigma$ while with more realistic interactions the separation is only $.6\sigma$. It turns out moreover that putting the "primary" single-particle energies to zero gives rise to much smaller widths and to ground-state intensities corresponding to very good symmetry, 95% or better in single representations; thus we may also say that the large admixing is due to the induced single-particle energies, and in particular, since $s_{1/2}$ carries little weight, to the spin-orbit splitting induced via the non-central interactions.

There are still further and more generally applicable ways of studying symmetries which also make use of the techniques described above. These involve decompositions of H into parts which preserve and violate the symmetry in question, and of appropriate measures for the parts; but we cannot discuss these now.

FINAL REMARKS

Even if we take a proper conservative view of the results derived so far by distribution methods, it seems clear that much may be learned and many simplicities exposed by the use of these methods. There remain many inter-

* In the informal remarks, of which this paper is an extension, an offhand guess was made that the admixing would be even stronger in that more representations might be involved and maximum-space-symmetry would contribute even less than the 40% or so now indicated.

esting technical problems, for example about the limit theorems and characteristic spectra, the invariants and other important operators associated with lattices of groups, and the recoupling algebras for these groups. And besides those indicated, there are other large domains of application including that of strength distributions for various excitations, and of an effective interaction theory which directly produces the significant low-order moments. The combination of symmetry and statistics seems to be a very promising one.

ACKNOWLEDGEMENTS

The author is indebted for discussions, for use of unpublished results, and in other ways, ro F. S. Chang, L. S. Hsu, P. E. Mugambi, J. C. Parikh. K. F. Ratcliff, T. H. Thio, F. C. Williams, and S. S. M. Wong.

REFERENCES

1. For a review and references see E. P. Wigner, *SIAM Review* **9**, 1 (1967)
2. E. P. Wigner, *Phys. Rev.* **51**, 106 (1937); see also F. Hund, *Z. Physik* **105**, 202 (1937)
3. J. B. French, in *Nuclear Structure*, A. Hossain *et al.*, editors, North-Holland, Amsterdam (1967). K. F. Ratcliff, *Phys. Rev.* (in press)
4. F. S. Chang, J. B. French, and T. H. Thio, *Annals of Physics* (in press)
5. J. Griffin, *Phys. Rev. Letters* **17**, 478 (1966)
6. J. C. Parikh, in "Contributions to the International Conference on Properties of Nuclear States", University of Montreal Press, Montreal (1969); and private communication
7. Y. Akiyama, A. Arima, and T. Sebe, *Nucl. Phys.* **A138**, 273 (1969)
8. E. C. Halbert, J. B. McGrory, B. H. Wildenthal, and S. P. Pandya, to be published in *Advances in Nuclear Physics*, vol. 4, Plenum Press, New York

DISCUSSION

E. P. Wigner I have one remark and one question. I once made a calculation similar to that reported on and, even though it is published in a mysterious place, perhaps I should mention it. It presents a situation as follows, on the dissolution of a definite isotopic spin state into the background. There is a nucleus, which has a definite m_T. If one neglects electrostatic interactions, one has somewhere high up a state with $T = m_T + 1$. If one now introduces the electrostatic interaction, a situation similar to the one which you describe occurs. It is similar but nevertheless very different because the electrostatic interaction will cause only a dissolution of the $T = m_T + 1$ level in the other levels nearby, with $T = m_T$. Only the matrix elements which are connecting the $T = m_T + 1$ level with nearby levels are effective and it is easy to calculate the curve, which you found to be a gaussian. In the case considered by me, it is back of the envelop calculation and it does not look like a Gaussian. It looks like a resonance curve $\{(E - E_0)^2 + \frac{1}{4}p^2\}^{-1}$. Therefore, it is very different from the curve you obtained but, of course, the model is very different. Is the reason for the difference that you have interactions all the way through?

J. B. French Yes. An important thing to remember is that I am dealing with quantities averaged over a space, and in fact a space corresponding to an irreducible representation of a group. When the operator involved is of low particle rank (a k-body operator has rank k) compared with the number of particles (or holes) defining the space then a Gaussian form results. This is not in conflict with other forms arising under different circumstances, for example when one does not make an appropriate average or an appropriate restriction of the particle rank.

E. P. Wigner It is very interesting to learn that the normal state of Si^{28} contains so much less of the lowest supermultiplet ($P = P' = P'' = 0$) than the normal states of Na^{22} and of Ne^{22} contain of their lowest supermultiplet. For these nuclei, as we learned from Dr. Elliott, the contribution of the lowest supermultiplet is 92.6 and 84.6 per cent respectively. Is the drop of these percentages due to the increased total number of states in the $(sd)^{12}$ configuration as compared with the $(sd)^6$ configuration? Could you tell us which other supermultiplets contribute significantly to the normal state of Si^{28} and how much?

J. B. French Yes. This I think is the main reason and in particular the fact that for $(ds)^{12}$ the "particle-hole" excitation of the lowest symmetry representation gives rise to a very large representation whose centre is not very far above the lowest centroid.

G. Ripka Could you tell us what happens to the moments of the Hamiltonian in the limit of a very large number of particles and in the limit of a very large number of single particle levels?

J. B. French The only general thing I know about this, is that, in the limit of a large number of particles with a large number of single particle states, then the width is dominated by that part of the interaction, which behaves like a one body interaction. This is extremely important. When I say "behaves like a one body interaction" (after all H is dominantly here a two body interaction) I mean: it has the unitary symmetry which is lower by one than the maximum which the space could support or in very crude terms: it behaves like a number dependent one body hamiltonian. This means that, in the limit of a very very large space, I think larger than the (ds) shell, one seems to recapture a reason or a justification for the kind of combinatorial counting that people do with level densities ignoring residual interactions.

G. Ripka Would your calculations give the result that the validity of $U4$ symmetry is better at the end of the $2S$–$1d$ shell, in nuclei such as ^{36}Ar, than in the middle of the shell, as in ^{28}Si for example?

J. B. French By the arguments which I have given I would be led to that conclusion. It is unfortunate in fact that I do not have here the results available. Some calculations of centroids and widths of $SU(4)$ representations were reported by J. C. Parikh at the Montreal Conference in 1969, but it is only quite recently that we have had sufficient confidence in the accuracy of the $SU(4)$ distributions to apply them in the way I have described.

I. Talmi With what kinds of interactions have these calculations been made?

J. B. French With the more or less "realistic" interactions as used for example by Halbert and her collaborators in the (d, s) shell.

SL3R Symmetry and Nuclear Rotational Structure

L. C. BIEDENHARN
Department of Physics,
Duke University, Durham,
North Carolina

THE WORK which I wish to report here is an outgrowth of a suggestion by Professor Gell-Mann, and has been carried out in collaboration with Dr. O. L. Weaver. Our detailed results have already appeared in Physics Letters[1]; accordingly, I shall not go so much into the technical details of the model as to discuss the viewpoints and motivations which underlie the work, as well as some of the objections to this approach.

Let me begin by emphasizing the very great change in our attitude toward symmetry techniques which stems from the fundamental work of Gell-Mann in particle physics. Previous to his work, the customary approach to symmetry consisted in treating the Hamiltonian for a system as having a *major* term, which respects a given symmetry group, and a *minor* term which disturbs this symmetry. The language "major" vs "minor" implies the existence of a small parameter whose vanishing leads to complete symmetry. The prototype for such a structure is the relativistic hydrogen atom. Spin and relativistic effects generate a fine structure that splits the $R4$ symmetry of the non-relativistic problem. The parameter of smallness here is the fine structure constant; this, in fact, led to the name.

Gell-Mann introduced a very different concept. Following Gell-Mann, we may say that we do not care if a symmetry breaking is large or small—only that the Hamiltonian be a function of certain *transition operators* which generate the symmetry—that is to say a multiplet characteristic of the symmetry maybe *split* by the Hamiltonian *but not mixed* with other multiplets.

One starts then with a set of operators that obey (the equal time) commutation relations characteristic of some algebra. These operators are identified with physical transition operators which—acting on a given state—use up most of their strength in transitions to a few nearby states. If one is lucky, the algebra may be such that (because of dynamics) the stationary, or quasi-stationary states, fall into a few (unitary) irreducible representations of the group. If so, this is said to be a "good symmetry". An example is the Elliott ($SU3$) model, in which the symmetry breaking ($Q:Q$ interactions) are not small, but $SU3$ is still a useful symmetry.

In all fairness, I should mention that this view had been partly developed earlier (1960) by Lipkin and Goshen[2], and even earlier (1955) by Tomonaga[3].

In addition to emphasizing the changed rôle played by symmetry, Gell-Mann also introduced a second idea: that of the algebra of currents. Because the commutation relations are kinematical statements, this algebraic structure is preserved independently of any symmetry breaking. Following the model by which the weak and electromagnetic currents were exploited, Gell-Mann considered the (symmetric) energy-momentum tensor, which couples to gravity. He showed that, in the quark model, the time-derivative of the quadrupole moment of the zero-zero component of this tensor, $\mathbf{Q}$, and the *orbital* angular momentum, $\mathbf{L}$, close on the algebra of $SL(3R)$.

$SL3R$, as I shall discuss in a moment, possesses as irreducible representations band structures characteristic of Regge trajectories in hadrons. This was the original motivation behind the introduction of $SL3R$, but it was suggested by Gell-Mann, that the same model should also be applicable to nuclei.

There is another interesting way to approach our problem. Consider a symmetric top. The three Euler angles α, β, γ are respectively the angles giving the (laboratory) orientation of the symmetry axis of the top (specified by the unit vector $\hat{r}$) and the rotation angle around this symmetry axis. The orbital angular momentum operator $\mathbf{L}$ is conjugate to the angles α and β of $\hat{r}$. Using algebraic language we may say that the kinetic energy—which is proportional to L^2—does not commute with the orientation vector $\hat{r}$. Instead we get a *new* vector defined by the commutator: $\mathbf{V} \equiv i[H, \hat{r}] = i[L^2, \hat{r}]$. The two vectors $\mathbf{L}$ and $\mathbf{V}$, together form a closed algebra. This is the algebra of $SL2C$, which is familiar as the algebra of the Lorentz group.

This group has unitary irreducible representations ("unirreps") which form infinite rotational bands, characterized by a quantum number K. The bands are: $L = K, K + 1, K + 2, \ldots$ This is the famous symmetric top model; the K quantum number is an eigenvalue of the rotation operator conjugate to the Euler angle γ. Note, however, that our construction actually implies $K = 0$. Only by abstracting the commutation relations from the model do we obtain the full set of K-bands.

Let us now generalize this approach. Instead of characterizing the symmetric top by the orientation vector $\hat{r}$, let us now characterize the system by a traceless symmetric tensor—the mass quadrupole moments—which we view as a primitive model of a "rotational" nucleus. The commutator of the mass quadrupole tensor with the Hamiltonian now defines a new tensor, **T**. That is: $\mathbf{T} \equiv i\,[H, Q] \neq 0$.

This new quadrupole operator, **T**, and the orbital angular momentum **L** now close on an algebra which is that of *SL3R*.

It would appear from all these considerations that the group *SL3R* should play a fundamental rôle in both hadron and nuclear physics. It is interesting, however, that there are other, very different, ways in which the same group *SL3R* may be introduced into nuclear structure. If we take, as a crude physical approximation, the energy of a nucleus to be primarily a function *only* of its volume—and to a first approximation independent of shape—then we find that the symmetry group of such a structure is that of volume conserving deformations and rotations[4]: this is again precisely the group *SL3R*. [In fact, an early introduction of this group as a nuclear symmetry was by Nataf[5], who, however, discarded *SL3R* as a useful symmetry since it is contrary to the usual techniques in which the symmetry breaking is small.] Even earlier, Tomonaga[6] remarked that the Heisenberg algebra of a mass point in three space is the algebra of $Sp(3, 3)$, which contains as subgroups not only *SL3R* but also the *SU*3 group introduced by Elliott. This bigger group, $Sp(3, 3)$, can be considered as a primitive model of a system having both rotational and vibrational excitations.[2] By taking a (Segal-Wigner) contraction of this algebra, one, in fact, arrives at the algebra characterizing Bohr's Hamiltonian, which has been applied extensively in nuclear structure discussions.

I might also add that there is considerable purely mathematical interest in *SL3R*. It is a surprisingly difficult group to work with.

Let us now turn to applying the symmetry *SL3R* to nuclear structure. We will first identify the electric quadrupole moment operator of a nucleus as proportional to the mass quadrupole operator.

The structure of *SL3R* and its representations limit the nuclei to which our analysis may be applied. Since *SL3R* involves only a quadrupole transition operator, one should first look for nuclei in which $E2$ transitions alone are of major significance. Furthermore, the unirreps, which we have studied, involve only integer angular momenta, so we limit ourselves to those nuclei for which spin does not play an essential rôle. The first ten or so states of a deformed even-even nucleus satisfy these criteria.

Now we must discuss the *SL3R* representations.[7,8]

(a) There are two distinct $K = 0$ band irreps:

$$0^+ : L = 0, 2, 4, \ldots$$

$$0^- : L = 1, 3, 5, \ldots.$$

These irreps are denoted by the (generalized) Young pattern labels $[pq0]$ with:

$$p = -\frac{3}{2} + i\eta,\ \eta = \text{real number}$$

$$q = 0.$$

The continuous variable η is related to the moment of inertia.

(b) There are irreps corresponding to single K-bands, $K \neq 0$. The angular momentum content is: $L = K, K + 1, \ldots$ The labels $[pq0]$ take the form:

$$p = \frac{K-3}{2} + i\eta,\ q = K.$$

Once again the continuous variable η is related to the moment of inertia.

(c) There exist a family of irreps having two continuous labels[7]. The most interesting for physical applications are the sub-set characterized by the finite subgroup V (= Vierergruppe) having representation labels $(++)$.

The angular momentum content for this latter case is found to consist of an infinite series of K-bands:

$$\begin{aligned} K &= 0^+ : L = 0, 2, 4, \ldots \\ K &= 2\ \ : L = 2, 3, 4, \ldots \\ K &= 4\ \ : L = 4, 5, 6, \ldots \\ &\cdots \qquad\qquad \cdots \end{aligned}$$

The extended pattern labels $[pq0]$ are now:

$$p = \frac{\sigma - 3}{2} + i\eta$$

$$q = \sigma > 0;\ \sigma, \eta \text{ real numbers.}$$

The resemblance of this family of irreps to solutions of the Bohr Hamiltonian is noteworthy.

All of our comparisons to experimental data are based on the assumption that even-even rotational nuclei belong to a single irrep of $SL3R$ of type (c), as given above.

Before we embark on this comparison to data, we would like to point out a very significant general result already implicit above. One notes that the symmetric top treated algebraically by means of *SL2C* possesses an irrep consisting of a single $K = 0$ band with L-content $L = 0, 1, 2, \ldots$ By contrast, the *SL3R* algebraic treatment *splits* this $K = 0$ band into two separate irreps: $K = 0^+$ (L even) and $K = 0^-$ (L odd). Ground state bands in even-even rotational nuclei are known to exhibit the *split* $K = 0^+$ band. We feel this fact is significant. Group theoretically one would not be too surprised if nature chose to *combine* two irreps; this would mean only that an extra operator existed which was not in the original algebra.

The splitting of an irrep into two separate pieces, of which only one is retained, is totally different: *this destroys the group as generated by the given set of operators.*

This observation implies a very strong result: nuclei considered as rotators do *not* have the symmetry of *SL2C*, but rather that of *SL3R*. Expressed differently—and more physically—the description of an intrinsic state of a nuclear rotator requires a *quadrupole*, not a *vector*.

Let us now turn to the application of the type (c) irreps to rotational nuclei. Our assumptions imply a formula for the reduced transition probability, $B(E2)$:

$$B(E2: LK \to L'K') = \frac{\text{Constant}}{E_\gamma^2} \frac{|\langle L'K' \| T \| LK \rangle|^2}{(2L+1)} \tag{1}$$

where the constant is not determined group theoretically, and the energy E_γ is *taken from experiment, not theory*. [We view *SL3R* as a set of algebraic constraints on the quadrupole transition operators, not as a symmetry of a Hamiltonian, so we cannot predict energies.] Using the results quoted in reference 1, we can obtain the interesting special case of (1):

$$\frac{B(E2: LK \to L_1 K-2)}{B(E2: LK \to L_2 K-2)} = \left(\frac{E_{\gamma_2}}{E_{\gamma_1}}\right)^2 \left|\frac{C^{2LL_1}_{-2K}}{C^{2LL_2}_{-2K}}\right|^2 . \tag{2}$$

This is a *no-parameter* formula for the ratio of two $\Delta K = 2$ transitions. We will compare it with data in a moment.

First we consider another special case of (2): $\Delta K = 0$ transitions. If the energy spectrum of the ground state band is *exactly* that of a rotator—so that $E_\gamma \propto 2L-1$—then the *SL3R* result for $B(E2)$ becomes identical with the result of the simplest rotator model. Deviations arise in *SL3R* only if the energy spectrum differs from the $L(L+1)$ rule.

There is much more data on the $\Delta K = 2$ transitions. (We interpret the "γ-vibrational band" as $K = 2$ in *SL3R* also.)

A sample of the data is presented below. No band mixing is included in the rotator result quoted.

Nucleus	$\frac{(L2 \to L'0)}{(L2 \to L''0)}$	*B*(*E*2) ratio		
		Experiment	*SL*(3, *R*)	Rotator
Gd^{154}	$\frac{22 \to 00}{22 \to 20}$	0.43 ± 0.04	0.54	0.70
	$\frac{22 \to 20}{22 \to 40}$	6.9 ± 2	10.3	20
	$\frac{32 \to 20}{32 \to 40}$	1.03 ± 0.10	1.41	2.50
	$\frac{42 \to 20}{42 \to 40}$	0.15 ± 0.06	0.21	0.34
Er^{168}	$\frac{22 \to 00}{22 \to 20}$	0.56 ± 0.02	0.57	0.70
	$\frac{22 \to 20}{22 \to 40}$	12 ± 4	11.3	20
	$\frac{32 \to 20}{32 \to 40}$	1.56 ± 0.04	1.50	2.50
	$\frac{42 \to 20}{42 \to 40}$	0.15	0.22	0.34

The relation of the *SL3R* formula for $\Delta K = 2$ transitions to the band mixing result in the rotator model is an interesting one. If the energy E_γ in the transition $L, K = 2 \to L', K = 0$ is assumed to be:

$$E_\gamma = A + BL(L+1) - BL'(L'+1)$$

then, using Eq. (1), we find:

$$\sqrt{B(E2)/|C_{2-2}^{L2L'}|^2} \sim \frac{1}{A - B\Delta} \approx \frac{1}{A}\left(1 + \frac{B}{A}\Delta\right)$$

where $\Delta = L(L+1) - L'(L'+1)$. This has the *form* of the Mikhailov plot: it is linear in Δ.

CONCLUDING REMARKS

There are several objections to this *SL3R* model that we have presented. First, it is in some ways not very ambitious. We have talked about $E2$ transitions but not about $M1$ or $E3$—in fact *cannot* discuss them. Nor do we predict energy spectra. These limitations are indeed built into the theory: we have presented an algebraic model for certain transition operators, not a model of nuclear structure.

A second, related, objection is that the model is hard to generalize. If more transition operators are to be discussed, the whole group structure changes. There is no room for small perturbations. The use of several irreps is possible, but unattractive.

We should also mention that spin* and isospin have not been included.

A fourth objection has been raised by A. Bohr: the energy weighted sum rules are not exhausted by the transitions within a single type (c) irrep. This is true—states of much higher energy need to be included in the sum rules. The *SL3R* model as we use it is essentially a low energy phenomenology in an algebraic setting. It is a striking characteristic of *SL3R* irreps that they contain at most one $L = 0$ or $L = 1$ state. Thus as soon as a second 0^+ state appears in the physical spectrum, or a 1^-, we must admit that *SL3R* has broken down: it can only describe the low lying states.

The value of the *SL3R* model seems to us to be that it *does* give a simple algebraic picture of the low lying states of a rotator; that it yields a simple prediction for interband transition rates; and that it thereby encourages more ambitious use in nuclear physics of Gell-Mann's fundamentally new approach to symmetry.

ACKNOWLEDGEMENTS

I am very much indebted to Dr. O. L. Weaver both for his collaboration in this work, as well as in preparing this communication. I would also like to acknowledge helpful discussions with Professors R. Y. Cusson, Y. Dothan, M. Gell-Mann, and Y. Ne'eman as well as a friendly correspondance with Professor A. Bohr on the question of sum rules.

REFERENCES

1. L. Weaver and L. C. Biedenharn, *Phys. Letters* **32B** (1970) 326
2. S. Goshen and H. J. Lipkin, *Ann. of Phys.* **6** (1959) 301
3. S. Tomonaga, *Prog. Theor. Phys.* **13** (1955) 467
4. R. Y. Cusson, *Nuclear Physics* **A 114** (1968) 289

5. R. Nataf, Orsay preprint (1965)
6. S. Tomonaga, *loc. cit.*
7. L. Weaver, Ph. D. Dissertation, Duke University, Durham, North Carolina, U.S.A. (1970)
8. L. Weaver and L. C. Biedenharn, Contrib. Intern. Conf. on Properties of Nuclear States (University of Montreal Press, 1969)

* Note added in proof:

The question as to including spin in *SL3R* has been answered. The group *SL3R* possesses unirreps having (only) half-integer angular momenta (L. C. Biedenharn, R. Y. Cusson, M. Y. Han, and O. L. Weaver, *Phys. Letters* **42**B (1972) 257–260).

DISCUSSION

A. Bohr One test of the applicability of the *SL3R* group to the description of low energy nuclear excitations is provided by the $E2$ oscillator sum rule $\sum_i (E_i - E_0) B(E2; 0 \to i)$. It is found that only 10% of the total strength goes to the rotational and low energy vibrational excitations; the rest is associated with high frequency excitations. Another sum rule, first suggested by Dothan and Gell-Mann, is obtained from the commutation relation $[T, T] \sim L$, where $T = \dot{Q}$. Evaluation the expectation value in the first excited $2+$ state and including on the left side only the rotational excitation as intermediate states, one obtains a relation between the moment of inertia and the quadrupole moment. This relation is violated by about a factor of two by the experimental data, which again implies the need to include high frequency transitions.

The experimental data on $E2$ transitions between the $K = 2$ and $K = 0$ bands quoted by Professor Biedenharn are similar to those illustrated in Fig. 3 of my report. In such plots, the experimental $E2$ amplitudes are always well represented by a straight line, and it is a question of the slope of this line. The analysis reported by Professor Biedenharn implies a definite value for this slope; this value is of the order of magnitude of the experimental values; the latter vary from case to case, typically by a factor of two. The slope can also be expressed in terms of the dependence of the moment of inertie on the deformation parameter γ associated with the $K = 2$ vibrational excitation. The empirical values correspond to a moment of inertia with a dependence on deformation intermediate between linear and quadratic.

L. C. Biedenharn Professor Bohr has correctly emphasized an important difficulty in the *SL3R* approach. One may indeed regard the fact that the *SL3R* sum rule is off by 50% as a serious defect. On the contrary though—if one bears in mind the tentative nature of the approach—one might feel that it is remarkable that *SL3R* works so very well! [In fact, the first sum rule Bohr quoted indicates that the low energy region (to which *SL3R* is appropriate) should be only a *small* part ($\approx$10%).]

Actually when we obtained our two parameter induced representations—which imply a doubly infinite array of levels—we were very worried about *oversaturating* the sum rules! (That the energy appears in denominator in Eq. (2) helps here.)

One further remark: if the energy spectrum is taken to be precisely rotational ($\sim J(J+1)$) then the $SL3R$ results for the $BE2$ value become exactly those of the simplest rotator model. If the spectrum is slightly non-rotational, the change in $BE2$ given by $SL3R$ tends toward the changes parametrized by the band mixing model. Our results, it should be recalled, are absolute and do not involve adjustable parameters.

The fact that these predictions of interband transition ratios fit very well with experiment, and the fact that the $SL3R$ sum rule is within 50%, we regard as triumphs of $SL3R$—at least to the extent that there may *really* exist true connections between $SL3R$ and nuclear physics.

M. Moshinsky I would like to make a comment on the foundations of the reasoning that Biedenharn presented.

In my opinion his work goes deeper than the references he has given to Gell-Mann and hadron physics. The basic point seems to be that the fundamental group in any dynamical system is the symplectic group $Sp(2N)$ when N is the number of degrees of freedom. This is the group of canonical transformations.

Any other group we want to introduce in the picture seems to be a subgroup of this fundamental group. We are then concerned with the unitary representation of this group and its subgroups in quantum mechanics.

L. C. Biedenharn Moshinsky is quite right and I regret that I did not emphasize enough that this was precisely the point that Tomonaga was making: that when one is dealing with the group of transformations that leave Heisenberg's relations invariant, one is on very firm grounds. The only question which then occurs, is: how many degrees of freedom do you put into the problem? In the quarks, you put in three, for there are three quarks in the hadron and you can arrive at $SL3R$ as a subgroup. But I would like to caution here and make a remark. The remark is that although it is perfectly true that you start with the Heisenberg group over these independent degrees of freedom, if you try to realize this space in the space of boson operators, you will quickly come into difficulties. This group and its realization in the space of boson operators is rather tricky. Louck and I call this "symplecton analysis"—which is a different topic and probably should not be discussed here.

H. J. Lipkin A few remarks regarding particle and nuclear physics. Particle physicists tend to avoid admitting that they can learn something from nuclear physics. For example, when they found that the $SU(3)$ group is the natural generalization to the $SU(2) \times U(1)$ of isospin and strangeness, they called it the "eight-fold way"—presumably because it took them eight years to find it. They could have found it immediately if they had listened to Prof. Racah's lectures on "Group Theory and Spectroscopy", but they were sure at that time that they could learn nothing useful from these lectures.

L. C. Biedenharn To be fair to our European collegues, this gap of eight years was not too critical because, if you will look, you will find that Michel was lecturing on a sort of travelling discussion through various universities in Europe around 61–62 in which he showed every possible generalization of "$SU(2) \times U(1)$" and arrived almost uniquely at $SU(3)$ before it was published, I believe.

H. J. Lipkin The dual role of a symmetry as an "approximate symmetry algebra" and as "transition operator algebra" as discussed by Dothan *et al.* is familiar to nuclear physicists in the example of isospin in nuclear beta decay. The isospin generators not only generate symmetry transformations under which the nuclear hamiltonian is approximately invariant, they also describe nuclear beta decay transitions. The operator whose matrix elements give the allowed Fermi transitions is just an isospin generator, and this is completely independent of the role of isospin as an approximate symmetry. The existence of *superallowed* transitions follows from the combination of *both* properties of isospin.

The case of isobaric analog states is similar. Here again the excitation of these states is described by an isospin generator, regardless of the validity of isospin as a symmetry.

One must be careful in assuming the *same* symmetries to be valid for both particle and nuclear physics, even though the fundamental interactions may be the same. Atomic and molecular physics both result from the electromagnetic interaction, but the symmetries of the *hydrogen atom* and of the *hydrogen molecule* are very different.

L. C. Biedenharn It is a very serious question as to how one goes from particle to nuclear physics. The simplest example is this: in particle physics one puts the lowest representation for the hadron into the $[3\dot{0}]$ of $SU(6)$, whereas the Wigner supermultiplet model puts it in quite a different representation. This is a clear hint that we should be very careful, and if I understand the matter, I think Prof. Lipkin is going to tell us that really for nuclear physics you often have the *Sakata* model in effect.

L. Radicati I hate to appear old fashioned but perhaps at my age I can afford that luxury. I still believe that there is a difference between good guys and bad guys and between good groups and bad groups. The fact that the

Hamiltonian is left invariant, almost invariant, by certain group of transformations is still important. Certainly there is a difference between isospin transformations and other transformations. Groups that leave no part of the hamiltonian invariant are probably not terribly useful. It is true there are commutation relations between operators which are not constants of the motion, which are nevertheless important. That we know since the Reiche-Kuhn's sum rules and we can get useful results out of them but in that case the sum rules must be satisfied. If it is not satisfied I believe the algebraic structure of these operators is probably not the right one, so we have to abandon it. Really groups that leave no part of the hamiltonian invariant, at least in some limit, we have then little experience about them and we are not used to that situation. Since particle physics has been dragged in the discussion, let us take the case of $SU(3) \times SU(3)$. It is probably not a good symmetry but nobody actually knows whether there is not a limit in which this symmetry is not satisfied. In fact there are people who believe that there is a limit in which this symmetry is a good one. And at any rate the commutation relations which I used to deduce the Adler-Weissberger sum rule are good because the sumrule is essentially satisfied.

L. C. Biedenharn I can only agree with Dr. Radicati: that is the reason I started by saying that this point of view is actually quite revolutionary in physics and we have to learn how to use it. There are deep connections that Gell-Mann has told us about namely the algebra of weak currents, and that includes not only the electromagnetic and weak currents but also gravitational currents. They simply exist, we must learn to use them. Now as to the sum-rules: your emphasis on dynamics is absolutely correct. If we knew how to treat dynamics we should do it; on the other hand if we do not know how to do it then perhaps the transition operator algebra (in the language of Ne'emann) is not a completely foolish thing to do. If of course the sum rule is badly violated, I agree with you that is shocking. But unfortunately our result is neither really bad nor all that good. I think that perhaps more data will either kill it or cure it and we must look to experiment. It is indeed a revolution in one's thinking but it is not completely stupid.

L. Radicati No, I don't think it is stupid at all. But I question the revolutionary part of this. The commutation relation between x and p can be used to get a sum rule which is well satisfied even though these operators are not constants of the motion at all. That the algebra generated by an integral over the space of the currents is a good algebra in some cases is true, and get the Adler-Weissberger sum rule from it. But that does not mean that necessarily $SU(3) \times SU(3)$ is a good symmetry. Nevertheless even in this case it is not at all impossible that there should be a limit where one can neglect pion mass and so on, and where these operators are essentially constants of the motion for the largest part of the hamiltonian.

W. Heisenberg I would like to express a general warning with respect to the use of the two concepts: "true approximate symmetry" and "number of degrees of freedom", which have played some rôle in the report and in the discussion. An exact symmetry of the underlying hamiltonian (or what corresponds to it) is of course well defined. But an approximate symmetry may refer only to some part of the spectrum, other symmetries may be a better approximation in other parts—it is probably not possible to say what a "true" approximate symmetry should be. One might think of defining the "degrees of freedom" by some operations. But then we have to keep in mind that the "number of degrees of freedom" is well defined in classical physics (number of coordinates and velocities necessary for defining the exact state of the system), it is not in quantum theory; since by stating just the energy of the stationary state we have already defined the system "completely".

E. P. Wigner About the number of degrees of freedom, I would completely agree with Dr. Heisenberg. When one defines approximate symmetries, one has to restrict one's attention, as a rule, to certain parts of the multitude of all possible states. For instance, $l - s$ coupling in atomic physics is a good approximation only for the low lying states. At high excitation, there are hundreds of states close together, and $l - s$ coupling will be a poor approximation because the spin-orbit forces will mix up everything. For low lying states, in many cases, it is a very good approximation. So that I think it is fair to say that, if one speaks of approximate symmetries, one should also specify for what part of the spectrum or for what states it is an approximate symmetry.

May I make another comment? I concur very much in what Dr. Radicati said. If a sum-rule seems to be violated strongly, that is a clear indication that the operator in question leads out of the region which you consider into other parts of the spectrum. Hence, even though the sum-rule may be very important, there are no symmetries. I mean, there is a sum-rule, as Dr. Radicati explained, for f. And for the sodium atom, it is a very useful sum-rule because the first transition already exhausts it. But if you go to states of higher excitation, it is not very useful.

B. R. Mottelson There is a point in the derivation of these relations which I have not been able to understand. While I can understand that the representation considered can contain the states corresponding to the rotational bands of an asymmetric rotor, I do not see how the K quantum number can be assigned to the stationary states, unless one makes some assumption about the Hamiltonian.

For example, the two spin 2 states are degenerate. Then you can take arbitrary linear combinations of them. Therefore the matrix elements for the ground state to the second 2^+ state, there must be a parameter in that, to describe the ratio of the $K = 0$ to the $K = 2$ part.

L. C. Biedenharn Professor Mottelson has made a very important point. He is quite correct. Although we do indeed claim that the introduction of $SL3R$ does not explicitly involve any particular Hamiltonian, when we come to comparing our results for $BE2$ to experiment we have implicitly made a dynamical assumption. That is, we identified the K-quantum number of our representations with the "physical" K-quantum number of the experimentally found bands. This is indeed a dynamical assumption.

I would like only to remark that the K-quantum number does not have a model-free definition experimentally; the situation is not really a great deal better theoretically.

It is at least consistent to use $SL3R$ symmetry itself to *define* the K-band quantum number. This subject is, however, not closed.

H. J. Lipkin I should like to stress that algebras which have no validity as approximate symmetries of a Hamiltonian can be very important and very useful in physics. The assumption that in nuclear beta decay, the operators describing allowed transitions with positive and negative electron emission respectively satisfy commutation relations like angular momenta is completely independent of the assumption that isospin is a good symmetry. One assumption refers to a property of weak interactions; the other to a property of strong interactions. A priori either may be right or wrong and the two can be tested separately by different experiments.

The Adler-Weissberger sum rule is an excellent example of a transition operator algebra which is not an approximate symmetry. The neutron and proton are not classified in any good approximation in a single irreducible representation of the $SU(3) \times SU(3)$ group, even though one may think of some limit where the $SU(3) \times SU(3)$ symmetry is valid; e.g. zero mass pions. The agreement of the sum rule with experiment involved the use of an infinite set of states; i.e. it was crucially dependent on those matrix elements which exist only because the symmetry is broken. The success shows that the operators describing these weak interactions satisfy $SU(3) \times SU(3)$ commutation relations, but that $SU(3) \times SU(3)$ is not a good symmetry of strong interactions.

L. Radicati I have been completely misunderstood. I never said that sum rules do not test an algebraic structure which is interesting. In fact I quoted the example of x and p and that is very interesting and I quoted Adler-Weissberger exactly in that sense. Of course the generators for these kind of transformations are not constants of the motion but you test their commutation relations. I only object to call that a symmetry; there is no symmetry whatever in the old fashioned sense when there is no part of the hamiltonian that is left invariant.

SU(3) Symmetry in Hypernuclear Physics

HARRY J. LIPKIN

The Weizmann Institute of Science
Department of Nuclear Physics
Rehovot, Israel

THIS WORK was done in collaboration with A. K. Kerman. A detailed account will appear in the volume of Annals of Physics[1] dedicated to the memory of Amos de-Shalit. It seems appropriate to recall the name of Amos de-Shalit at this meeting on the first anniversary of his death, since he would have been very welcome at this meeting if he were still with us.

The subject of hypernuclear physics is somewhere between nuclear physics and particle physics and has been generally ignored on both sides. Particle physicists think that hypernuclear physics is a dull subject because it is similar to nuclear physics. Nuclear physicists think that hypernuclear physics is dull because so far it only resembles the dullest aspects of nuclear physics.

The reason for this dullness is that nearly all of the experimental information deals with ground state configurations of hypernuclei. These are simply described as a Λ particle added to nucleus.[2,3] There is no Pauli principle preventing the Λ from going down into the lowest $1s$ orbit and staying there. Thus the system is described as a Λ added to an ordinary nucleus and moving in a potential well. Analysis of the data can give parameters of the well, but nothing much more interesting. This same sort of analysis is done in nuclear physics with much more interesting and varied experimental data. Thus most nuclear physicists prefer doing more exciting nuclear physics rather than this uninteresting hypernuclear physics.

I should like to suggest that hypernuclear physics can now be "excited to an exciting state" by studying the excited states of hypernuclei. About two years ago such interesting hypernuclear excited states were found in photographic emulsions by Drs. Davis and Sacton of this university.[4] The experiment had been suggested six years ago[5,6] and results predicted. However, at that time the experiment was not very feasible and nobody paid attention to the suggestion. When the experiments were actually performed the predictions were completely forgotten.

Table I shows the experimental results of Davis and Sacton for negative kaons stopped in carbon and nitrogen. They selected events corresponding to the reactions

$$K^- + C^{12} \to \pi^- + {}_\Lambda B^{11} + p, \tag{1a}$$

$$K^- + N^{14} \to \pi^- + {}_\Lambda C^{13} + p. \tag{1b}$$

The energy spectrum of the outgoing pions shows a very definite peak in a small energy region. This indicates that the reaction proceeds via a two-body final state containing a pion and an excited state of the hypernucleus ${}_\Lambda C^{12}$ or ${}_\Lambda N^{14}$. The excited state then decays by proton emission. The excitation energy of the hypernuclear state[3] is found from the data to be about 10 MeV.

TABLE I 1a. Kinetic energies of the π^- mesons in the reactions[4] $K^- + C^{12} \to \pi^- + H^1 + {}_\Lambda B^{11}$ and $K^- + N^{14} \to \pi^- + H^1 + {}_\Lambda C^{13}$

T_π (MeV)			
$K^- + C^{12} \to \pi^- + H^1 + {}_\Lambda B^{11}$		$K^- + N^{14} \to \pi^- + H^1 + {}_\Lambda C^{13}$	
156.7	147.2	173.3	161.2
157.2	139.2	171.9	
156.7	145.0	169.2	
156.5	123.3	168.8	
154.2	94.3	167.8	
156.2	79.5	167.0	
157.5	67.2	167.0	
	66.8		

1b. Parameters of the Λ-hypernuclear continuum states[3] reported by Davis and Sacton[4] from the work of the European K^- Collaboration. B^*_Λ denotes the Λ binding energy in this state, and E* the excitation energy of the state relative to the ground state of this hypernucleus; the E* in brackets is based on the assumption $B_\Lambda = 13.3$ MeV for ${}_\Lambda N^{14}$, an extrapolation only since this B_Λ value has not yet been established. The parameters are uncertain especially because of the subjective element involved in deciding which events belong to the resonance peak and which are background events

Species	B^*_Λ (MeV)	E* (MeV)	Γ (MeV)	Q (MeV)
${}_\Lambda C^{12*}$	0.4($\pm$0.3)	10.7	0.9	1.5 for $(p + {}_\Lambda B^{11})$
${}_\Lambda N^{14*}$	2.9($\pm$0.3)	($\sim$10.4)	1.8	6.5 for $(p + {}_\Lambda C^{13})$

The natural explanation given by the hypernuclear physicists who are used to a model in which the Λ is added to a nucleus is that the Λ is excited from an s-orbit to a p-orbit in the same potential.[3] Anyone who knows nuclear physics sees that this is the wrong way to look at it, because exciting a particle from the s shell to the p shell gives a particle-hole excitation. Such particle-hole excitations have a tendency to be collective. In the case of $_{\Lambda}C^{12}$ excitations can be produced either by raising a Λ or a nucleon from the s-shell to the p-shell, and these two types of excitations should be mixed. This kind of excitation is very similar to collective particle-hole excitations appearing in nuclei, particularly in the isobaric analog resonances.

Let us now consider more precisely the expected properties of such excited states. The analysis uses a number of interesting ideas from both particle physics and nuclear physics but in a rather unconventional way. From nuclear physics we use the concepts of collective states and isobaric analog states.[7,8] From particle physics we use $SU(3)$ symmetry.[9,10,11]

The reactions (1) can be considered as "strangeness exchange reactions" on single neutrons in the target nucleus, in which a neutron is transformed into a Λ.

$$K^- + n \rightarrow \pi^- + \Lambda. \tag{2}$$

This is very reminiscent of the charge exchange reactions which produce isobaric analog states, where a neutron in a nucleus is changed into a proton. In the hypernuclear case the neutron is changed into a Λ instead of a proton, and a similar kind of analog state can be expected. We call such a state a "strangeness analog state."

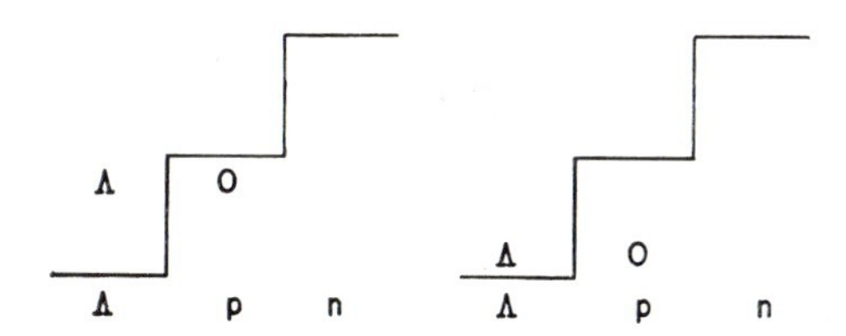

FIGURE 1 Two examples of Λ-particle proton-hole states which go into a strangeness analog state. Since the average potential for the Λ has the same shape as that for a proton all such states have the same energy in first approximation

Figure 1 shows possible components for this strangeness analog state in a picture similar to those used for the isobaric analog states. It is drawn for a nucleus with a neutron excess and containing no Λ's, and the excitations shown change a proton into a Λ in the same orbit. The operators which change a neutron into a proton are called isospin. Similar operators can be defined which change neutrons into Λ's or protons into Λ's. These are called U-spin and V-spin respectively.[10] The action of these three different kinds of spins on neutrons, protons and Λ's is indicated in Fig. 2 and defines the well-known $SU(3)$ algebra. However it is not the $SU(3)$ algebra called "the eight-fold way" used by the particle physicists (see the discussion after Biedenharn's talk).

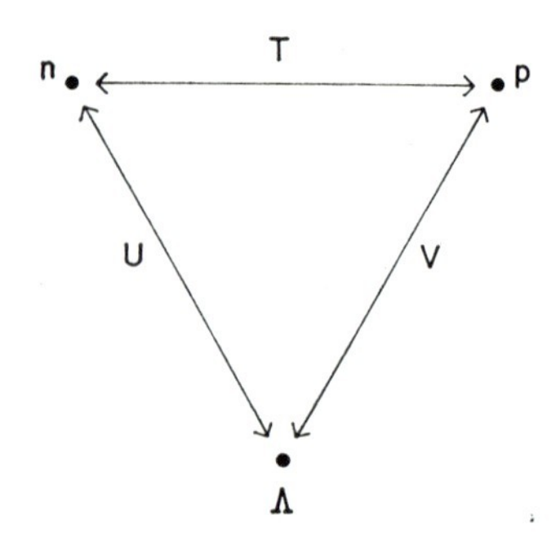

FIGURE 2 Action of isospin, U spin and V spin operators on npΛ triplet

There is no reason a priori why the same symmetries should appear in both particle and nuclear physics. The symmetries of the hydrogen atom and hydrogen molecule are very different even though their dynamics follow from exactly the same fundamental interaction. However there is a peculiar correspondence between the symmetries of particle and nuclear physics. First there is isospin which seems to be just as good in both cases. Then there are other symmetry groups used in both cases but with a peculiar difference. The $SU(4)$ Wigner supermultiplet group is just as good in nuclear as in particle physics, but the nucleon is classified in different representations in the two cases. In particle physics the nucleon is put together with the Δ (3 – 3 resonance) in the 20 dimensional supermultiplet whereas in nuclear physics it is put in a four dimensional supermultiplet containing only the four charge and spin states of the nucleon.[11] Of course the excitation energy of the Δ is so large that we know that a nucleus contains only nucleons and does not contain deltas.

The case of $SU(3)$ symmetry is similar. The octet model used by particle physicists places the nucleon together with the Λ, Σ and Ξ in an eight-dimensional multiplet. However in hypernuclei only nucleons and Λ's are relevant because the $\Lambda - \Sigma$ mass difference is 80 MeV. This is a small number on the scale of particle physics and can be neglected as a first approximation. However a Σ in a nucleus will get rid of 80 MeV excitation very fast and will not remain a Σ in the nucleus very long, just like a Δ will not remain in a nucleus. It is therefore silly to consider states of hypernuclei which contain linear combinations of Λ's and Σ's.

In discussing hypernuclei what is really needed is the old-fashioned version of $SU(3)$, originally proposed by Sakata[13] who suggested that all hadrons were built from the $np\Lambda$ triplet. This was later shown not to be relevant for hadron physics where the octet model described the baryons. However the Sakata triplet model is relevant to hypernuclear physics because hypernuclei are really built from the $np\Lambda$ triplet.

Before we consider the validity of the Sakata $SU(3)$ symmetry, it is useful to examine the two aspects of symmetry algebras mentioned in Biedenharn's discussion. They can be both approximate symmetry algebras and transition operator algebras[12] in the terminology of Dothan *et al.*[14] In the reactions (1) a neutron is changed into a Λ and this transition is described formally by operating on the target nucleus with a U-spin operator. Even if U-spin is not an approximate symmetry of the nuclear Hamiltonian, it still describes the transitions. There is also the question of what is meant by "good or bad" approximate symmetry. When a symmetry is not exact, whether it is good or bad depends on the particular phenomena considered. For example, for hadrons or light nuclei, isospin is a very good symmetry, but for a neutron star, isospin is no symmetry at all.

In the hypernuclear case we have an $SU(3)$ algebra which does not seem to be a good symmetry and in general it is not good. But for the particular cases relevant to the reactions (1) we shall see that it is not so bad.

Let us now examine this question of how good the symmetry is by writing a nuclear Hamiltonian and separating it into a term which is invariant under the Sakata $SU(3)$ symmetry and a symmetry-breaking term

$$H = H_{\text{nuc}}(\Lambda N \equiv np) + \sum_{ij=\Lambda N} \Delta V_{ij}^{\Lambda N}. \tag{3}$$

The first term $H_{\text{nuc}}(\Lambda N \equiv np)$ is the entire nuclear Hamiltonian but has the ΛN interaction set equal to the np interaction. This term gives an exact description of all non-strange nuclear states and in the approximation where we neglect electromagnetic effects which violate isospin symmetry, it is also invariant under the Sakata $SU(3)$ transformations.

The states produced in the strangeness exchange reactions (1) on nuclear targets are hypernuclei containing only a single Λ. To describe these states we add to the Sakata-symmetric Hamiltonian a term which describes the difference $\Delta V^{\Lambda N}$ between the ΛN and np interactions. We need not consider the $\Lambda - \Lambda$ interaction as there is never more than a single Λ in the states under consideration. The second term in Eq. (3) adds this difference to all ΛN pairs.

All ordinary nuclei are exact eigenstates of the unperturbed Hamiltonian. We are thus better off than in the case of isobaric analog states, where the target nuclei used in an experiment are not exact isospin eigenstates and mixing must be considered. The reason for this difference is that ordinary nuclei have a certain nucleon excess, but still contain both neutrons and protons. In the framework of the Sakata model where nucleons and Λ's are considered together, they have an *extreme* "nucleon excess", as they have no Λ's at all. This is like the case of a nucleus containing only six neutrons, which would be a good isospin eigenstate with $T = 3$ with no possibility of mixing in $T = 4$ or $T = 2$ no matter how badly isospin is broken. In the same way those particular nuclei containing no strange particles are automatically good eigenstates of the Sakata model. So far we have neglected isospin breaking, but even if isospin symmetry breaking is included and nuclei are no longer good isospin states, they are *still* good U spin and V spin eigenstates because they do not contain any Λ's, and it is the U spin and V spin which is relevant to the production of strangeness analog states.

Thus the initial nuclear state in a strangeness exchange experiment is a good eigenstate of U spin and V spin and the experiment itself itself is described by acting on this target state with a U spin or V spin generator. The final state created in the experiment is thus produced by acting on a U spin or V spin eigenstate with a U spin or V spin generator. It is thus also a good U spin or V spin eigenstate.

We must now consider whether this state is also a good eigenstate of the Hamiltonian. This immediately raises the question of what is meant by a "good" eigenfunction of the Hamiltonian? We know that because of the symmetry breaking, this state lies in the continuum just like the isobaric analog states. It will therefore decay as a result of the symmetry-breaking interactions. If the symmetry-breaking interactions are strong, the state will decay very fast and have a large width, but if they are weak the state will decay very slowly and have a narrow width. If we wish to know whether we are really about to open a new chapter in hypernuclear physics, the question is whether the width is reasonably small, so that the state can be detected experimentally as a resonance, or whether it is spread out over the whole energy spectrum and cannot be observed. Even though we have no detailed model for such states we can get some general idea from what we already know about nuclear physics. We first note that the symmetry-breaking interaction is like a nuclear two-body interaction. Since the nuclear shell model works the two-body interaction can be broken up into two pieces, one a contribution to the average field and the other a residual interaction

$$\sum_{ij=\Lambda N} \Delta V_{ij}^{\Lambda N} = \sum \Delta U_i^{\Lambda N} + \sum_{ij=\Lambda N} \Delta V_{ij}^{\Lambda N}\,(\text{res.}) \tag{4}$$

where $\Delta U_i^{\Lambda N}$ is the average symmetry-breaking field seen by the particles and $\Delta V_{ij}^{\Lambda N}$(res.) is the difference between the Λ-nucleon and the neutron-proton residual interactions. This shell model picture gives an indication that symmetry-breaking can be small in a nucleus even though it is large at the elementary two-body level where the Λ-nucleon interaction is only about half of the nucleon-nucleon interaction and what is neglected is the same order as what is kept. In a complex nucleus it is not the two-body interaction which is important but the interaction with the rest of the nucleus of the particular nucleon which is changed into a Λ.

Thus we need the difference between the nucleon-*nucleus* interaction and the Λ-*nucleus* interaction.

The interaction of the odd particle with the nucleus is described to a first approximation by a potential well. The range of this well does not depend upon the strength of the forces, but upon the size of the nucleus. The only effect of the strength of the forces is to change the depth of the well. The gross features of the wave functions of a particle moving in a particular well are determined by the angular momentum and by the radius of the well. The depth of the well affects only the tail of the wave function. Thus, even though Λ-nucleon and nucleon-nucleon interactions are quite different, the wave functions of Λ's and nucleons in the field of a given nucleus should have quite a big overlap.[5]

This argument is expressed formally in Eq. (4). The first term on the right-hand side is the change in the average field seen by a particle when one nucleon is changed into a Λ. The second term is the change in the residual two-body interaction. It happens that in nuclear physics we already know of excited states in the continuum which have measurable widths and where the interactions responsible for these widths are just of the two types appearing in Eq. (4). The change in the average field is found in isobaric analog states. There it is the difference between the average field seen by a proton and by a neutron and gives rise to an escape width, Although comparing the hypernuclear and isobaric cases appears to be comparing a strong interaction with a much weaker electromagnetic interaction, this is not really true. The strong interaction is of short range and the electromagnetic interaction is long range. Thus if strong interactions are of order unity and electromagnetic interactions are of order 1/137, the relevant parameters for nuclei are 1 and $Z/137$. These are much more comparable and indicate that the contribution of the average field to the width of a strangeness analog state should be "not too different" from widths of isobaric analog states. It is not serious if they are one or two orders of magnitude larger, since the isobaric analog states themselves are quite narrow and an increase in two orders of magnitude still leaves widths that are small enough to be observable.

The residual interaction is a strong two-body interaction. However since it is only the difference between the nucleon-nucleon and the Λ-nucleon interactions it is weaker than the residual two-body interaction which breaks the shell model and is responsible for the decays of collective particle-hole excitations like the nuclear giant dipole resonance. Since the strangeness analog states are also collective particle-hole excitations, this argument suggests that the widths of these states should be comparable to giant dipole widths and therefore measurable. Of course these arguments are very crude and at best only qualitative but I think they are sufficient to suggest that it is worth while to do more experiments and to do more serious calculations as data become available. A further question is what excitation energy is expected for this analog state, and whether the observed value of 10 MeV is reasonable. Again by making use of things known from nuclear spectra a qualitative estimate can be made by the following argument. The parent of the analog state differs from the analog state only by having a nucleon instead of the Λ, but otherwise has the same wave function. The ground state of the hypernucleus differs from the parent of the analog state by having one less nucleon in the highest orbit and a Λ in the lowest s orbit. The ground state of the hypernucleus can therefore be transformed into the strangeness analog state in the following three steps: 1) removing the Λ from the s orbit, 2) adding a nucleon into the highest orbit to make the parent of the strangeness analog state, and 3) changing a nucleon into a Λ without changing the wave function to obtain the strangeness analog state. The excitation energy is the sum of the changes in binding energy in the above three steps. (The effects of the Λ-nucleon mass difference cancel in the difference in energy between the analog state and the hypernuclear ground state and is therefore disregarded.)

Figure 3 shows plots of the energy levels relevant to this calculation for two cases. The removal of the Λ from the lowest s state of the hypernucleus costs the Λ binding energy B_Λ. The addition of the nucleon in the outermost unfilled orbit gains the nucleon separation energy B_n for the parent nucleus. The transformation of the nucleon into the Λ without changing the wave function loses an amount proportional to the difference ΔV_n between the depths of the Λ and nucleon wells. In C^{12}, (Fig. 3a) the neutron separation energy is 19 MeV, and the Λ binding energy is around 11 MeV for hypernuclei in this mass region.[3] If we insert the experimental value of 10 MeV for the excitation energy of the analog state we obtain 18 MeV for the difference between the effective Λ and nucleon well depths. This is not far from the value given for the well depth by analysis of hypernuclear spectra[2] and can be considered as support for our suggestion that the observed continuum state in C^{12} is indeed a strangeness analog resonance.

The same calculation can be done on a heavy nucleus. Here we are interested in guessing the positions of the analog states, and it is very different from the case of ^{12}C. In heavy nuclei you do not need 18 MeV to remove a

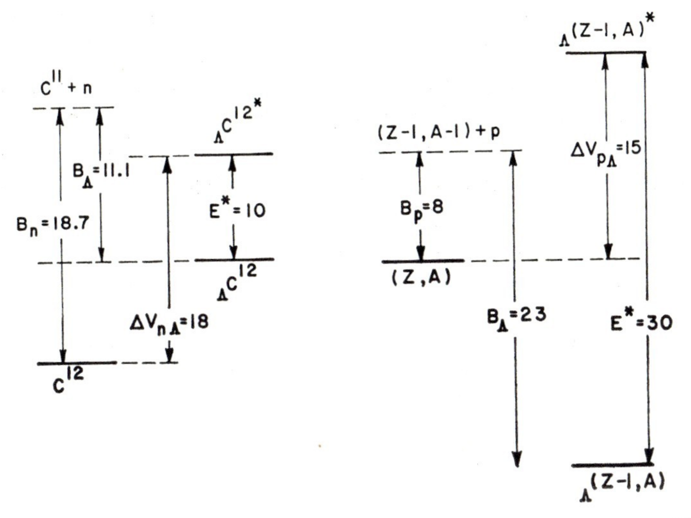

FIGURE 3 Energy level diagram for strangeness analog states a) in C^{12} b) in a heavy nucleus

nucleon but only about 8 and the binding energy of a Λ is much bigger in a heavier nucleus because the Λ still goes down into the bottom *s*-shell. If we put in 15 MeV for the difference in well depth and 23 MeV for the Λ-binding energy, we obtain a excitation energy of about 30 MeV. This is not an unreasonable place to look for collective particle-hole excitations.

In this case I have considered analog states produced by changing a Λ into a proton rather than those produced by changing a Λ into a neutron. This is because of isospin considerations. For ^{12}C there is no problem because ^{12}C has isospin zero and changing either a neutron or a proton into a Λ which has zero isospin makes a nucleon hole state with $T = 1/2$. In a heavy nucleus with a neutron excess and isospin T, removing a proton makes a proton hole state with isospin $T + 1/2$ which is still a good isospin. However removing a neutron makes a neutron hole state which is not an isospin eigenstate but a mixture of $T - 1/2$ and $T + 1/2$. Such a state would not be observed as a single resonance but would be split into two isospin components as shown in Fig. 4. Starting from a parent state $|\pi\rangle$ we make the " V-spin analog" state $|S_p\rangle$ which has a proton changed into a Λ. But the analog state produced by changing a neutron into a Λ is split into two pieces, one denoted by $|S_>\rangle$ having a greater isospin and one denoted by $|S_<\rangle$ having a lower isospin. We estimate the excitation energy of $|S_p\rangle$ by the argument of Fig. 3. This also gives the energy of $|S_>\rangle$, because $|S_>\rangle$ is just the isobaric analog of $|S_p\rangle$ and is shifted by the Coulomb energy. We can calculate the energy of $|S_<\rangle$ because this differs from $|S_>\rangle$ only by the symmetry energy. It is also possible to calculate the isospin admixture in the collective state; i.e., how the excitation strength is divided between the two isospin eigenstates. This is a simple exercise in SU_3 algebra. It can be done by using U and V spins as angular momenta, and using nothing more complicated than ordinary angular momentum algebra. The results show that for nuclei which have an appreciable neutron excess nearly all the strength goes into the lower states.

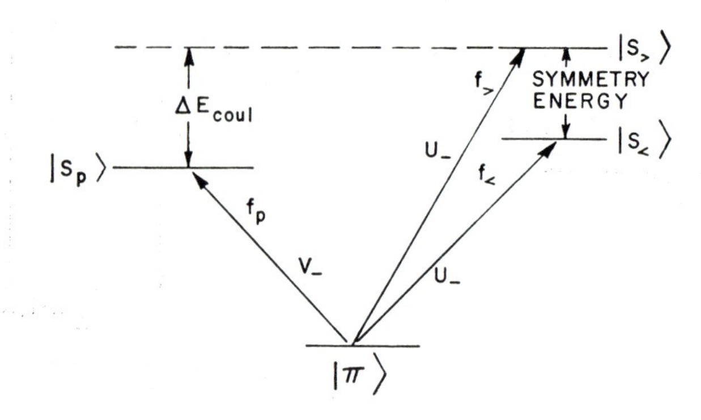

FIGURE 4 Schematic representation of the strangeness analog states and the transitions from the parent

I conclude by pointing out that the study of these continuum excited states of hypernuclei suggest a much more exciting period in hypernuclear physics, even if what I have said about analog states may not turn out to be correct. The ground states of hypernuclei beyond $A = 4$ all have five particles in the s-shell and therefore cannot be compared with ordinary nuclear states which never have this configuration. However, excited hypernuclear states with only four baryons in the s shell can be compared with existing nuclei having identical configurations. Thus one can consider what happens when a nucleon is changed into a Λ in a particular state, keeping the same configuration and study directly the difference between ΛN and NN interactions. In this way it should be possible to learn many new things both about nuclear and hypernuclear structure and about the Λ-nucleon interaction.

REFERENCES

1. A. K. Kerman and H. J. Lipkin, "Strangeness Analog Resonances", *Annals of Physics* **66**, 73B (1971)
2. A. R. Bodmer, *High Energy Physics and Nuclear Structure* (ed. G. Alexander, North Holland, Amsterdam, 1967) p. 60
3. R. H. Dalitz, *Proceedings of International Conference on Hypernuclear Physics* (ed. A. R. Bodmer and L. A. Hyman, Argonne National Laboratory, 1969) p. 708

4. D. H. Davis and J. Sacton, *Proceedings of International Conference on Hypernuclear Physics* (ed. A. R. Bodmer and L. A. Hyman, Argonne National Laboratory, 1969) p. 159
5. H. J. Lipkin, *Phys. Rev. Lett.* **14**, 18 (1965)
6. H. Feshbach and A. Kerman, in *Preludes in Theoretical Physics* (eds. A. de-Shalit, H. Feshbach and L. Van Hove, North Holland, Amsterdam, 1966) p. 260
7. J. D. Anderson, C. Wong, and J. W. McClure, *Phys. Rev.* **126**, 2170 (1962)
8. A. K. Kerman, in *Nuclear Isospin* (eds. J. D. Anderson, S. D. Bloom, J. Cerny and W. W. True, Academic Press, New York and London, 1969) p. 315
9. M. Gell-Mann and Y. Ne'eman, *The Eightfold Way*, W. A. Benjamin New York (1964)
10. H. J. Lipkin, *Lie Groups for Pedestrians* (North Holland, Amsterdam, 1965)
11. F. J. Dyson, *Symmetry Groups*, W. A. Benjamin (1966)
12. Y. Ne'eman, *Algebraic Theory of Particle Physics*, W. A. Benjamin (1967)
13. S. Sakata, *Prog. Theo. Phys.* **16**, 686 (1956)
14. Y. Dothan, M. Gell-Mann, and Y. Ne'eman, *Phys. Letters* **17**, 148 (1965) reprinted in ref. 11

Remarks on Wigner's Supermultiplet Theory on its 34th Anniversary

L. A. RADICATI
Columbia University, New York, N.Y.

1 INTRODUCTION

It is rare that a paper which is thirty-four years old should be discussed at a physics conference: at that age a paper has either been forgotten or has become part of the common knowledge and need no longer be discussed. Wigner's supermultiplet theory [Wigner 1937a] had a different fate. It has certainly not been forgotten, but has often been regarded more as a mathematical *tour de force* than as a realistic approximate description of nuclei.

At the origin of the reluctance to accept the supermultiplet theory there might have been the mathematical difficulties of the language used by Wigner, even though he later gave a simplified version of it [Wigner and Feenberg 1941] which, however, probably due to the war conditions, never became widely known. However, I do not think that mathematical difficulties would alone have prevented people from accepting the theory. The main reason for its slow acceptance was, I think, that it was written too early, perhaps more than twenty years too early. In 1936, when Wigner presented his theory at the Harvard bicentennial conference, many of the things that we now know about nuclear forces, about weak interactions, and about the relations between currents and symmetries were essentially unknown. In some prophetic way they were anticipated by Wigner.

Equipped with all this knowledge, we can today look at his theory in a different perspective. The validity of the isospin group is of course no longer questioned, and what Wigner calls the second approximation [Wigner and Feenberg 1941] now appears to be the natural starting point for every treatment of the nuclear forces. Similarly, the relations between the hadronic vector current coupled to the electromagnetic and lepton fields and the generators of the isospin group is now well established by the experiments which have demonstrated the conservation of the vector current.*

We now also know the form of the weak interactions, and this is precisely the one which best fits the supermultiplet theory. The approximate symmetry between vector and axial-vector currents of the $(V - A)$ interaction leads in a natural way to extending the $U(2)$ symmetry generated by the vector current to a larger group, $SU(4)$, generated by both the vector and the axial-vector current. However, since the symmetry is rather badly broken, we expect that the transition from $U(2)$ to $SU(4)$ corresponds to a much less accurate description of nuclei.

To check the validity of Wigner's predictions we have today many more data than were available in 1936. The spins, parities and isotopic spins of a large number of levels are now measured even at fairly high excitations. Furthermore, we know the rates of many transitions—weak and electromagnetic—between these levels, and we can compare different excitation modes like photo-excitation, μ-capture and electron excitation, so that the theory can be tested also on its dynamical predictions.

2 THE $SU(4)$ ALGEBRA

The physics of subatomic systems is characterized by the appearance of dynamical symmetries—to use the classification introduced by Wigner [1964]—which play no role at the atomic or molecular scale. These symmetries are connected with the interactions which determine the stability and the transitions of subatomic systems: the weak interactions, which cause transitions with charge transfer, and the strong, short range, interactions. The electromagnetic interactions continue, of course, to play their usual role.

Feynman and Gell-Mann [1958] were the first to state explicitly the connection between the vector currents coupled to the non-hadronic fields—the electromagnetic and the lepton fields—and the generators of the isotopic spin group.† If we extend their point of view to include the axial-vector current, Wigner's supermultiplet theory appears, in the static limit, as the natural generalization of isospin invariance.

---

* For a review of the present experimental evidence for CVC, see Moskowski and Wu (1966).

† A precise statement of this assumption can be found in a review article by Lee and Wu [1965]. The consequences of CVC for nuclear physics have been discussed by Wigner [1957] soon after Feynman and Gell-Mann's theory was proposed.

Let $j_\alpha^{em}(x) = j_\alpha^0(x) + j_\alpha^3(x)$ be the hadronic electromagnetic current operator which is a function of $x = (\mathbf{x}, t)$ and contains an isoscalar and an isovector part. The charged hadronic currents coupled to the leptonic field will be denoted by:

$$h_\alpha^\pm(x) = j_\alpha^\pm(x) + g_\alpha^\pm(x) = (j_\alpha^1 \pm ij_\alpha^2) + (g_\alpha^1 \pm ig_\alpha^2), \tag{1}$$

where $j_a^\pm(x)$ and $g_a^\pm(x)$ are the vector and axial vector components respectively. In the definition (1) we have not included the part of h_a which causes transitions with hypercharge change since they are irrelevant for nuclear physics. Therefore the coupling constant which appears in the coupling of $h^\pm$ to the lepton current will contain the cosine of Cabibbo's angle.

According to Feynman and Gell-Mann, the generators T^a $(a = 1, 2, 3)$ and Y of the symmetry group $U(2)$ of the strong interactions are determined by the electromagnetic and charged currents defined above through the relations:

$$T^\alpha = \int d^3x\, j_0^a(x) \tag{2}$$

$$Y = 2 \int d^3x\, j_0^0. \tag{3}$$

T^a are the isotopic spin operators, and Y is the hypercharge. The approximate constancy of T^a and Y is a consequence of the approximate conservation of the currents j_a^0 and j_a^a which is violated only by the electromagnetic and weak interactions.

The axial current plays, in the Feynman-Gell-Mann assumption, a role entirely different from the vector current. The two charged axial currents, $g_\alpha^\pm$, are the components of an isovector operator whose neutral component, g_α^3, is presumably only coupled to itself in non-leptonic, hypercharge-conserving, weak interactions. The axial current is not, however, in this model connected with any symmetry.

Since our discussion is essentially confined to transitions between nuclear states it is legitimate, as a first approximation, to take into account only the contribution to the hadronic current $h_a^\pm$ arising from the free nucleons. In this way we neglect exchange currents, but this is probably a comparatively mild approximation.

Under this approximation all the hadronic currents which we have defined so far are of the form:

$$\psi^+(x)\, \Lambda \psi(x), \tag{4}$$

where $\psi(x)$ is the nucleon field operator and Λ is an appropriate matrix which is the tensor product of an element of the Dirac algebra by one of the $U(2)$

algebra. The space integrals of the currents (4) are operators which depend upon time:

$$Q(\Lambda; t) = \int d^3x\, \psi^+ \Lambda \psi. \tag{5}$$

Their equal-time commutators are

$$[Q(\Lambda; t), Q(\Lambda'; t)] = \int d^3x\, \psi^+ [\Lambda, \Lambda'] \psi = Q([\Lambda, \Lambda']; t). \tag{6}$$

From the hadronic currents which we have introduced so far (with the addition of g_a^3) one constructs 28 operators $Q(\Lambda; t)$. By means of the commutator (6) one gets four more operators, i.e., altogether a set of 32 operators which give at each time t a representation of the algebra of $U(4) \times U(4)$ on the Hilbert space of the hadronic states. The four additional operators needed to complete the basis of the algebra are easily seen to be the integral over space of an axial isoscalar current $g_a^0(x)$.

We do not know any situation in which all the elements of this large algebra could be considered, even approximately, as constants of the motion and therefore as the generators of an approximate symmetry group. However, two subalgebras, besides the $U(2)$ generated by the time components of the vector current, are particularly significant:

(i) The chiral algebra $SU(2) \times SU(2)$ (a subalgebra of the $SU(3) \times SU(3)$ algebra considered by Gell-Mann [1962]) whose elements are the isotopic spin operators T^a and the odd (under space reflection) operators:

$$X^a = \int d^3x\, g_0^a. \tag{7}$$

The relations between the operators X^a and T^a implied by this algebra have led to establishing a number of interesting results, the most important of which is the Adler-Weissberger sum rule [Adler 1965; Weissberger 1965] for the axial vector coupling constant renormalization. However, it is doubtful that the chiral algebra will play an important role in the interpretation of nuclear phenomena, except perhaps those due to pion-capture. Indeed the operators X^a seem to have non-vanishing matrix elements essentially only between states whose energy difference is considerably greater than those encountered in nuclear physics.

(ii) The $U(4)$ algebra generated by the even (under space reflection) operators Y, T^a and by the integrals of the axial current space components

$$Y_k^a = \int d^3x\, g_k^a \tag{8}$$

$$S_k = \int d^3x\, g_k^0. \tag{9}$$

The equal-time commutation relations satisfied by these operators are

$$
\begin{aligned}
[T^a, T^b] &= i\varepsilon^{abc}T^c \\
[S_i, S_j] &= i\varepsilon_{ijk}S_k \\
[T^a, S_i] &= 0 \\
[T^a, Y_i^b] &= i\varepsilon^{abc}Y_i^b \\
[S_i, Y_j^a] &= i\varepsilon_{ijk}Y_k^a \\
[Y_i^a, Y_j^b] &= i(\delta^{ab}\varepsilon_{ijk}S_k + \delta_{ij}\varepsilon^{abc}T^c)
\end{aligned}
\tag{10}
$$

whereas Y commutes with all the other operators.

As long as we are allowed to neglect the contribution to the hadron currents from other fields besides the nucleon field, the equal-time commutation relations (10) are exact independently of whether the operators involved are constants of the motion. It may even be that the commutation relations continue to be valid also if one takes into account the contributions from other fields, as is the case for the $U(2)$ subalgebra which contains Y and T^a.

The connection of the elements of the $SU(4)$ algebra with the observable currents coupled to the non-hadronic fields is perhaps the most significant aspect of Wigner's theory. To quote Dyson [1966] "a group is useful in physics if its generators are physically observable quantities whether or not it is a symmetry group". This connection was realized by Wigner at a very early stage [Wigner 1939] even though the uncertainty of the weak interaction Lagrangian prevented him from stating it in this form.

3 $SU(4)$ AS AN APPROXIMATE NUCLEAR SYMMETRY

In order that the elements of the $SU(4)$ algebra (10) may be considered as the generators of a unitary representation of the $SU(4)$ group on the Hilbert space of hadronic states, we must find a meaningful limit in which they are constants of the motion. There is no problem for the elements of the $U(2)$ algebra generated by Y and T^a which are constants of the motion if one neglects the weak and electromagnetic interactions.

The constancy of the other elements which arise from the axial currents is not dependent upon the vanishing of the axial vector divergence as would be the case for the odd parity generators of the chiral $SU(2) \times SU(2)$ group. The operators Y_k^a and S_k are time independent for nucleons at rest and we thus expect that the frequency of their time evolution for a nucleus will be of the order of a few MeV.

To check the validity of the $SU(4)$ classification scheme which follows from assuming that all the elements of the algebra are constants of the motion, one must first of all verify if the separation between states belonging to different supermultiplets (the representation spaces of $SU(4)$ on Hilbert space) is greater than the separation between states within the same supermultiplet. As a second test, one must see if the transitions which would be forbidden if $SU(4)$ were an exact symmetry are significantly slower than the allowed ones.

(i) The separation between supermultiplets. The evidence for the validity of the first condition comes mainly from the Wigner mass formula [Wigner 1937b, Wigner and Feenberg 1941] which is now supported by many more data than at the time when it was first proposed [Franzini and Radicati 1963; Bourdet, Maguin and Partensky 1968]. The mass formula states that the excitation energy of a supermultiplet in a nucleus of mass number A is a linear function of the eigenvalues of the second order Casimir operator C_2 of $SU(4)$. To compare the predictions of the mass formula with the empirical data one must assign to each state the three quantum numbers $[PP'P'']$ on which C_2 depends quadratically. According to Wigner's assumption we can set for the ground state $P = |T_3|$ where T_3 is the eigenvalue of the third component of the isospin and choose for P' and P'' the lowest values compatible with the value of P. In this way C_2 becomes a function of only T_3 and the quantity

$$R(T_3) = \frac{E(A_1C_2(T_3)) - E(A_1C_2(T_3 - 2))}{E(A_1C_2(T_3 - 1)) - E(A_1C_2(T_3 - 2))}$$

is independent of A. Here $E(A_1C_2(T_3))$ is the energy of the level of mass number A belonging to the supermultiplet $[P, P', P'']$. The empirical data analyzed by Franzini and Radicati [1963] show that $R(T_3)$ remains indeed constant even for fairly large values of A (see Fig. 1). If one takes into account the Casimir operators of higher degree, one can write down an improved mass formula which is in agreement with the experimental data even for very heavy nuclei. We can thus conclude that the $SU(4)$ invariant forces must play a significant role in determining the average stability of nuclear systems. This of course does not mean that each nuclear state will appear at exactly the energy predicted by the mass formula. The $SU(4)$-violating forces which remove the degeneracy of each supermultiplet necessarily complicate the picture. On the average, however, the position and the order of the supermultiplets appear to be those predicted by Wigner.

(ii) Allowed β-transitions. We now turn to the second problem, namely the rates of the transitions which are forbidden according to $SU(4)$. From the definitions (2) and (8), it follows that all allowed Fermi and Gamow-Teller transitions must occur only between states belonging to the same super-

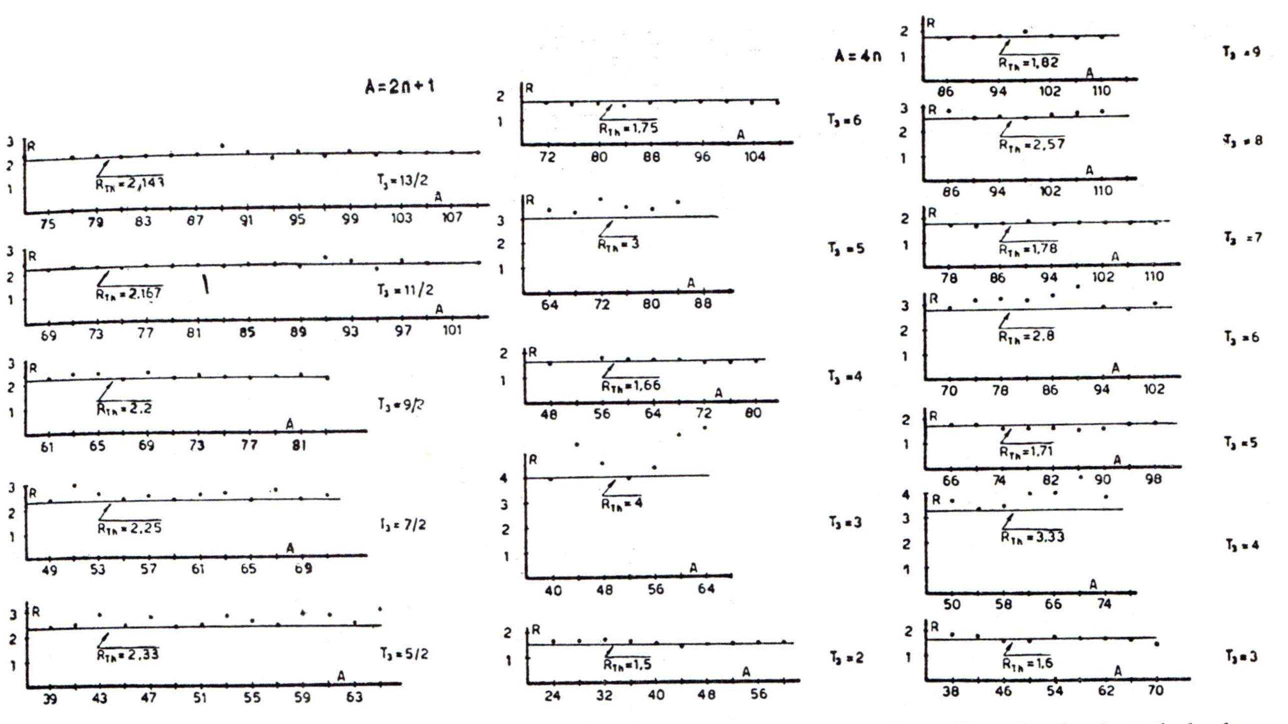

FIGURE 1 The ratio $R(T_3)$ for nuclei with (a) odd-A, (b) $A = 4n$ and (c) $A = 4n + 2$. The lines give the theoretical values computed according to Wigner's prescriptions

multiplet [Wigner 1939]. Of course, since nuclear levels are not pure $SU(4)$ states, we expect also allowed transitions between different supermultiplets. However, their strength should be reduced as they can go only through the $SU(4)$ mixing which, if the theory has any validity, should not be too large.

To compare this prediction with the empirical evidence, we must be able to assign to each state its $SU(4)$ quantum numbers $[PP'P'']$. This is often possible on the basis of the isotopic spin assignment if one assumes the order of excitation of different supermultiplets given by Wigner [1937b]. However, Wigner's rule only determines the $SU(4)$ quantum numbers of the ground state and little is known of the excited states.

For the isobars $A = 4n$ and $A = 4n \pm 1$ the excitation energy of the first excited supermultiplet is determined, according to Wigner's rule, by the appearance of the first level with an isospin higher than the ground state. This of course is not necessarily always true, although it is presumably a reasonable guideline. One may thus tentatively assume that in the $A = 4n$ isobars, all β-transitions which lead to a level below the first $T = 1$ level are transitions between two different supermultiplets. The transitions within the same supermultiplet are those occurring between the two components $T_3 = 1$ and $T_3 = 0$ of the $T = 1$ level.

Five β^+ transitions (see Table 1) with $\log ft \leq 3.44$ between members of the first excited supermultiplet are now accurately measured for $A = 4n \leq 40$ [Armini, Sunier, Richardson 1968]. The theoretical interpretation of the ft values in terms of $SU(4)$ has been discussed by Bosterli and Feenberg [1955] who obtain for the square of the Gamow-Teller matrix element for ^{24}Al the value

$$|M_{GT}|^2 \leqq 1/10.$$

This gives $3.44 \leqq \log ft \leqq 3.49$, which is compatible with the experimental value.

It is interesting to remark that the transition to the level at 8.43, 3^+ in ^{24}Mg has an ft considerably smaller than the other allowed transitions in this

TABLE I Allowed β^+ transitions between states of [110] supermultiplet

INITIAL		FINAL			Log ft.
Nucleus	J^π	Nucleus	Energy	J^π	
^{8}B	2^+	^{8}Be	16.67	2^+	2.9
^{24}Al	4^+	^{24}Mg	9.516	4^+	3.44
^{28}P	3^+	^{28}Si	9.32	3^+	3.35
^{32}Cl	1^+	^{32}S	6.99	1^+	3.43
^{40}Sc	4^-	^{40}Ca	7.67	4^-	3.34

nucleus ($\log ft = 4$) (see Table II). It is not unreasonable to speculate that this state too might belong to the first excited multiplet (it could be the level $T = 0$, $S = 1$) or at least contain a large fraction of it. A similar interpretation may hold for the transition to the 8.57, 3^+ level in $^{28}S_i$.

All the other allowed transitions in the $A = 4n$ isobars occur between the first excited and the ground supermultiplet. This is consistent with their *ft*-values, which are collected in Table II. Apart from the two transitions already mentioned in the decays of ^{24}Al and ^{28}P, low *ft* values appear only in the $A = 12$ isobars. This may suggest a higher degree of impurity in this system than in the other cases.

In the odd isobars we have a situation which is in principle similar to the one discussed for the $A = 4n$ isobars. Here, too, the energy of the first excited supermultiplet can be inferred from the position of the first level whose iso-

TABLE II Allowed β-transition between different supermultiplets in $A = 4n$ nuclei

INITIAL		FINAL			Log ft.
Nucleus	J^π	Nucleus	Energy	J^π	
^{8}Li	2^+	^{8}Be	2.90	2^+	5.6
^{12}B	1^+	^{12}C	0	0^+	4.1
			4.43	2^+	5.1
			7.66	0^+	4.2
^{12}N	1^+	^{12}C	0	0^+	4.1
			4.43	2^+	5.2
			7.66	0^+	4.4
^{16}N	2^-	^{16}O	6.13	3^-	4.4
			7.11	1^-	5.1
^{20}F	2^+	^{20}Ne	1.63	2^+	5.0
^{24}Al	4^+	^{24}Mg	4.12	4^+	5.98
			6.01	4^+	5.98
			8.43	3^+	4.00
^{28}P	3^+	^{28}Si	1.78	2^+	4.95
			4.62	4^+	5.30
			6.28	3^+	4.44
			8.57	3^+	4.13
^{32}Cl	1^+	^{32}S	0	0^+	6.7
			2.23	2^+	4.61
			4.28	2^+	5.3
			4.69	1^+	4.85
			5.55	2^+	4.31
^{40}Sc	4^-	^{40}Ca	3.74	3^-	4.75
			4.49	5^-	4.71
			5.62	4^-	4.40

spin is one unit higher than the ground state isospin. One finds that all transitions between the two members of the $T = 1/2$ ground level have small $ft(\log ft \leqq 3.8)$.

There are several transitions which *could* be between members of the lower supermultiplet because both the initial and the final state are below the first excited supermultiplet. Some of them have a small ft but some have a $\log ft$ as high as five.

Most of the transitions of the type

$$(A, Z = \tfrac{1}{2} A - \tfrac{3}{2}) \rightarrow (A, Z = \tfrac{1}{2} A - \tfrac{1}{2}) + e^- + \bar{\nu}$$

occur between different supermultiplets. All have $\log ft$ 4.5 except for two transitions in $A = 13$, one in $A = 15$ and one in $A = 17$.

The situation for the $A = 4n + 2$ isobars is more complicated. First it is not easy to identify the position of the first excited supermultiplet since its highest T is 1 as in the case of the lowest one. On the basis of the Wigner mass formula, one finds that the excitation energy is considerably lower than the energy of the first excited supermultiplet of the two adjacent $A = 4n$ isobars. We can thus expect a larger mixing.

Nevertheless, up to $A = 26$ we encounter large values of the GT-matrix element only for $A = 14$ and $A = 22$ (for a discussion of the transitions in $A = 14$ see Rose, Hanesser and Warburton 1968). The corresponding transitions between different supermultiplets have, as before, $\log ft \geqq 4.4$. Beyond $A = 26$, the $G - T$ transitions become rather slow but in this region the mixing is expected to be already rather large.

Table III summarizes the preceding discussions. The transitions for the $A = 4n$ isobars are those of Table I and Table II. For the odd isobars, the transitions in columns 2 and 3 are between states with the same T. For the $A = 4n + 2$ isobars, we have excluded from the average the 'pathological' transitions in $A = 14, 22$. From these data it appears that on the average the squares of the matrix elements of the "forbidden" (in the sense of the supermultiplet model) transitions are reduced by about 1/100 compared to the "allowed" ones. This corresponds to an approximately 10% mixing between different supermultiplets. One should remark that the $SU(4)$ symmetry is already broken in the simplest supermultiplet, i.e. the nucleon. Indeed from the commutation relations (10) we deduce

$$[Y_k^+, Y_k^-] = 2T^3. \tag{11}$$

By taking the matrix element of (11) on a proton state, we would deduce $g_A = G_V/G_A = \pm 1$ instead of the observed value $g_A = -1.20$. (G_V and G_A are the vector and axial vector coupling constants.)

TABLE III Average ft for Allowed Transitions

	Within Supermultiplet		Outside Supermultiplet	
	Number of Transitions	$\langle ft\rangle/10^5$	Number of Transitions	$\langle ft\rangle/10^5$
$A = 4n \leqq 40$	5	0.02	24	1.2
$A = 4n \pm 1 \leqq 43$	21	0.04	29	4.26
$A = 4n + 2 \leqq 26$	10	0.02	4	8.5

To apply Eq. (11) to complex nuclei, one should therefore renormalize the single particle matrix elements of Y_k^a to take into account the fact that each nucleon is not a pure $SU(4)$ state. This can be done by putting:

$$\left\langle f \middle| \int \mathrm{d}^3x\, g_k^a \middle| i \right\rangle = g_A \langle f_0 \,|Y_k^a|\, i_0\rangle \tag{12}$$

where $f\rangle$ and $i\rangle$ are the final and initial nuclear states made up of physical nucleons and $f_0\rangle$ and $i_0\rangle$ are the states made up of (unphysical) nucleons which belong to the $[\frac{1}{2}\,\frac{1}{2}\,\frac{1}{2}]$ supermultiplet.

The sum rule which follows from Eq. (11) can be tested, after applying the correction (12), in other cases beside the nucleons. Preliminary results [Phan Tri Nang and Rosa-Clot 1970] are in reasonable agreement with the predictions which follow from saturating the sum rule with states which belong to the same supermultiplet.

It seems thus fair to conclude that:

(i) The position of the supermultiplets is in good agreement with the prediction of Wigner's mass formula.

(ii) The amplitude for super-allowed transitions is given by the matrix elements of the space integrals of the weak currents, i.e. by the generators of $SU(4)$. The group $U(4)$ can therefore be considered as an approximate symmetry for nuclear systems.

4 μ-CAPTURE

μ-capture was first discussed from the point of view of $SU(4)$ by Foldy and Walecka [1964]. It has recently been re-examined by Krüger and van Leuven [1969] and by Cannata, Leonardi and Rosa-Clot [1970]. The interest of μ-capture for our subject lies in the fact that, at least in some cases, it gives a way to study excited supermultiplets or, more generally, to study transitions with larger momentum transfer than the allowed transitions investigated by Wigner.

The simplest interpretation of the definitions (2) and (8) is obtained by assuming that $j_0^a(x)$ and $g_k^a(x)$ are components of a tensor operator which transforms like the adjoint representation of $SU(4)$. From this assumption one can derive a set of relations between the matrix elements of $j_0^a(x)$ and $g_k^a(x)$, i.e. a set of relations between transitions induced by the moments of the vector and axial vector currents.

Foldy and Walecka have investigated the consequences of this assumption for nuclei with $A = 4n$ and $T = 0$. Their ground state is invariant under $SU(4)$ transformations, i.e., in Wigner's notations, it belongs to the [000] representation. After capturing a μ-meson the nucleus will be left in a state which belongs to the 15-dimensional adjoint representation $[110] = (1, 0) \oplus (0, 1) \oplus (1, 1)$ ((T, S) are the representations of $SU(2) \times SU(2)$).

If $SU(4)$ is valid, the transition cannot be a monopole transition since the generators of $SU(4)$ annihilate [000]. It will thus predominantly be a dipole transition leading to negative parity states. The average excitation in μ-capture on these nuclei ($4 \leqq A \leqq 40$) is about 20–25 MeV, which is of the same order as the peak energy of the giant $E1$ absorption.

From the point of view of $SU(4)$, this is not a coincidence. Indeed, the amplitude for $E1$ absorption on a $T = 0$ nucleus is proportional to the matrix element of the operator ($u = 1, 2, 3$)

$$D_u^3 = \int \mathrm{d}^3x \, x_u j_0^3. \tag{13}$$

Similarly, the first forbidden contribution of the vector current to μ-absorption is given by the matrix element of

$$D_u^- = \int \mathrm{d}^3x \, x_u (j_0^1 - i j_0^2). \tag{14}$$

Finally, the first forbidden contribution of the axial current is obtained from:

$$G_{ku}^- = \int \mathrm{d}^3x \, x_u (g_k^1 - i g_k^2). \tag{15}$$

The operators D_u^a, G_{ku}^a and G_{ku}^0 constructed as in (15) from the isoscalar axial current, are the 15 independent components of an $SU(4)$ tensor operator and have therefore the same reduced matrix elements. In particular, we can get the relation between the electric dipole matrix element and the operators which appear in μ-capture with the use of the commutators:

$$[T^-, D_u^3] = D_u^- \tag{16}$$

$$[Y_k^-, D_u^3] = G_{ku}^-. \tag{17}$$

A picture of the excitation of the [110] supermultiplet is illustrated in Fig. 2 (not drawn to scale) where the states are supposed to be split by $SU(4)$ breaking forces. The ordering of the levels has been taken to be the same as the one calculated by De Shalit and Walecka [1965] for the case of He^4.

A further splitting of the levels (illustrated in Fig. 3) is introduced by spin-orbit forces. The three $L = 1$ levels with $J = 0^-, 1^-, 2^-$ are excited respectively by the three components of G^-_{ku}:

$$G^-_{ku} = \delta_{ku} P^- + \varepsilon_{kut} M^-_t + Q^-_{ku} \tag{18}$$

where

$$\begin{aligned} P^- &= \tfrac{1}{3} \int d^3x\, x_t g^-_t \\ M^-_t &= \tfrac{1}{2} \varepsilon_{trs} \int d^3x (x_r g^-_s - x_s g^-_r) \\ Q^-_{ku} &= \tfrac{1}{2} \int d^3x [x_u g^-_k + x_k g^-_u - \tfrac{2}{3} \delta_{ku}(x_s g^-_s)]. \end{aligned} \tag{19}$$

D^-_u will of course only excite the $J = 1^-$ level.

Let us write the μ-capture rate in the usual way (neglecting retardation effects and using natural units, $\hbar = c = 1$)

$$\Gamma = \frac{m_\mu^2}{2\pi^2} R(Z\alpha m_\mu)^3 G^2 |\mathcal{M}|^2 \tag{20}$$

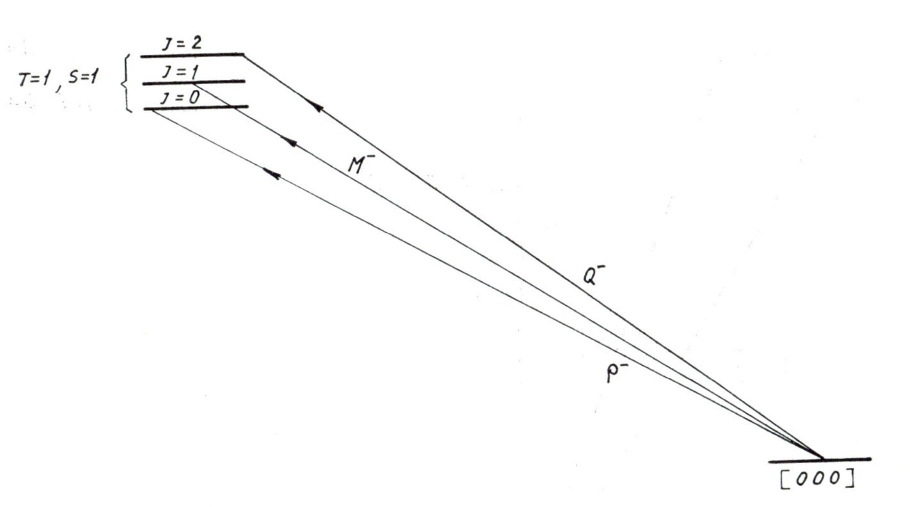

FIGURE 2 μ-capture and γ $E1$ excitation in $A = 4n$ nuclei. Spin orbit splitting not included

where m_μ is the μ-mexon mass, α the fine structure constant, Z the atomic number, G the Fermi constant, R a numerical factor of order one, and

$$|\mathcal{M}|^2 = g_V^2 M_V^2 + 3g_A^2 M_A^2 + (g_P^2 - 2g_P g_A) M_P^2. \tag{21}$$

In this expression the renormalization effect discussed at the end of the last section has been taken into account by separating out the breaking of $SU(4)$ on the single nucleon. The coupling constants g_V, g_A and g_P which are momentum-dependent have for μ-capture the approximate values:

$$g_V \simeq 1.01, \quad g_A \simeq -1.55, \quad g_P \simeq -0.58.$$

In the $SU(4)$ limit the three matrix elements M_V, M_A and M_P are equal. The part arising from the vector current can be directly obtained from the photoabsorption cross-section $\sigma_\gamma(E)$

$$M_V^2 = \frac{m_\mu^2}{2\pi^2\alpha}\left(\frac{E_m}{m_\mu}\right)^4 \int_0^{E_m} \mathrm{d}E \, \frac{(E_m - E)^4}{E_m^4} \, \frac{\sigma_\gamma(E)}{E} |F|^2 \tag{22}$$

so that Γ is a function of E_m which is of the order of 100 MeV, of $\sigma_\gamma(E)$ and of the elastic electromagnetic form factor F.

In this idealized picture, the μ-capture rate can be directly deduced from $E1$ photoabsorption. Actually, the breaking of $SU(4)$ introduces a difference between the three matrix elements which appear in Eq. (21). Cannata,

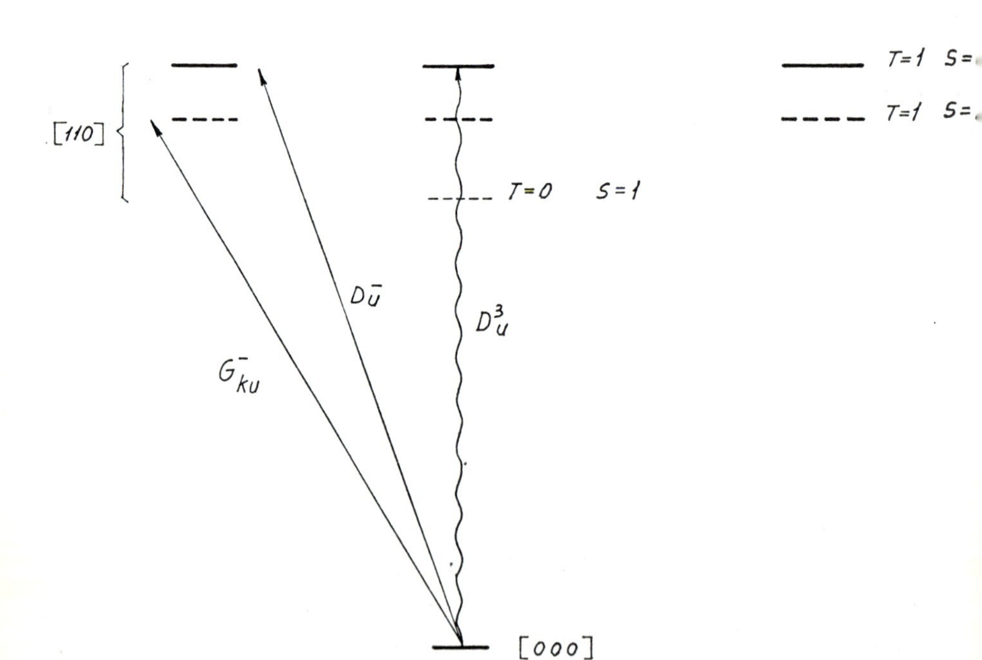

FIGURE 3 μ-capture in $A = 4n$ nuclei. The $T = 1$, $S = 1$ level is split by spin orbit forces

Leonardi and Rosa-Clot [1970] have estimated the effect of the $SU(4)$ breaking and have shown that:

$$\frac{M_A}{M_V} \sim 1 - 2\frac{\Delta\omega}{E_m - \omega_F} \tag{23}$$

where $\Delta\omega$ is the separation between the two $T = 1$ levels with $S = 1$ and $S = 0$, and ω_F is the peak energy for photoabsorption.

In Table IV the experimental capture rates (column 2) are compared with Foldy and Walecka's calculations (column 3) and with those of Cannata, Leonardi and Rosa-Clot (column 4). There seems to be little doubt that the relation implied by $SU(4)$ between the axial and the vector current is borne out by these data which, however, also show the need to take into account $SU(4)$ breaking if one requires an accuracy greater than 15–20%.

TABLE IV μ-Capture Rates in $A = 4n$ Nuclei

Nucleus	Experimental $10^5\ s^{-1}$	Calculated		Energy Separation MeV	$\|M_A\|^2/\|M_V\|^2$
		Without Level Splitting	With Level Splitting		
^{40}Ca	25.5 ± 0.5	28.3	25.2	7	0.73
^{16}O	0.97 ± 0.03	0.95	0.98	3	0.86
^{12}C	0.30 ± 0.08	0.26	0.28	2	0.91
^{4}He	336 ± 76	249	300	0.9	1

μ-capture can, of course, be studied also on nuclei whose ground state is not an $SU(4)$ scalar. In this case, transitions between states within the same supermultiplet are not forbidden by $SU(4)$ and have indeed been observed. For example, capture on ^6Li and ^{11}B has been observed [Deutsch *et al.* 1968a, b] to lead to the ground state of ^6He and to the first excited state of ^{11}Be. Here the capture rate can be related to the electric form factor of ^6He or ^{11}B by using the commutation relation:

$$[Y_k^+, g_k^-] = 2j_0^3. \tag{24}$$

The matrix element of (24) between states of ^6He with momenta p and p' yields the relation:

$$\sum_q \langle {}^6He(p')\,|g_{-q}^-|\,{}^6Li(p);\, q\rangle = -\frac{6}{\sqrt{2}}\langle {}^6He(p')\,|j_0^3|\,He(p)\rangle \tag{25}$$

where $q = -1, 0, 1$ are the values of the third component of the ^{6}Li angular momentum. If, instead of (24), one considers the commutator

$$[Y^+_+, g^-_-] = 4(j^3_0 + g^0_3) \tag{24}$$

one can derive a relation between the amplitude for μ-capture on ^{6}Li leading to ^{6}He and the matrix element of g^0_3 on the ^{6}Li ground state. In the non-relativistic limit the latter is measured by the magnetic form factor of ^{6}Li. It is thus possible to express the μ-capture rate in terms of this factor [Krüger and van Leuven 1969]. Several rates for capture into well-defined low lying states have how been accurately measured in Louvain [Deutsch *et al.* 1968a, 1968b]. It will therefore soon be possible to test the predictions of the supermultiplet theory for values of the momentum transfer considerably higher than those accessible in β-decay. Although departures from $SU(4)$ symmetry are here to be expected, the agreement which obtains for the $A = 4n$ nuclei seems to indicate that the violation will not be larger than 10–15%.

5 CONCLUSION

The aim of this talk was not to present a complete analysis of the experimental evidence in support of Wigner's supermultiplet theory. In particular I have not discussed the problem of the magnetic moments whose relevance to the $SU(4)$ model was realized by Wigner from the very beginning [Margenau and Wigner 1939]. Recently the connection between the magnetic moments and the Gamow-Teller matrix elements has been reanalyzed by Leonardi and Rosa-Clot [Leonardi and Rosa-Clot 1969, 1970] who have shown that, under plausible assumptions, the expectation values of the operator Y^a_k and S_k are in close agreement.

More important is the omission of a justification for the validity of the $SU(4)$ symmetry in nuclei. The reason for leaving out this topic is that I do not know any better justification than those suggested long ago by Wigner [Wigner 1937b; Wigner and Feenberg 1941]. To make Wigner's arguments more precise and quantitative, though probably not impossible today with our better knowledge of nuclear forces, would require a considerable calculational effort. I will thus limit myself to quote once more Dyson [1966] according to whom

> The Heisenberg force does not mix supermultiplets much because (i) the force has roughly the same space dependence as the Majorana force and (ii) it is effectively proportional to T^2 which is a function of the generators of $SU(4)$!

This, of course, is not a complete explanation and does not provide us with a justification for not undertaking detailed calculations.

The main point that I wanted to make in this talk was to illustrate the deep connection which in Wigner's theory exists between the symmetry of the strong interactions and the form of the hadron-lepton coupling. This connection could not be fully understood at the time of Wigner's writing and his theory was thus essentially an approximate description of nuclear systems. Today the symmetry between spin and isotopic spin inherent in the supermultiplet theory appears as a consequence of the almost symmetrical role played, in the static limit, by the vector and axial vector currents. Of course, the two currents are not entirely equivalent and we thus cannot expect to deduce from their analogy an exact symmetry for nuclei. However the validity of the mass formula, the ft values for the $SU(4)$-forbidden transition and the good agreement obtained in the prediction of the μ-capture rate, are, in my opinion, a convincing proof that $SU(4)$ is on its 34th anniversary even more relevant to the description of nuclei and their transition than it was at its birth.

REFERENCES

Adler, S. L. (1965) *Phys. Rev. Letters* **14**, 1051
Armini, A. J., Sunier, J. W., and Richardson, J. R. (1968) *Phys. Rev.* **165**, 1194
Bosterli, M. and Feenberg, E. (1955) *Phys. Rev.* **97**, 736
Bourdet, G., Maguin, C., and Partensky, A. (1968) *N. Cimento* **54 B**, 1
Cannata, F., Leonardi, R., and Rosa-Clot, M. (1970) to be published
DeShalit, A. and Walecka, J. D. (1965) *Phys. Rev.* **147**, 763
Deutsch, J. P., Grenachs, L., Igo-Kemenes, P., Lipnick, P., and Macq, P. C. (1968a) *Phys. Letters* **26 B**, 315
Deutsch, J. P., Grenachs, L., Lehman, J., Lipnick, P., and Macq, P. C. (1968b) *Phys. Letters* **28 B**, 178
Dyson, F. J. (1966) *Symmetry Groups in Nuclear and Particle Physics*, Benjamin, New York
Feynman, R. P. and Gell-Mann, M. (1958) *Phys. Rev.* **109**, 193
Foldy, L. L. and Walecka, J. D. (1964) *N. Cimento* **34**, 1026
Franzini, P. and Radicati, L. A. (1963) *Phys. Rev. Letters* **6**, 322
Gell-Mann, M. (1962) *Phys. Rev.* **125**, 1067
Krüger, F. and van Leuven, P. (1969) *Phys. Letters* **28 B**, 623
Lee, T. D. and Wu, C. S. (1965) *Ann. Rev. Nucl. Science* **15**, 381
Leonardi, R. and Rosa-Clot, M. (1969) *N. Cim. Lett.* **1**, 829
Leonardi, R. and Rosa-Clot, M. (1970) *Phys. Rev. Lett.* **24**, 407
Margenau, H. and Wigner, E. P. (1939) *Phys. Rev.* **58**, 103
Moszkowski, A. A. and Wu, C. S. (1966) *Beta Decay*, Interscience Publishers, New York.
Pham Tri Nang and Rosa-Clot, M. (1970) unpublished
Rose, H. J., Hanesser, O., and Warburton, E. K. (1968) *Rev. Mod. Phys.* **40**, 591
Weisberger, W. I. (1965) *Phys. Rev. Lett.* **14**, 1047
Wigner, E. P. (1937a) *Phys. Rev.* **51**, 106

Wigner, E. P. (1937b) *Phys. Rev.* **51**, 447
Wigner, E. P. and Feenberg, E. (1941) *Rep. Prog. in Physics*, Phys. Soc., London **8**, 274
Wigner, E. P. (1939) *Phys. Rev.* **56**, 519
Wigner, E. P. (1957) *Proc. of the Robert A. Welch Foundation Conference on Chemical Research. I. The Nucleus*, p. 67
Wigner, E. P. (1964) *Proc. Nat. Acad. Sci.* (*U.S.A.*) **51**, 965

DISCUSSION

B. R. Mottelson I would like to comment first on the evidence that has been adduced for the validity of the U_4 symmetry. It seems difficult to conclude from the approximate validity of the U_4 mass formula to the applicability of the symmetry classification since this mass formula can also follow from much more general models in which the U_4 symmetry is not at all obeyed. Similarly the evidence on β-decay selection rules that has been refered to by Dr. Radicati, while consistent with U_4 symmetry in some cases, leaves considerable uncertainly concerning the quantitative validity of this symmetry; this feature is illustrated by the simple model that I discussed in my talk earlier this week.

My second comment concerns the multiplet of states produced in μ-capture as discussed by Dr. Radicati. It seems to me that the the assumption that the main capture will go to single states (for each LSJ), goes considerably beyond the original assumption of U_4 symmetry. Indeed such collective states can only be generated if there are appropriate interactions present to organize the many different one-particle configuration into a collective mode of corresponding symmetry.

L. Radicati I do not claim that the approximate validity of the mass formula (which on the average is remarkably good) is a proof of the validity of the $SU(4)$ symmetry. It is however consistent with $SU(4)$ and it is probably the easiest way to explain why the $T = 1$, $T = 2$ ... states appear just where they do. The same is true for the values of the β-decay matrix elements.

As for the second point raised by Dr. Mottelson it seems to me that the fact that the sum rule is approximately exhausted by single states is a good evidence of the validity of the $SU(4)$ algebra generated by the integrals of the currents.

E. P. Wigner I would like to say a few words about the subject of our discussion but before doing so let me admit that much of what Dr. Radicati said was new to me and that I learned a great deal from his paper.

As to the validity of the supermultiplet symmetry, this seems to me a rather complex question. If we understand it in its initial form, it includes the postulate that the orbital angular momentum L is a good quantum number and we know, from the success of the usual shell model, that this is only rarely the case. On the other hand, if we consider it as a permutational symmetry of the space-coordinates of the nucleons, (and hence also of their spin and isospin coordinates), there is reason to believe that it has, at least for the low lying states, considerable validity. Dr. Elliott mentioned a few cases in which the total wave function consists, to an extent of about 90 per cent and even more, of the proper permutational symmetry, and Dr. Radicati brought out other pieces of evidence to render it probable that this is a rather general situation.

In view of the strength of the spin-orbit and tensor forces, this may appear surprising. The situation may be, however, as follows. If we start with a Hamiltonian in which the spin-orbit and tensor forces are omitted, the states with different permutational symmetry are rather far from each other or, at least, the states with permutational symmetries different from that of the ground state, are rather far therefrom. Therefore, the introduction of the spin-orbit and tensor forces will, at least in the first two approximations, cause only little admixture of the higher supermultiplets to the wave function of the normal state.

This does not apply to states with the same permutational symmetry but different L. These may be quite close and a mixing of two states such as $^2P_{3/2}$ and $^2D_{3/2}$, for instance, is not unlikely. It is also clear that, at higher excitations, the mixing will be quite strong—we know, for instance from the experiments of Tollefsrud and Jolivetti which I just happened to read, that at excitations of 15 MeV or so even the isospin impurities become appreciable.

There is, however, a point at which the conclusion arrived at by Dr. Radicati may be correct but the argument is, in my opinion, not conclusive. I am referring to the argument for the validity of the supermultiplet symmetry, arrived at from β decay data. As Dr. Radicati pointed out, according to the supermultiplet theory, all s transitions are forbidden, except those between members of the same supermultiplet. And, in fact, they are, almost without exception, about a hundred times slower than would correspond to a s transition within a supermultiplet. Incidentally, this is in conflict with the $j - j$ coupling shell model and I wish this were discussed a bit more in detail.

To return to our subject, however, it should be admited that the β decaying nucleus is, almost without exception, in its normal state and its supermultiplet quantum numbers $(PP'P'')$ are something like $P = T_3$, $P' = 0$ or 1, $P'' = 0$ in case of even mass number. Let me specialize on this case and, for the sake of concreteness, assume $P' = 1$. The normal state of the nucleus to which it decays is then either in the $P = T_3 - 1$, $P' = P'' = 0$ state (in case of electron emission) or the $P = T_3 + 1$, $P' = P'' = 0$ state (in case of positron emission). Let me consider the former case.

The small magnitude of the β decay matrix element then means that the normal state of the daughter nucleus has little admixture of the $P = T_3$, $P' = 1$, $P'' = 0$ supermultiplets. It *may* have, however, a significant admixture of $P = T_3 - 1$, $P' = P'' = 1$ multiplet, for instance. It is unreasonable to assume this if the energy of the latter multiplet is just about as high as that of the $P = T_3$, $P' = 1$, P'' multiplet and this is the care, approximately, for low T_3, i.e. for very light nuclei. If we talk, however, let us say, $\varphi T_3 = 4$ for the parent nucleus, the usual formula, for the energy, so nicely verified by Dr. Radicati, gives

$$(P + 2)^2 + (P' + 1)^2 + P'' = 5^2 + 1^2 + 0^2 = 26$$

for the normal state, $6^2 + 2^2 + 0^2 = 40$ for the supermultiplet of the parent which, according to the β decay data, in only little admixed to the normal state, but only $5^2 + 2^2 + 1^2 = 30$ for the supermultiplet which may be admixed to the normal state without this showing up in the β decay matrix element. The energy separation of this is $(30 - 26)/(40 - 26) = 4/14$ times lower than for the supermultiplet which does not show up.

The situation is quite similar in the other cases, such as positron emission, or odd A, etc. It is not impossible, therefore, that of the three supermultiplet quantum numbers P, P', P'', only the first has the validity proved by Dr. Radicati. At least this may be true if P is not too small, if it is, let us say, 3 or more. For low P, on the other hand, his argument is, it seems to me, valid. Of course, as all other non rigorous quantum numbers, P, P', P'' lose validity at higher excitations—this was discussed before and is a general rule.

L. Radicati Yes, one case I did not mention is the $A = 4n + 2$ nuclei. I was careful not to mention it. But since you raise the point of these hidden supermultiplets, there is a supermultiplet that one really cannot identify because it has the same T and S as the ground state supermultiplet and my impression is that: it is for example lower than the corresponding first excited supermultiplet in the two adjacent $A = 4n$ nuclei. In this case, I think there is evidence that the mixing is not so low. Let me state explicitely that *I* am not claiming that $SU(4)$ is only slightly broken. It seems to me however that there are still traces of the symmetry in the sense that for example the first excited $T = 2$ state is always found in the region where the supermultiplet mass formula tells you to do.

I have also tried to stress the connection between the supermultiplet theory and the electromagnetic and weak currents. Wigner, of course, came to supemultiplets in an entirely different way. He never mentioned currents because there were no currents in those days. But *I* think that today, that we know the existence of vector and axial vector currents $SU(4)$ appears to us as the natural symmetry of non relativistic nucleon physics.

C. E. Brown The point about permutation symmetry raised by Prof. Wigner seems to me to be the major one. Supermultiplets with lower spatial symmetry have fewer relative even-state interactions. The nucleon-nucleon force is known to be strongly attractive in the relative s-state; small, or repulsive in the relative p-states. Thus, the supermultiplets with high spatial symmetry come much lower in energy, and a gap develops.

With respect to the work of Foldy and Walecka, I would remark that these states were known quite well in O^{16} for some time before their considerations, and that they and the associated states obtained by rotating in isospin, formed the "15" in $SU(4)$. It was not felt, however, that this point of view was very helpful in getting into the real physics of the situation as, for example, the reason why only part of the dipole sum rule resides in the dipole state in O^{16}, etc.

L. Radicati The main point in the Foldy and Walecka calculation is not in to relate the states of N^{16} to those of O^{16} by an isospin rotation as in relating the matrix elements of the moments of the vector and axial vector currents. This relation depends upon the validity of the $SU(4)$ algebra.

R. Brout Professor Radicati, would you please indicate the degree of agreement between calculations of magnetic moment and Gamow Teller matrix elements in the SU_4 limit in favorable cases.

L. Radicati The magnetic moment operator contains a part which is related by an $SU(4)$ transformation to the operator Y and a part which is independent of it. The latter has to be calculated with some model. Therefore the comparison between the theoretical values and the experimental one is not a model independent proof of $SU(4)$.

P. Macq We think a dynamical evidence for supermultiplet theory predictions is given by Van Leuven calculations; V.L. links the μ-capture rate inside the ^{6}Li ground state supermultiplet with the elastic scattering results. With the assumption $G_P/G_A = 8$, V.L. found Γ-capture $= 1500\ \text{sec}^{-1}$ to be compared to the experimental value $1600 \begin{matrix} +\ 330 \\ -\ 120 \end{matrix}$.

L. Radicati I think here too, the interpretation of the results is not totally model independent, because this excitation does not only depend upon σ. There are the other parts which also contribute to the magnetic moment and you have to make assumptions about those. I believe the excitation discussed by Dr. Macq is what Dr. Mottelson would call a monopole excitation, namely one induced by Y_k^a, plus the other parts which are more difficult to calculate. Is that a correct interpretation of a monopole oscillation?

B. R. Mottelson I prefer to keep the multipole order as a description of the tensor character under total rotation.

So, these will be pseudovector, therefore I will label these by the quantum numbers $\lambda = 1, \varkappa = 0, \sigma = 1, \tau = 0$. They are really vector fields.

L. Radicati Would you call a monopole the pseudo scolar part of the first moment $G_{k\mu}^a = \int d^3x \, x_\mu g_k^a$

B. R. Mottelson That is one of the spin dependent monopole but what has been discussed the other day was spin independent.

D. R. Inglis To add to the commemorative nature of this session tracing the latter-day consequences of Prof. Wigner's 1936 idea, it is perhaps appropriate to introduce some of the simple physics of that time. Attempts were being made to understand nuclei in terms of a shell model, but spin-orbit coupling was either ignored or thought to be very weak, perhaps as weak as the Thomas coupling. The two-nucleon interactions were those that go by the names of Heisenberg, Wigner, Majorana and Bartlett, empirically replacing what now might be considered to be the interplay of a repulsive core and symmetry in relative coordinates. Experimental data were scarce but we did have the table of stable elements and to these crude data Professor Wigner applied his elegant theory explaining why a given neutron excess first appears at very nearly the same mass number for nuclei of the categories $4n$, $4n + 1$ and $4n + 3$. To obtain this result he made the drastic "long range" assumption that nucleon interactions have no radial dependence. This oversimplification made valid an independent-particle shell-model approximation with which it was possible to obtain the same physical result with very simple arithmetic, suggesting that deriving this result from Prof. Wigner's rather high-powered supermultiplet treatment was something like shooting a mosquito with an elephant gun. When at that time I displayed my lack of appreciation of Prof. Wigner's forsight into the future importance of symmetry in nuclear physics by pointing out this feature of his application of the theory, perhaps in a critical tone, he answered with his usual polite modesty "Well, I thought there might be some interest in the formulation itself".

P. Kramer With respect to the use of supermultiplet theory I would like to make a distinction between two approaches: (a) the attempt to describe a ground or excited state of a nucleus by a pure supermultiplet and (b) the use of the supermultiplet scheme to provide a labeling scheme for nuclear states.

Even if the first approach is not successful, the second one may still be useful. One may compare the situation with the use of orbital angular momentum: it is good to have a state of definite angular momentum, but for more complex configurations we do have well-defined procedures for handling superpositions of such states. With respect to the supermultiplet labels, Prof. Elliott has pointed out that we get a very rapid convergency if the states are expanded in terms of the available supermultiplets. What is required, therefore, is a careful classification of the states, interactions and transitions in terms of the supermultiplet scheme along with the study of the corresponding algebraic coefficients. I think that the recent progress in supermultiplet theory came along these lines.

B. R. Mottelson I would like to return to the question of the proper role of U_4 symmetry in our thinking about nuclear structure. It seems to me that there remain important questions for us to understand in this connection since the strong spin orbit forces clearly tell us that this symmetry is not at all satisfied if we take a single slater determinant in $j - j$ coupling. However the strong exchange force γ are a fact of nature and they are striving to impose this symmetry on the system. We are then faced with the problem of understanding the correlations produced, the configuration mixings implied, and in general the proper characterization of the modifications implied by these forces that are attempting to impose U_4 symmetry.

L. Radicati I am not really answering directly your question in quantitative terms.

I think nobody really understands why the currents that you can build with the nucleons and which are coupled to these fields which have nothing to do with hadrons, should be the generators of the symmetry.

It seems to be a very deep idea. I am sure Dr. Wilkinson will tell us about the point in the case of isotopic spin.

Parity and Time Reversal in Nuclear Physics

by HANS FRAUENFELDER

Department of Physics
University of Illinois at Urbana-Champaign
Illinois

I WILL TRY to give a short review treating the status of parity ($\mathscr{P}$) and time-reversal ($\mathscr{T}$) invariance in nuclear physics. I had hoped to listen to a discussion of these problems by Lobashov or by Blin-Stoyle. In their absence I will do my best to sketch the present situation.

PARITY VIOLATION

The *history* of the problem of parity violation in nuclear physics can be traced back to 1932 when Heisenberg[1] introduced spinless electrons to account for the exchange forces between nucleons in a nucleus. Pauli's lecture on the neutrino at the 1933 Solvay meeting[2] led Fermi to his beta decay theory.[3] Heisenberg then suggested that there may be a connection between the Fermi electron-neutrino field and the forces between neutrons and protons.[4] Calculations by Tamm and by Iwanenko,[5] however, indicated that the force produced by such an exchange is too weak by about a factor of 10^{12}.

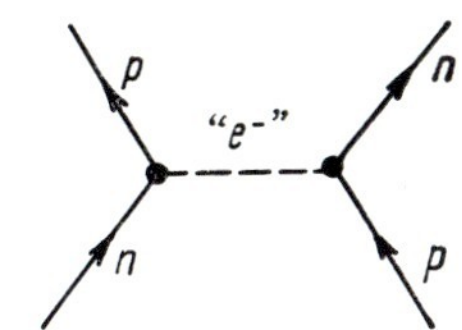

Heisenberg's spinless electron

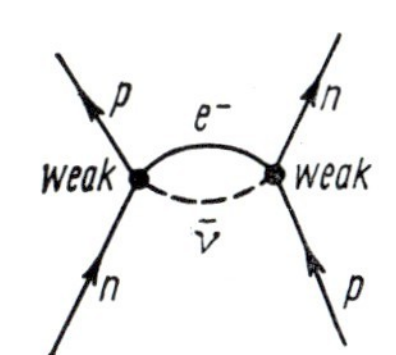

Exchange of an electron-neutrino pair

FIGURE 1

Nevertheless the weak force contributes to the interaction between two nucleons, even if this contribution is extremely small. Late in 1934, Yukawa made the next step by introducing heavy quantas, mesons, to explain the observed strong (hadronic) forces.[6] These mesons have weak interactions also; they can therefore be emitted hadronically and absorbed weakly or vice versa. Such processes are weaker than the hadronic forces "only" by about a factor 10^6, but it would still be hopeless to discover their contribution to nuclear forces were it not for one fact: Weak interactions do not con-

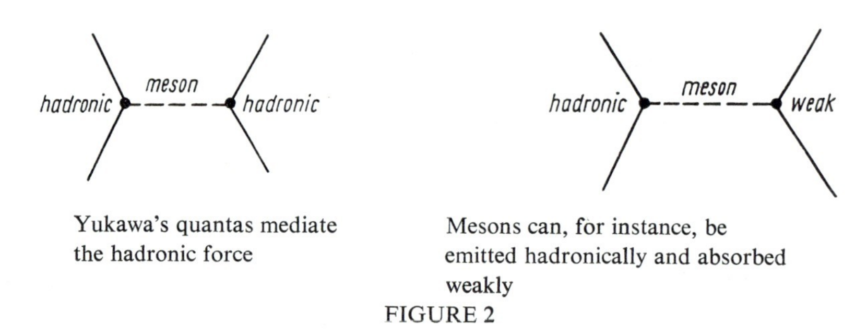

Yukawa's quantas mediate the hadronic force

Mesons can, for instance, be emitted hadronically and absorbed weakly

FIGURE 2

serve parity. In 1956, Lee and Yang suggested this possibility[7] and it was verified experimentally soon afterwards.[8] Parity non conservation provides a signature that makes it possible to detect a small weak contribution to the nuclear interaction.

Most theories of weak interactions that include parity violation also lead to a parity-violating contribution to the hadronic forces. In order to make realistic estimates, a specific form is needed and the current-current interaction with a conserved vector current[9] makes such estimates feasible.[10] Calculations that take into account all aspects of nuclear structure still remain to be done.*

Experimentally, the parity-non-conserving contribution to the nuclear force has been observed in a number of ways.[11,12] Probably the simplest experiment involves the observation of the circular polarization of nuclear gamma rays. The basic idea is straightforward. Consider unpolarized nuclei. Photons emitted in a transition between states of well-defined parity then cannot be circularly polarized: A circular polarization appears if the transition contains electric and magnetic multipole radiation of the same order, $E(L)$ and $M(L)$. These, however, have opposite parity; circularly polarized photons therefore indicate that at least one of the states, $|m\rangle$, involved in the transition has mixed parity and is described by

$$|m\rangle = |\text{reg}\rangle + F\,|\text{irreg}\rangle. \tag{1}$$

* For details and for additional references see Refs. 11 and 12.

Here, |reg⟩ and |irreg⟩ have the same spin, but opposite parity. F is very small so that the state $|m\rangle$ has predominantly the regular parity. A hypothetical example is shown in Fig. 3. The upper state has spin 2 and the predominant parity is negative. With $F = 0$, only magnetic quadrupole radiation is emitted; $F \neq 0$ permits a small admixture of electric quadrupole

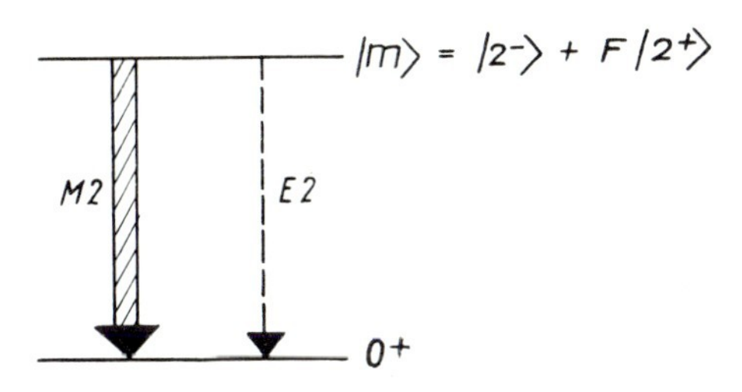

FIGURE 3 A hypothetical example

radiation. The circular polarization is given by

$$P_\gamma = \frac{2\langle 0|M2|m\rangle\langle 0|E2|m\rangle}{\langle 0|M2|m\rangle^2 + \langle 0|E2|m\rangle^2}, \tag{2}$$

where all matrix elements have been chosen to be real. Since F is very small, P_γ becomes in a good approximation

$$P_\gamma = 2F\frac{\langle 0|M2|2^-\rangle\langle 0|E2|2^+\rangle}{\langle 0|M2|2^-\rangle^2} = 2F\frac{\langle 0|E2|2^+\rangle}{\langle 0|M2|2^-\rangle},$$

or

$$P_\gamma = 2FR, \tag{3}$$

where R is the ratio of matrix elements of multipole operators of opposite parity.

To search for parity violation, a transition has to be chosen where the admixture of irregular parity and the ratio R are as large as possible. An example of such a transition occurs in ^{181}Ta which has a $7/2+$ ground state and a $5/2+$ excited state at 482 keV. The $M1$ transition $5/2+ \rightarrow 7/2+$ is strongly hindered. If the $5/2+$ state contains a small admixture of negative parity, then an irregular $E1$ transition can occur and it can interfere with the hindered $M1$ decay. Indeed, using an ingenious detecting system, Lobashov and coworkers[13] found a polarization of $P_\gamma = -(6 \pm 1) \times 10^{-6}$. Since experimental physicists at this meeting have rarity value, it may be appropriate to stress the enormous difficulties inherent in such an experiment. Polarization analyzers have an efficiency of only a few per cent. In order to measure a polarization to one part in 10^6, count rate or current has to be measured to nearly one part in 10^8.

An example where the polarization is larger than in ^{181}Ta is ^{180}Hf. The essential levels are indicated in Fig. 4. Two facts make the 501 keV $8^- \rightarrow 6^+$

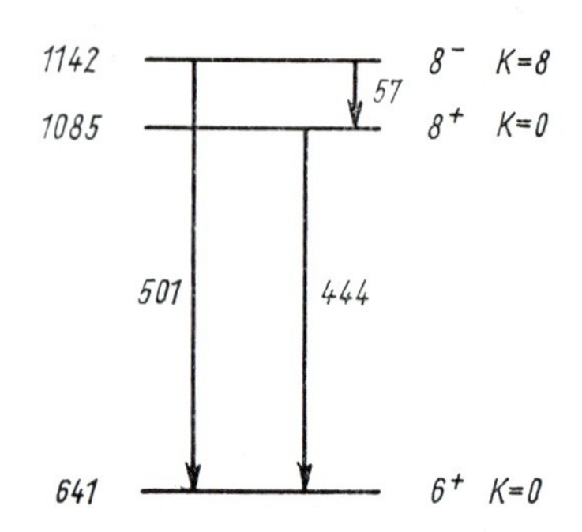

FIGURE 4 Decay of ^{180}Hf.Only the levels of interest are shown. Energies are given in keV

transition a good candidate for parity studies: (1) The upper state (8^-) lies close to a state with the same spin but opposite parity. In first-order perturbation theory, the admixture parameter F is given by

$$F = \frac{\langle 8^+ |H_w| 8^- \rangle}{\Delta E}, \quad \Delta E = 47 \text{ keV}, \tag{4}$$

where H_w is the parity-violating weak interaction. (2) The transition is highly K forbidden. The regular transition between the 8^- and the 6^+ state can occur through $M2$ und $E3$ and these transitions are suppressed by factors 5×10^{14} and 1.6×10^9, respectively, with respect to the Weisskopf estimate. The irregular $E2$ transition, however, is admixed from the 8^+, $K = 0$, state and is therefore no K forbidden. It has an excellent chance to compete with the regular transitions and to interfere with the $M2$ component. In fact, Eq. (3), taking into account the $E3$ component, gives

$$P_\gamma \approx 10^8 F. \tag{5}$$

F is usually of the order of 10^{-7} and P_γ would then be large indeed. However, the matrix element in Eq. (4) is also K forbidden and F is expected to be reduced.[10] P_γ for the 501 keV transition in ^{180}Hf has been measured,[14] with the result

$$P_\gamma = -(2.8 \pm 0.45) \times 10^{-3}, \tag{6}$$

corresponding to $F \approx 3 \times 10^{-11}$. Unfortunately no theoretical calculation exists as yet with which this number can be compared.

The present situation can be summarized as follows[11]: Parity violation in nuclear forces is well established. New experiments, such as for instance with ^{3}He-^{4}He dilution refrigerators[15] promise considerably more insight in the next few years. Unfortunately, theory and experiment have not yet really met. Successful experiments have been performed with heavy nuclei, where the theory is not yet well developed. I hope that theoreticians will attack the difficult question of the computation of F in heavy nuclei and that experimentalists will be able to solve the equally difficult problem of measuring parity violation in light nuclei, for instance in the reaction $np \rightleftharpoons d\gamma$. Experiments with light nuclei may also allow a separation of the isovector part of the parity-violating force from the isoscalar and isotensor part.

TIME REVERSAL VIOLATION

The situation concerning time-reversal violation in nulear physics is very different from that of parity violation. Experimentally, time reversal violation has only been found in the neutral kaon system. Theoretically it is not clear which interaction must be blamed for the violation and it is therefore doubtful whether effects can be expected in nuclear physics.

In 1964, it was found that the long-lived neutral kaon, K_L^0, can decay into two pions.[16] Such a decay violates conservation of the combined parity, $\mathscr{CP}$, and it leads to three interrelated questions:

(1) Does $\mathscr{CP}$ violation imply a $\mathscr{T}$ violation?

(2) Which interaction is responsible for the $\mathscr{CP}$ violation?

(3) Can evidence for $\mathscr{CP}$ or $\mathscr{T}$ violation be found outside the neutral kaon system?

In the six years since the fall of $\mathscr{CP}$, an enormous amount of experimental and theoretical work has gone into attempts to solve these three problems.

The first question has been answered within the last two years. By 1964 the $\mathscr{TCP}$ theorem was well established theoretically and experimentally.[17] It was therefore assumed that the $\mathscr{CP}$ violation implied a $\mathscr{T}$ violation. To test this assumption, the relations among the decay parameters of the neutral kaons were examined without assuming $\mathscr{CP}$, $\mathscr{T}$ or $\mathscr{TCP}$ symmetry. This examination leads to the conclusion that $\mathscr{TCP}$ symmetry is valid but that both $\mathscr{T}$ *and* $\mathscr{CP}$ are violated.[18]

The answer to the second question is not yet clear. After the experimental discovery of $\mathscr{CP}$ violation, several types of theories were put forward to explain the unexpected.[19] We will only consider two of these here. Wolfenstein introduced a *superweak* interaction that satisfies a selection rule $|\Delta Y| = 2$, where Y is the hypercharge.[20] Berstein, Feinberg, and Lee assumed that the violation occurs in the *electromagnetic* force.[21] The consequences of these two theories are very different. If $\mathscr{T}$ and $\mathscr{CP}$ violations occur in the electromagnetic interaction, then it is not excluded that observable time-reversal violation effects appear in nuclear decays. The theory, however, makes no definite predictions concerning the parameters occurring in the decay of the neutral kaons. The superweak interaction, on the other hand, would be about 10^7 times weaker than the ordinary weak one and it would not lead to observable effects in decays other than those of the neutral kaons. It enters the physics of neutral kaons through the introduction of off-diagonal elements in the mass matrix. These in turn produce a non-vanishing amplitude for the decay $K_L^0 \to 2\pi$. Any hope of seeing effects of such a super weak interaction in hadronic or electromagnetic processes is remote indeed. As a compensation, the superweak interaction leads to definite predictions for the decay parameters of the neutral kaons.[22] In order to discuss these predictions, a few definitions have to be introduced. Consider the decays

$$K_L^0 \to \pi^+\pi^-, \quad K_L^0 \to \pi^0\pi^0,$$
$$K_S^0 \to \pi^+\pi^-, \quad K_S^0 \to \pi^0\pi^0. \tag{7}$$

The following ratios of matrix elements of the transition operator T are then defined

$$\eta_{+-} = \frac{\langle\pi^+\pi^- |T| K_L\rangle}{\langle\pi^+\pi^- |T| K_S\rangle}, \quad \eta_{00} = \frac{\langle\pi^0\pi^0 |T| K_L\rangle}{\langle\pi^0\pi^0 |T| K_S\rangle}. \tag{8}$$

The superweak interaction predicts, with $\eta = |\eta|\, e^{i\varphi}$:

$$|\eta_{+-}| = |\eta_{00}| \tag{9}$$

$$\varphi_{+-} = \varphi_{00} = \arctan\,(2\,\Delta m \tau_S) = (43.1 \pm 0.4)^\circ. \tag{10}$$

Here Δm is the mass splitting between K_S and K_L, and τ_S the mean life of K_S.

Even for the "non-K^0" experimental physicist, the values of the parameters η_{+-} and η_{00} are important. If they do not obey the relations (9) and (10) then a search for $\mathscr{T}$ violation in nuclear and for $\mathscr{CP}$ and $\mathscr{T}$ violation in particle physics outside the K^0 system is a reasonable gamble. If, on the other hand, the relations (9) and (10) are satisfied then the decision is much more agonizing, as I will point out again below. The situation has changed considerably in the past three years. At the end of 1967, the values were[23]

$$|\eta_{+-}| = (2.00 \pm 0.06) \times 10^{-3}, \quad |\eta_{00}| = (4.2 \pm 0.3) \times 10^{-3},$$
$$\varphi_{+-} = (65 \pm 11)^\circ. \tag{11}$$

In 1967, the superweak interaction appeared to be ruled out. Now, three years later, more of the delicate experiments have been performed and the present results are[24]

$$|\eta_{+-}| = (1.92 \pm 0.05) \times 10^{-3}, \quad |\eta_{00}| = (2.0 \pm 0.2) \times 10^{-3}$$
$$\varphi_{+-} = (44 \pm 5)^0, \quad \varphi_{00} = (51 \pm 30)^0. \quad (12)$$

The superweak interaction no longer appears to be ruled out. However it must be pointed out that the value for $|\eta_{00}|$ is not as trustworthy as the quoted error makes it appear. Some experiments give values much higher than 2×10^{-3} and the cause for the discrepancy has not yet been found. A number of additional experiments are in progress and, hopefully, the situation will soon be cleared up.

A great deal of effort has been expended to answer the third question. After the fall of parity, a number of time-reversal tests were performed, but the limits obtained on the irregular amplitude $\mathscr{F}_T$ was only a few per cent. After 1965 efforts were renewed when the situation looked as in Eq. (11) and when Lee's proposal[21] made tests attractive. Old techniques were improved and new ones were devised. The various experiments (detailed balance, neutron dipole moment, Mössbauer effect) have been reviewed by Henley.[11] The result is unambiguous: In none of the tests outside the K^0 system has a $\mathscr{T}$ violation been seen and the present limit is $\mathscr{F}_T < 3 \times 10^{-3}$.

At the present time every experimental physicist who is involved in time-reversal tests stands before a difficult choice. It has taken a few years of very hard work to reduce the limit on $\mathscr{F}_T$ by one order of magnitude. Without a brilliant new approach, the next order-of-magnitude improvement will be even harder and probably take longer. This thought alone is no deterrent. However, the data given in Eq. (12) suggest that the $\mathscr{CP}$ violation is due to a superweak interaction. If this suggestion is confirmed, then all present searches in nuclear physics are hopeless. I therefore feel that it is best to wait till the values of η_{+-} and η_{00} are better known before committing a great deal of time, equipment, and enthusiasm to improve the limit on $\mathscr{F}_T$. It is probably better to turn the question around and ask: Is there a place outside the neutral kaon system where a superweak interaction would show up in some other way?

REFERENCES

1. W. Heisenberg, *Z. Physik* **77** (1932) 1.
2. W. Pauli, in *Noyaux Atomiques*, Proc. of Solvay Congress, Brussels, 1933, p. 324.
3. E. Fermi, *Z. Physik* **88** (1934) 161, Ric. Sci. **2**, Part 12 (1933).
4. W. Heisenberg, Lectures at the Cavendish Laboratory, Cambridge, 1934. Unpublished.

5. I. Tamm, *Nature* **133** (1934) 981; D. Iwanenko, *Nature* **133** (1934) 981; H. A. Bethe and R. F. Bacher, *Rev. Mod. Phys.* **8** (1936) 201; See also G. Feinberg and J. Sucher, *Phys. Rev.* **166** (1968) 1638.
6. H. Yukawa, *Proc. Phys. Mat. Soc. Japan* (3) **17** (1935) 48.
7. T. D. Lee and C. N. Yang, *Phys. Rev.* **104** (1956) 254.
8. C. S. Wu, E. Ambler, R. W. Hayward, D. D. Hoppes, and R. P. Hudson, *Phys. Rev.* **105** (1957) 1413.
9. R. P. Feynman and M. Gell-Mann, *Phys. Rev.* **109** (1958) 193.
10. R. J. Blin-Stoyle, *Phys. Rev.* **118** (1960) 1605; F. C. Michel, *Phys. Rev.* **133** (1964) B 329.
11. E. M. Henley, *Ann. Rev. Nucl. Sci.* **19** (1969) 367.
12. W. D. Hamilton, *Progr. Nucl. Phys.* **10** (1969) 1.
13. V. M. Lobashov, V. A. Nazarenko, L. F. Saenko, L. M. Smotritskii, and G. I. Kharkevich, *Soviet Phys. JETP Letters* **5** (1967) 59, *Phys. Letters* **25**B (1967) 104.
14. B. Jenschke and P. Bock, *Phys. Letters* **31B**, 65 (1970)
15. W. P. Pratt, Jr., R. I. Schermer, J. R. Sites, and W. A. Steyert, *Phys. Rev. C*, **1499** (1970)
16. J. H. Christenson, J. W. Cronin, V. L. Fitch, and R. Turlay, *Phys. Rev. Letters* **13**, 138 (1964); A. Abashian, R. J. Abrams, D. W. Carpenter, G. P. Fisher, B. M. K. Nefkens, and J. H. Smith, *Phys. Rev. Letters* **13**, 243 (1964).
17. L. B. Okun, Comm. *Nucl. Part. Phys.* **2**, 116 (1968)
18. R. C. Casella, *Phys. Rev. Letters* **21**, 1128 (1968), **22**, 554 (1969); K. R. Schubert, B. Wolff, J. C. Chollet, J. M. Gaillard, M. R. Jane, T. J. Ratcliffe, and J. P. Repellin, *Phys. Letters* **31B**, 662 (1970)
19. R. E. Marshak, Riadzuddin, and C. P. Ryan, *Theory of Weak Interactions in Particle Physics* (Wiley-Interscience, New York, 1969, p. 646)
20. L. Wolfenstein, *Phys. Rev. Letters* **13**, 562 (1964)
21. J. Bernstein, G. Feinberg, and T. D. Lee, *Phys. Rev.* **139**, B1650 (1965); T. D. Lee, *Phys. Rev.* **140**, B959, B967 (1965)
22. T. D. Lee and L. Wolfenstein, *Phys. Rev.* **138**, B1490 (1965)
23. A. H. Rosenfeld *et al.*, *Rev. Mod. Phys.* **40**, 77 (1968)
24. M. Roos *et al.*, *Phys. Letters* **33B**, Number 1 (August, 1970); M. Cullen *et al.*, *Phys. Letters* **32B**, 523 (1970); S. H. Aronson *et al.*, *Phys. Rev. Letters* **25**, 1057 (1970); I. A. Budagov *et al.*, *Phys. Rev. D*, **2**, 815 (1970); J. C. Chollet *et al.*, *Phys. Letters* **31B**, 658 (1970)

DISCUSSION

D. H. Wilkinson Parity violation may be sought as a selection rule effect in which case the order of magnitude of the effect will be F^2 rather than F where F is the relative amplitude of the parity-irregular component of the wave-function. The 8.88 MeV state of ^{16}O is of $I^{\pi} = 2^{-}$ and may be populated by the beta decay of ^{16}N. The state is unstable against alpha-particle decay to $^{12}C_{g.s.}$ and this decay will go through the F-components of the nuclear wavefunctions only. Many groups over the last decade have sought this parity-violating decay and it now appears that it has been found with good confidence by Wäffler. He quotes $\Gamma_{\alpha} = (2.0 \pm 0.8) \times 10^{-10}$ eV which is indeed some 10^{14} times less than what might be expected for a parity-allowed decay. This figure is close to estimates made by Gari and Henley on the basis of the Michel parity-non-conserving NN interaction and reasonable nuclear wavefunctions.

L. Radicati The experiment that Wilkinson mentioned is very interesting. The 8.88 MeV is presumably a $T = 0$ state. Now it is of the greatest importance to distinguish the $T = 0$ from the $T = 1$ parity violation because this may settle the question of how the hypercharge conserving, parity violating interactions arise. If they come from the coupling of only charged currents or if they involve neutral currents, the ratio between the $T = 1$ and $T = 0$ is completely different. In one case it is much stronger than in the other. This distinction can be made in the electromagnetic transitions.

E. P. Wigner Is the theory of two subsequent transitions sufficiently well established so that one can be sure that the emission of the α-particle with such a low probability is a violation of parity?

D. H. Wilkinson The 8.88 MeV state has gamma-decay as its chief mode of deexcitation with a width of about 10^{-2} eV and so is quite well defined. Other processes such as the intervention of the atomic electrons in the decay, which could similate the effect of mixed parity in the nuclear wave functions, have been considered by Blin-Stoyle and have been found to be negligibly small. The effect of Coulomb excitation in the beta-decay process has also been considered by Spiers and has been found both to be too small and to lead to an alpha-particle spectrum very different from that found.

L. Radicati This is to answer partly to what Dr. Wigner said.

The theory of the parity mixing as Frauenfelder pointed out before is still not at all under control. What is the exact value of the parity mixing that is expected from weak interaction, is really very unclear. There are two sorts of uncertainty: the effect he mentioned which is certainly important and, I think is not under control and the other one, i.e. the amount of predicted mixing which is also not precisely known.

E. P. Wigner I presume the theory is right. I should not really question it, but I also have an uneasy feeling because the theory of successive disintegrations, if there are many other levels present, is not very clear mathematically.

D. H. Wilkinson Perhaps, I should just add one more comment on all this. In the case of the O^{16} decay there have been several calculations on what one might expect for the parity violating width on the basis of nuclear models of course, a careful one by Henley and another careful one by Gari both using the now traditional parity violating nuclear potential, which has already been refered to, by Michel and they both do indeed give about 2×10^{-10} eV. Unfortunately there is a very recent paper by Tadić claiming that there is a mistake in the original Michel calculation, such that the γ-ray experiments Dr. Frauenfelder spoke about, should give an effect of the opposite sign from what has been obtained experimentally. I am only reporting something that I have seen very briefly and in a preprint form; my knowledge is rather casual.

Isobaric Analogue Symmetry

D. H. WILKINSON
Nuclear Physics Laboratory, Oxford

1 INTRODUCTION

The last Solvay conference on nuclear structure was held in 1933. The neutron had been discovered in 1932 and in the same year Heisenberg had borrowed the formalism of the Pauli spin matrices to distinguish it from the proton, recognizing that the two particles were sufficiently similar to warrant being regarded merely as alternative states of the same particle. But Heisenberg stopped short of charge independence: guided by the analogy with the H_2, H_2^+ and H^+H^+ systems he saw the nuclear force as strongest for the *np* system, weaker for the *nn* system and zero for the *pp* system. Thus although Heisenberg introduced the formalism that led to the isospin quantum number his forces themselves did not lead to that quantum number.

We may at this point recognize another important nuclear event associated with the epoch of the last Solvay conference and of importance for our deliberations today, namely the invention of the neutrino. Pauli invented the neutrino in 1931 and by the time of the 1933 conference, consequent upon the discovery of the neutron, the neutrino was almost as firmly established as if he had discovered it rather than invented it; its ultimate direct detection by Reines and Cowan 20 years later was a tremendous tour de force but surprised no one. It is, incidentally, amusing to speculate what could have been the effect on the world of science had these neutrino experiments of 1953 been accurate enough, as later ones were, to show that the statistical weight of the neutrino is merely one rather than two.

To return to the nuclear force: Shortly after the Solvay conference realization began to grow that the nuclear force must be at least approximately charge independent; the full hypothesis seems to have been first put forward explicitly by Young in 1935 although Guggenheim was suspecting it a year earlier. The rapid development of nucleon scattering and other experiments and their analysis, in 1935 and 1936, made the hypothesis quantitatively more and more likely. In 1937, following work by Breit, Feenberg, Bartlett, Cassen and Condon, and detailed consideration of Coulomb energies, Wigner brought out the full isospin concept, clearly separated from the less-absolute supermultiplet theory. Immediately afterwards the application to beta-decay was made and it was, in particular, recognized that if isospin is a good quantum number Fermi transitions take place only between members of an isospin multiplet, i.e., between isobaric analogue states. The isospin selection rules for nuclear reactions were enunciated by Oppenheimer and Serber in 1938 who recognized at the same time the first isobaric analogue resonance and the likely magnitude of the relaxation of isospin by the Coulomb force.

So within 5 years of the 1933 Solvay conference all the essentials were there and had been quantitatively discussed: charge independence in the elementary interactions, isospin and its impurity, the Coulomb energy as a signal of charge independence, isospin selection rules for nuclear reactions and beta-decay and the existence of isobaric analogue resonances. There has, of course, been some progress since then. Charge independence and isospin lie behind, explicitly or implicitly, most of what we do in nuclear structure physics and have almost become an industry in their own right, particularly since the discovery of isobaric analogue resonances in their modern form. But isospin is only permissive and must not be overdramatized. It is well to remember Wigner's words "Isotopic spin is not a key to nuclear structure which will one time unlock its secrets ... it chooses itself what information it is willing to provide and this is far from all that one might desire".

It is only when one supplements isospin with other notions that one really moves towards a theory of nuclear structure and this is what other speakers at this 1970 Solvay conference will be doing. It is my own task to examine isobaric analogue symmetry proper, that is to restrict myself to a discussion of isospin pure and simple (or perhaps slightly impure and then not so simple), and to try to quantify this necessary basis for the more extended symmetries that lead into nuclear structure. I shall, therefore, very largely restrict myself to the simpler situations and not, for example, attempt any significant discussion of isobaric analogue resonances in the heavier elements. These resonances are fascinating: they have presented a great challenge, initially to understand why they are so narrow, then to understand why they are so broad; they also offer a great quarry of spectroscopic information that as yet very largely remains to be exploited; but they teach us little if anything about the underlying nuclear symmetry that they represent. We have come rather to understand the isobaric analogue resonances from the symmetry point of view in terms of what we already know and do not seriously expect to learn much from them about the charge dependence of the nuclear force. For that we must turn to simpler systems and this is what I shall do. I will look at the simpler places where charge independence and symmetry are manifest and where charge dependence or asymmetry might be detected and ask for the degree to which we might feel that the situation is quantitatively established. I shall speak entirely from the experimental viewpoint in an attempt to establish facts and shall present little in the way of theoretical discussion except in the form of a few assertions.

My treatment will necessarily be somewhat eclectic since the subject is vast. I shall in no way attempt to introduce any of the topics pedagogically and have selected them to illuminate a wide range of matter all of which bears on the question of the quantitative validity of the charge independence and isospin concepts. This means that I have ignored much fascinating material and many phenomena that relate to reaction mechanism and spectroscopic detail rather than to the basic question of analogue symmetry itself.

2 CHARGE INDEPENDENCE AND SYMMETRY IN SIMPLE SYSTEMS

Charge symmetry says that, in corresponding states, the *nn* and *pp* nuclear forces are the same. Charge independence says that, in addition, the *np* force is the same as these two. These symmetries, if true for the nuclear force, would remain approximately true for the total force between nucleons because the electric interactions, characterized by the fine structure constant α, are much weaker than the nuclear force. But, of course, the electric interactions are always present: their subtraction, to get at the dominant nuclear component, must always be theoretical and so associated with model-dependent uncertainty and their ineluctable presence means that the symmetries in nature must always be approximate only. But the smallness of α means that it makes sense to hypothesize exact symmetries and to study their relaxation by the electromagnetic interactions, hoping thereby to reveal or limit any additional breakdown that would be due to a charge dependence or asymmetry of the nuclear force itself.

Charge independence and symmetry should be investigated for all spectroscopic states of the *NN* system for which the answer has relevance for the nuclear structure or other problem in hand. However, accurate comparisons in any state are hard to come by and attention for quantitative purposes has scarcely passed beyond the low-energy aspects of the $^1S_0$ state, enjoyed by all three systems *pp*, *np* and *nn*. At low energy we make comparisons of the scattering lengths a_{NN} and effective ranges r_{NN}. Proton-proton measurements are themselves highly accurate but their correction for electromagnetic effects is somewhat model dependent. Henley (1969) in a systematic measurement allowing for these uncertainties derives:

$$a_{pp} = -(17.3 \pm 0.2) \text{ fm}$$
$$r_{pp} = 2.83 \pm 0.03 \text{ fm}$$

which values are corrected for direct Coulomb and also for vacuum polarization effects and for magnetic interactions.

For the *np* system Henley (1969) quotes:

$$a_{np} = -(23.515 \pm 0.013) \text{ fm}$$
$$r_{np} = 2.76 \pm 0.07 \text{ fm}$$

where again the magnetic interaction has been taken out.

The large scattering length in the ${}^1S_0$ state means that this quantity is a very sensitive index of departures from equality of the *pp* and *np* forces.

Using for the potential a Yukawa form with a hard core of 0.388 fm Henley (1969) then finds:

$$\bar{V}_{np} - \bar{V}_{pp} = (2.1 \pm 0.5)\%.$$

The data and analyses are not sufficiently accurate to establish a significant difference between the *np* and *pp* effective ranges.

Charge independence is clearly broken but charge symmetry remains to be questioned. The direct determination of *nn* scattering parameters is obviously very difficult. The closest approach so far has come through detailed studies of the reaction:

$$\pi^- + d \to 2n + \gamma$$

where the kinematical spectrum of the final state is affected by the *nn* interaction. The best measurements (Haddock *et al.* 1965; Nygren 1968) give $a_{nn} = -(18.42 \pm 1.53)$ fm where the error includes estimated uncertainties in the application of the Watson-Migdal formalism. A more recent measurement (Butler *et al.* 1968) by the same reaction gives $a_{nn} = -13.1\,^{+2.4}_{-3.4}$ fm. It is difficult to combine these results: a conventional weighting with an appropriately increased error gives:

$$a_{nn} = -(16.9 \pm 2) \text{ fm}.$$

This figure is consistent with the a_{pp} quoted above. It may also be directly compared with a somewhat different prediction also containing all the necessary electromagnetic corrections as between the two systems, derived from *pp* data (Miller *et al.* 1969):

$$a_{nn} \text{ from } pp = -(17.55 \pm 0.10) \text{ fm}.$$

This latter analysis, also consistent with the experimental result, treats 5 different detailed potentials of very different type, each of which gives a "exact" account of $^1S_0$ *pp* scattering up to 330 MeV. The error given for the above prediction of a_{nn} from *pp* scattering covers the range of values generated by the 5 potentials and may be taken as the best current estimate of the model-dependence of the prediction.

Comparison of the *nn* and *pp* data on the same basis as the above $pp - np$ comparison gives:

$$\bar{V}_{nn} - \bar{V}_{pp} = -(0.2 \pm 0.7)\%.$$

The comparisons made so far, which establish charge symmetry within 1% and show a 2% departure from charge independence, are relatively unambiguous in that it is unlikely that model-dependence and uncertainties about reaction mechanisms will cause the stated errors to be exceeded. The question of the *nn* interactions is, however, so important that any alternative approach to it must be explored, no matter how hazardous. Such an alternative approach is to study systems that have as their final state two neutrons plus other strongly interacting particles. Thus the reaction:

$$n + d \rightarrow p + 2n$$

will be affected by the final state *nn* interaction and an analysis that took adequate account of the reaction mechanism would reveal that interaction quantitatively. Studies in those kinematical conditions that leave the two neutrons with low relative momentum are indicated. The uncertainties are minimized if corresponding reactions involving known final state interactions are studied as a parallel exercise—for example in the above case the mirror reaction

$$p + d \rightarrow n + 2p$$

could be analyzed to reveal the known *pp* interaction. Very many such hopeful studies have been made using a range of light-nucleus reactions. A particularly careful and complete set is that of Gross *et al.* (1970a) who examined ^{3}He(^{3}He, α) *pp*, ^{3}He(t, α) *np*, ^{3}H(t, α) *nn* observing the alpha-particle spectra at small angles and analyzing the data in terms of the interactions in the spectator *NN* systems. In the first two reactions these authors find $a_{pp} = -(7.52 \pm 0.22)$ fm (including the Coulomb interaction and to be compared with $a_{np} = -(7.79 \pm 0.01)$ fm as directly determined from *pp* scattering) and $a_{pp} = -(21.5 \pm 2.3)$ fm. The good agreement on these known values of a_{NN} emboldens the authors to analyze the final reaction in terms of a_{nn} finding $a_{nn} = -(16.96 \pm 0.51)$ fm which agrees well with the above-quoted expectation derived from a_{pp}. But it must be emphasized that studies such as these that measure only one particle in the final state are not under sufficiently close kinematical surveillance for us to be confident that a particular reaction mechanism is predominating; great caution must be exercised in extracting the NN interaction using the Watson-Migdal formalism. This conclusion must remain valid even if, as in the case quoted, the spectrum of the observed particle is well-fitted by the final-state interaction theory over a wide range of relative momentum for the NN system and if the right answer is given for known final state interactions. By luck one may hit on acceptable conditions but this cannot be known. A valuable critique of such approaches is given by Larson *et al.* (1970). The situation is much improved if the final state is fully determined kinematically by a coincidence experiment so that conditions favoring particular reaction mechanisms can indeed be selected by the experimenter and so that quantitative estimates may be made of the contribution from other reaction mechanisms by systematically moving away from the kinematically-favored region. An illuminating study of these matters for the reaction:

$$p + d \rightarrow n + 2p$$

has been made by Boyd *et al.* (1969) who show that it is indeed possible, under controlled kinematical conditions, to separate out the part of the reaction mechanism amenable to the Watson-Migdal treatment of the final np state. They find, in a convincing manner, $a_{np} = -(23.8 \pm 0.5)$ fm which may be compared with the direct experimental 23.71 fm (note that the 23.52 fm quoted above has been corrected for certain small effects such as the magnetic interaction). This experiment clearly points the (arduous) way to a more reliable determination of a_{nn} using all-strongly-interacting particles in the final state but the appropriate experiment has not yet been carried out with adequate precision to add materially to our knowledge of a_{nn}. Zeitnitz *et al.* (1970) have, however, carried out a preliminary kinematically complete study of ^{2}H(n, 2n) p finding $a_{nn} = -16.4 \begin{smallmatrix} +2.6 \\ -2.9 \end{smallmatrix}$ fm which agrees with the other values.

We must therefore leave the charge symmetry/independence question at this stage insofar as NN approaches go and look at other evidence. In doing so, however, we may note that some relaxation of charge-independence is to be expected if only on account of the 3.4% difference in masses of the charged and neutral pions. It is not, at this time, possible to make any quantitatively-meaningful estimate of the differences in the NN forces to be expected on account of the differences (largely unknown apart from the pion) between the masses and couplings of the charged and neutral mesons that mediate them but it appears (Henley 1969) that the observed 2% difference between the np and pp interactions is wholly reasonable. The origins of a possible breakdown of charge symmetry are more subtle and lie, for example, in the isospin mixing of the mediating mesons—the $T = 1\,\pi^\circ$ will carry a small amplitude of the $T = 0\,\eta^\circ$, the $T = 1\,\varrho^\circ$ will contain a little of the $T = 0\,\phi^\circ$ and ω° and so on. Such admixtures would break charge symmetry in the *NN* force but are very difficult to extimate reliably. It seems (Henley 1969) as though differences between a_{nn} and a_{pp} as large as 1 fm could be due to this cause, the *nn* force being slightly the stronger. This would correspond to a difference in the $\bar{V}_{NN}$ values of about 0.35%.

Evidence on charge independence is also available from simple systems involving pions. The reactions:

$$p + d \begin{array}{l} \to {}^3\text{H} + \pi^+ \\ \to {}^3\text{He} + \pi^0 \end{array}$$

should have cross sections in the ratio 2 : 1 if charge-independence holds in its simplest form. The experimental ratio at $E_p = 591$ MeV is 2.13 ± 0.06 (Harting *et al.* 1960). The theoretical ratio, taking into account electromagnetic effects, is 2.20 ± 0.07 (Köhler 1960). (We may note for future reference that a large part of the departure of the theoretical ratio from 2 is due to the differential effect of the Coulomb force on the structure of ^{3}H and ^{3}He.) Without an explicit calculational model it is difficult to quantify these results in terms of symmetry breaking but if we assume charge independence in the pion interactions we may very quickly estimate the effect of a departure from charge symmetry in the nuclear system with the result $\Delta \bar{V}_{NN} \leqq 1\%$.

The experimental weakness of the reaction

$$d + d \to {}^4\text{He} + \pi^0$$

which is forbidden by charge symmetry leads (Henley 1969) to $\Delta \bar{V}_{NN} \leqq 0.5\%$. But we are here at the mercy of the model that we make of the reaction "if the nuclear force were not charge-symmetric" and so we have no way of assessing our real confidence in the number that emerges from the analysis.

We leave this study of simple systems recapitulating that charge symmetry is established to 1% or better and that charge-independence is broken by (an understandable) 2% or so but remembering that these are somewhat vague statements referring to the overall interaction strength and that only $^1S_0$ states have come seriously into question. Higher energy *NN* scattering, referring to other spectroscopic states, may show some departure from charge independence also but the conclusions are there scarcely quantitatively significant.

3 CHARGE SYMMETRY AND INDEPENDENCE IN COULOMB ENERGIES

The fact that the mass differences between mirror nuclei such as ^{13}N–^{13}C and between even mass nuclei such as ^{10}Be–^{10}B that belong to the same partition could wholly, or largely, respectively be accounted for by the differing Coulomb energies was a vital early clue that led Feenberg and Wigner in 1937 to adopt full charge independence as a tenet of faith and that gave rise to the full isospin concept. Now that we know rather accurately the charge distribution of nuclei from electron scattering and muonic X-ray work and the gross shell structure to some fair degree, at least in the lighter nuclei, from knock-out and pickup reactions it may be felt that the analysis of Coulomb energies should constitute a sharp test of charge symmetry and independence. This hope, however, is not yet fully realized because of the great complication of the problem at its detailed level, simple as it is conceptually and at the primitive level of argument restricted to classically-charged spheres. This is not to say that much interesting information, both detailed and gross, may not, in principle, be got from a study of Coulomb energies (see e.g. Jähnecke 1969; Nolen and Schiffer 1969). Detailed information may be hoped for from changes in Coulomb energies as between different excited states of the same A-value concerning configurational changes and rearrangements or as between different isotopes concerning pairing and related effects. Gross information as to the distribution of the neutron excess may be obtained by studying analogue state displacements in heavy nuclei. But even here little really quantitative progress has been made that has resulted in the acquisition of hard information. The situation always remains extremely complex. For example when configurational or isotopic information is sought it appears that other effects, obviously connected with the changes in the individual particle wave functions brought about by the different relationships of the nuclear forces in the new configuration or isotope and that themselves therefore entrain changes in the Coulomb energy, are often as big as or bigger than the direct effect expected from, say, the change of configuration of the nucleon in question. Again, when analogue state displacements in heavy nuclei are examined with an eye to determining the radial distribution of the neutron excess effects of state displacement due to coupling to the continuum and due to nucleon correlations are uncertain and of sufficient possible importance to bring about qualitative changes in the conclusions. In all cases it is vitally important to take proper account of the residual interactions among the valence nucleons that determine the parentage spectrum and to the effective binding of the nucleon(s) whose Coulomb energy is being considered. In other than the most ideal single-particle unique-parent situation, that never exists in practice, it is utterly inadequate to represent the nucleon in question by a wavefunction corresponding to the separation energy: if the parentage spectrum is concentrated at high excitation with essentially no parentage to the ground state of the parent nucleus the bulk of the single particle wavefunction proper for use in computing the

Coulomb energy will drop off correspondingly sharply even though in the extreme asymptotic region its fall off corresponds to the separation energy; these considerations are particularly clear for states in the nucleon continuum whose nucleon decay is isospin forbidden but they apply everywhere. The gravity of these considerations is illustrated by the fact that even for the *p*-shell nuclei the trivial error of only 2 MeV in locating the effective centroid of the parentage spectrum can make a difference of 0.1 MeV to the Coulomb energy.

The only computations that take quantitative account of the radial form of the nucleon configurations as determined by an experimental partial parentage spectrum supplemented by a semi-realistic theory of the residual parentage are in the 1*p*-shell (Wilkinson and Hay 1966). For the odd-*A* mirror nuclei agreement between theory and experiment indicates that charge symmetry holds in the nucleon-nucleus, and so presumably also in the *NN*, interaction to 0.7% or better. There is some indication of discrepancy in the even-*A* cases that could indicate an *np* strength some 2% greater than the *nn* or *pp* strength (assuming charge symmetry). These indications are then precisely in line with those that we have earlier seen, in section 2, to come from the elementary interaction data.

A further test of charge independence from Coulomb energies follows from a study of the isobaric multiplet mass relation:

$$M(T_z) = a + bT_z + cT_z^2$$

which is discussed in section 6. The Coulomb contributions, b^c, c^c to the coefficients b, c are related only through models (see e.g. Jähnecke 1969) but for the special case $T = A/2$ we have the rigorous relationship:

$$b^c = (A - 1)\, c^c$$

which may immediately be seen by setting equal the Coulomb contribution to the mass for $T_z = \frac{A}{2} (Z = 0)$ and $T_z = \frac{A}{2} - 1 (Z = 1)$. This relationship is only approximate, but cannot be far wrong for $T \neq A/2$. Analysis of the 1*p*-shell (Wilkinson 1964) shows that this relationship is obeyed such that the necessary charge-dependent nuclear supplementation corresponds to a force of order only 1% of the charge-independent force. As Garvey (1969a) has correctly cautioned, great store is not to be placed on this because, for example, the Coulomb vector interaction in second order may strongly modify c^c (see section 6) but at least this study is consonant with the other approaches to charge independence.

Coulomb energies are indeed very tricky to discuss quantitatively on an individual basis from the point of view of charge independence although their systematic trends may illuminate systematic nuclear changes that are of no interest to us in the present context. The best illustration of these very real difficulties is provided by the ^{3}He–^{3}H system. The wavefunctions for this system have been the object of exhaustive theoretical study for many years and the electron elastic scattering form factors are experimentally well known; a detailed interim report has been given by Delves and Phillips (1969). From the quantitative point of view our knowledge of the $A = 3$ wavefunctions far surpass that of the wavefunctions of more complex nuclei. On the other hand the uncertainties in going from the electron-scattering form factor to the Coulomb energy are in some ways reduced for the more complex nuclei because there the antisymmetization imposes firmer constraints than for $A = 3$. It is not possible, at this time, to compute the Coulomb energy of ^{3}He with a confidence of better than 0.1 MeV or so. There may be a significant tendency, emphasized especially by Okamoto and Lucas (1967), for the calculated Coulomb energy to underestimate the binding energy difference by as much as 0.05–0.2 MeV; this could indicate an *nn* interaction about 1% stronger than the *pp*, but this is by no means established yet.

4 CHARGE INDEPENDENCE IN FORBIDDEN FERMI BETA-DECAY

In a completely charge-independent situation Fermi beta-decay takes place only within isospin multiplets. This rule will be relaxed by charge-dependent effects: the Coulomb interaction and any charge-dependence of the nuclear force. Reliable and extensive information on forbidden Fermi matrix elements comes form Bloom and his colleagues (Bloom 1964 and later work) in their studies of $\beta - \gamma$ (circular polarization) correlations. The analysis of these data (Yap 1967) seems to demand, in addition to the Coulomb perturbation, a charge dependence to the nuclear force of magnitude, in relation to the charge-independent part, given by $s \approx 1\%$; $q \approx 3\%$ where the s and q parts have the form $\boldsymbol{\sigma}_i \cdot \boldsymbol{\sigma}_j T_{ij}$ and T_{ij} respectively ($T_{ij} = t_{zi}t_{zj} - \frac{1}{3}\,\mathbf{t}_i\mathbf{t}_j$). These results are consistent with those of sections 2 and 3.

This analysis assumes the validity of CVC and that other second forbidden effects are not large. These assumptions are probably acceptable.

5 SUMMARY OF EVIDENCE AS TO CHARGE INDEPENDENCE AND SYMMETRY

We may summarize the evidence referred to in sections 2, 3 and 4 by saying that all lines of enquiry indicate, or are consistent with, a charge dependence of the nuclear force of order 2% in $^1S_0$ states, the *np* interaction being stronger than the *nn/pp*, while charge symmetry is established to 1% or better, there being a little, weak, evidence that the *nn* force may be slightly stronger than the *pp* in $^1S_0$ states.

This well-established breakdown of charge independence of the nuclear force in the percentage order means that this charge dependence, as well as the effects of the Coulomb force, must be considered in computations of such matters as isospin mixing if serious quantitative results are sought. Neglect of the nuclear charge dependence would not appear to be justified unless a special case can be made in particular circumstances.

6 ISOBARIC MULTIPLET MASSES

Charge independence is certainly good enough to allow us to define the isospin of complex nuclei and to identify isobaric multiplets. If any departures from charge independence, the electromagnetic interactions included, may be treated as first order perturbations of a two-body nature we derive the isobaric multiplet mass equation:

$$M(T_z) = a + bT_z + cT_z^2$$

where the b and c terms that give the charge-dependence of the mass derive from the isovector and second order isotensor components of the charge-dependent perturbation respectively. From the form of the matrix element we should expect the c coefficient to be much smaller than the b, and so it is experimentally, by a factor of order 10 or more, for most of the 8 cases of $T = 3/2$ isospin quartets that have been investigated in detail (Cerny 1968). (For the classical case, b exceeds c by roughly a factor of A.)

It is of interest to inquire into the accuracy with which this equation is followed. This is indicated in Fig. 1 for the three cases, $A = 9$, 13 and 21, that have been investigated with greatest accuracy. The analysis depends on recent measurements by Mendelson *et al.* (1970) and by Trentelman *et al.* (1970). A least squares fitting has been made to the parabolic mass equation; the figure indicates the departure of the experimental mass from the parabolic fit. Only for $A = 9$ (with $\chi^2 = 4.5$) may the departure be significant. We may conclude with some confidence that, for these cases, the coefficient d of a possible dT_z^3 term in the above equation is less than 10 keV (as against about 200 keV for c); this directly fixes a limit to the diagonal matrix element of any interaction that transforms as a third order tensor in isospin space.

Beyond this formal remark about the third order tensor interaction care is necessary in discussing the remarkably high precision with which the

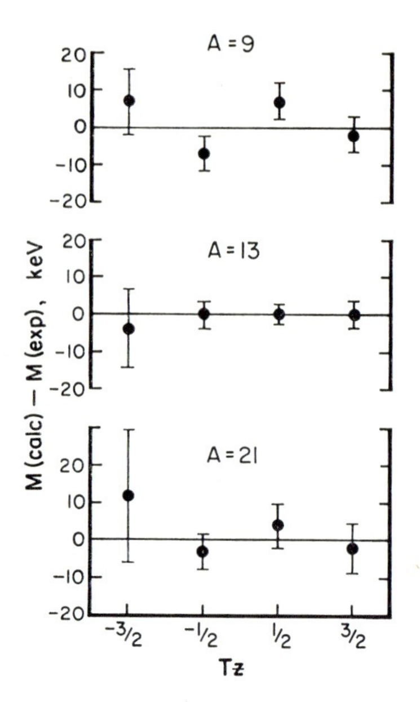

FIGURE 1 Least-squares fitting of the masses of the $A = 9$, 13 and 21 $T = \frac{3}{2}$ quartets to the isobaric multiplet mass equation

parabolic law is followed. It certainly tells us nothing in first order about the charge-dependence of the two body nuclear force for such a charge-dependence itself generates the parabolic form. It should also be noted that the formula should still hold, approximately at least, even though isospin-allowed decay may take place to the continuum provided that the wave-functions of the several members of the multiplet in the outgoing charged-particle channels are sufficiently similar.

To discuss the T_z^3 coefficient we may go to second order perturbation theory where the charge-dependent forces will mix isospins and where we find coefficients of the form:

$$d \approx \sum_i \frac{\langle \alpha_0 T \| H^1 \| \alpha_i T_i \rangle \langle \alpha_i T_i \| H^2 \| \alpha_0 T \rangle}{E_0 - E_i}$$

Here H^1 and H^2 are the isovector and second order isotensor perturbations respectively. Now, as already remarked, H^2 has small matrix elements; the H^1 matrix element above divided by $E_0 - E_i$, is a measure of the isospin mixing amplitude. As we shall see later, the isospin mixing appears to be 10^{-3} or so by intensity so we expect the d coefficient to be crudely of order less than a few per cent of the c coefficient, namely a few keV or less as we have seen it to be. The chief effects of isospin mixing are found theoretically in modifications to the b and c coefficients but they are not so found experimentally, of course, because experimentally they are simply subsumed in the total b and c coefficients.

An additional effect that must be considered is external mixing in the sense that is of great importance for discussion of the isobaric analogue resonance phenomena in heavier nuclei (see e.g. Lane 1969; Robson 1969). It does not seem likely that this is of great significance for the present cases of isobaric multiplets in light elements.

It must be emphasized, as has been done by Garvey (1969b), that the internal mixing, second order perturbation theory, approach relates the d coefficient to isospin mixing and so studies of the isospin purity of the multiplet are of importance. If significant d coefficients are associated with very high isospin purity then we must, in the absence of adequate external mixing, conclude that we have evidence for a charge-dependent (many-body) force transforming like H^3.

An alternative approach to the problem of the d coefficient is the non-perturbative model of Henley and Lacey (1969). $T = 3/2$ quartets were studied on a 3-valence nucleon model, including a $t \cdot T$ term in the nuclear potential, in which the $T_z = \pm 1/2$ members were of mixed $T = 3/2, 1/2$ isospin. The model rather satisfactorily reproduces the empirical relationship between the b and c coefficients and gives values for the d coefficients of order 1 keV or less.

The conclusion, from either approach, is that the superficially-surprising accuracy of the mass equation is indeed to be expected in a closer view. But we must repeat the remark that we have really not learnt much from this since, on the one hand, the parabolic form is itself very stable against perturbations while, on the other, its experimental coefficients are not susceptible of further breaking down as to their origin in terms of the various bits of the overall (charge dependent) Hamiltonian.

7 COMPARISONS WITHIN ISOBARIC MULTIPLETS

For a completely charge-independent system the different members of an isobaric multiplet are distinguished only by their T_z values. It is important to enquire whether the relaxation of charge independence by Coulomb or other charge-dependent forces appreciably modifies this zero-order expectation. We will examine two ways in which this has been investigated experimentally.

Consider reactions such as:

$$A + a \begin{array}{l} \to B + b \\ \to B' + b' \end{array}$$

where B, B' are members of an isospin multiplet (also b, b') and where the isospins are such that only a single isospin coupling is possible ($T_A + T_a = T_B - T_b$). Under these circumstances, if the nuclear wavefunctions are indeed identical within the multiplets, all aspects of the two reactions must be identical apart from kinematical and Coulomb-associated "DWBA" factors. The cross sections for the two reactions should stand simply in the ratio of the squares of the Clebsch-Gordan coefficients in the isospin.

Consider specifically the case $a = p$; $b = t$, $b' = {}^3$He. We then have:

$$\frac{(\mathrm{d}\sigma/\mathrm{d}\Omega)_{p,t}}{(\mathrm{d}\sigma/\mathrm{d}\Omega)_{p,{}^3\mathrm{He}}} = \frac{k_t}{k_{^3\mathrm{He}}}\frac{2}{2T_B - 1}$$

where the k are the respective wavenumbers and where we have ignored the "DWBA" effects.

So if comparisons are made between (p, t) and $(p, {}^3\mathrm{He})$ reactions leading to states of $T_B = T_A + 1$ under conditions where the awkward "DWBA" factors are small or negligible we can test this simple relationship and so obtain some quantitative measure of the similarity between the B, B' wavefunctions.

Not only should the differential cross sections be related by the above simple factor but also any other properties of the reactions, for example the polarizations involved, should be the same. Both these expectations are dramatically borne out in Figs. 2 and 3 which compare the differential cross sections and analyzing powers of the reactions ${}^{16}\mathrm{O}(p, t)\ {}^{14}\mathrm{O}_{\mathrm{g.s.}}$ and ${}^{16}\mathrm{O}(p, {}^3\mathrm{He})\ {}^{14}\mathrm{N}_{2.31\ \mathrm{MeV}}$ for an incident energy $E_p = 49.5$ MeV (Nelson *et al.* 1970). In the comparison of the differential cross sections the above-expected factor has been inserted.

A large number of such $(p, t) - (p, {}^3\mathrm{He})$ comparisons has been made by Hardy *et al.* (1969) also at $E_p \approx 50$ MeV where "DWBA" uncertainties are small. These data, integrated over angle are shown in Fig. 4 which also shows the T_B values and the expectation of the above formula based on isospin invariance. A direct weighted average of the ratios between the experimental and theoretical cross section ratios gives for the 13 cases the omnibus result:

$$\frac{[\sigma(p, t)/\sigma(p, {}^3\mathrm{He})]_{\mathrm{exp}}}{[\sigma(p, t)/\sigma(p, {}^3\mathrm{He})]_{\mathrm{theo}}} = 1.026 \pm 0.029.$$

It is not possible to give a very direct interpretation to this number but we may, in the very crudest way, set:

$$2\alpha \approx \varepsilon\sqrt{N}$$

where α is a measure of the isospin mixing amplitude, ε the departure of the above ratio from unity, and N the number of observations involved. This then

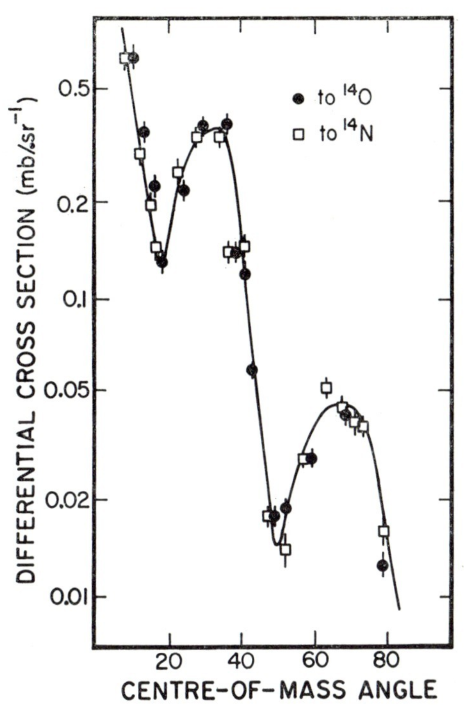

FIGURE 2 Differential cross sections for the reactions $^{16}O(p, t)\,^{14}O$ and $^{16}O(p, {}^3He)\,^{14}N_{2.31\,MeV}$ for $E_p = 49.5$ MeV

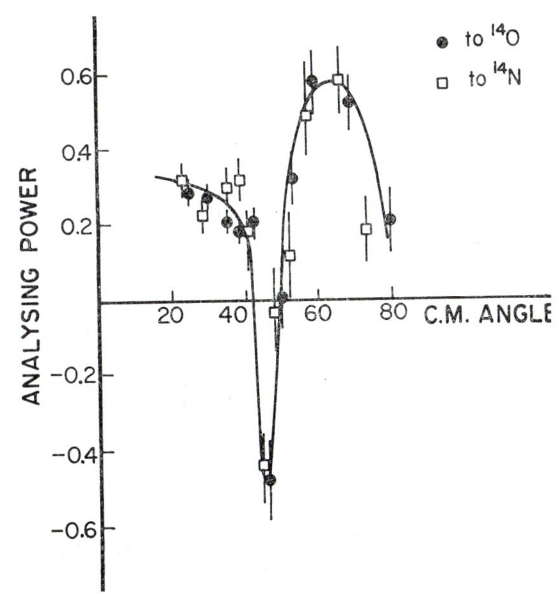

FIGURE 3 Analyzing powers of the two reactions of Fig. 2

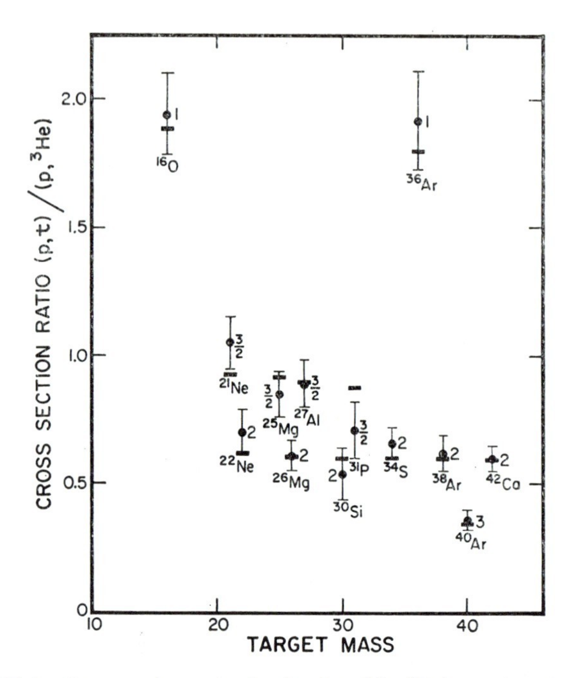

FIGURE 4 Cross section ratios for (p, t) and $(p, {}^3\text{He})$ reactions leading to members of isospin multiplets where unique isospin couplings are involved. The target nuclei are named below the experimental points; the isospins of the final states are shown to the side of the experimental points. The heavy horizontal bars show the theoretical ratios

suggests that $\alpha \leqq 0.06$ so that the analogue states are adulterated by less than a few parts per thousand by intensity. We may note that this figure also applies to the differences, as between the analogue states, in the mixing of other states of the same isospin which is something not susceptible of approach through selection rules; the present information, although crude, therefore becomes additionally valuable.

Another approach to the closeness with which analogues resemble each other is via the Barshay-Temmer (1964) theorem. Consider the reaction:

$$A + B \rightarrow C + C'$$

where C, C' are analogues and where $A + B$ is of definite isospin. In this case nothing in the strong interactions through which the reaction goes distinguishes C fom C' and so their angular distributions must be symmetrical fore-and-aft. This is illustrated in Fig. 5 for the reaction ${}^{12}\text{C}({}^{14}\text{N}, {}^{13}\text{C})\,{}^{13}\text{N}$ at $E_N = 78$ MeV (von Oertzen *et al.* 1969) leading to the ground states. The fore-aft differential cross sections are identical within 2.5% over the angular

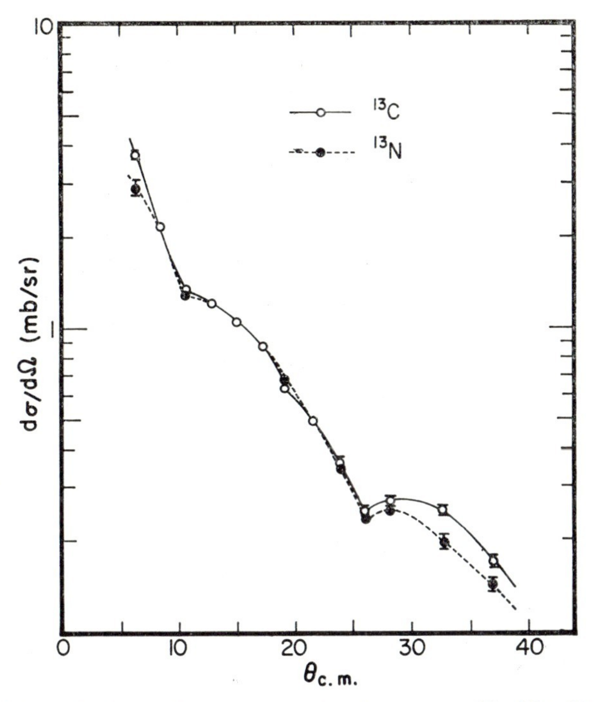

FIGURE 5 Differential cross section for the reaction $^{12}C(^{14}N, ^{13}T)\,^{13}N$ leading to the ground states. (Identity of angular distribution for the two product nuclei in the forward hemisphere implies reaction symmetry about 90°)

range (8–30°) in which there is good experimental confidence; integrating over this range we have, for ^{13}C:

$$\int_{8^\circ}^{30^\circ} \text{fore} \Big/ \int_{8^\circ}^{30^\circ} \text{aft} = 1.014 \pm 0.015.$$

Such a reaction as this, as is suggested by the form of the angular distribution, goes largely by surface transfer; this permits us to analyze the fore-aft symmetry in terms of the difference between the amplitudes of the corresponding nucleon wavefunction in the analogues (n in ^{13}C; p in ^{13}N) at the transfer radius (about 3.9 fm) we find:

$$(\psi_n - \psi_p)/\psi \leqq 0.8\%.$$

This is in accord with the earlier estimate from the $(p, t) - (p, {}^3He)$ comparison.

A particularly interesting reaction in this context is:

$$^4He + d \rightarrow {}^3H + {}^3He.$$

This has been the object of two accurate studies. In the first (Kim Sin Nam *et al.* 1970) tritons of energy 1 to 1.5 MeV were used and symmetry about 90° was established to better than 1.5% [the angular distributions being approximately of the form $1 + 0.7\,P_2(\cos\theta)$]. In the second (Gross *et al.* 1970b) alpha-particles of 82 MeV were used; the angular distributions were deeply modulated and showed departures from 90° symmetry of up to 10% while the differential cross section itself varied by a factor of about 5. The former measurements appear to limit the magnitude of the isospin-forbidden amplitudes to about 2%. The latter measurements appear to constitute a clear violation of simple charge symmetry; it is rather unlikely that Coulomb-associated "DWBA" type corrections in themselves can be significant at the quite-high energies involved. An *nn-pp* force difference of the percentage order would seem to be called for if the blame is laid to the ^{3}He–^{3}H difference. This is indeed of the order of strength of the Coulomb force and we have seen in section 2 how the Coulomb-induced structural difference between ^{3}He and ^{3}H was enough largely to account for the several per cent effect in their production in the $p + d$ pion-producing reaction. One must, of course, also consider the possibility that the reaction mechanism is inducing the difference. It is by no means clear that this cannot be so. The intermediate "^{6}Li", even though short lived and the reaction a direct one, might be expected to show effective $T = 1$ amplitudes of order 1/137 (which would be greatly increased if the system were longer-lived) and this could be adequate to produce the observed effect. There is also the "fast" mechanism proposed by Noble (1967) which would seek to lay the blame on the indirect induction of a $T = 1$ component into the deuteron consequent upon the Coulomb distortion of its wavefunction; the effect of the distortion is not the direct Coulomb induction of the $T = 1$ component but is rather consequent upon the Coulomb polarization, to expose the neutron and proton of the deuteron to different strengths of the (charge-independent) nuclear spin-orbit interaction and so to produce a differential spin-flip probability between the two.

It is not clear why, on either view, the effect should change much as between the two experiments. The centre of mass excitations differ (16 MeV in the first and 27 MeV in the second) but not greatly and it is difficult to see how the low centre-of-mass energies of the ^{3}He, ^{3}H in the first experiment could bring about a symmetry that is lacking in the second experiment, short of the unlikely chance that a single compound nucleus state dominates the cross section. In summary, it cannot be said that this rather-large violation of the Barshay-Temmer theorem obviously demands more than the Coulomb force for its understanding.

If isospin multiplets are of pure isospin then certain other predictions follow concerning their dynamical properties. One simple consequence of isospin conservation is that corresponding $\Delta T = \pm 1$ transitions in mirror nuclei have identical strengths. It is unfortunately not easy to test this prediction but it has been rather crudely checked in the $A = 13$ system by Adelberger *et al.* (1969) who compare the ground state radiative widths of the lowest $T = 3/2$ states of ^{13}C (15.10 MeV) and ^{13}N (15.07 MeV) finding 25 ± 7 and 27 ± 5 eV respectively. The states in question are of very pure isospin (see later) so the equality of the radiative strengths is to be expected to a much higher degree of precision than is here seen experimentally but the demonstration remains a welcome one.

It should be noted that such equality of gamma-transition strengths for mirror transitions is only expected for the case, as usually assumed, that the electromagnetic current has only isoscalar and isovector components and so can induce only $\Delta T = 0$ and $\Delta T = \pm 1$ transitions. A possible isotensor component could be sought directly in $\Delta T = 2$ transitions but it can also be sought in an inequality between such $T = 3/2 \rightarrow T = 1/2$ mirror transitions as we have just considered in $A = 13$. The ratio of such transition strengths is (Blin-Stoyle 1969a)

$$1 - 4\left(\frac{3}{5}\right)^{1/2} R$$

where R is the ratio of the transition amplitudes due to the isotensor and isovector parts of the electromagnetic interaction. From the example, in $A = 13$ it can now be said that $R \leqq 0.15$. This approach to an isotensor electromagnetic interaction may be better than the direct search for $\Delta T = 2$ transitions since both approaches are obscured by isospin impurities and there should perhaps be less of these for the $T = 3/2$ states than for the $T = 2$ (or $T = 5/2$) states since the former are somewhat lower-lying.

It may be useful to emphasize that it is only in restricted circumstances that charge independence (symmetry) demands equality of radiative strengths in mirror nuclei: for all $\Delta T = \pm 1$ transitions and, in the case of $E1$ only, for $\Delta T = 0$ also. In other cases there may be differences, perhaps large (see e.g., Warburton and Weneser 1969). This we illustrate with reference to $E2$ transitions where, very obviously, differences across an isobaric multiplet are to be expected. Consider a triad of $E2$ transitions between analogue $T = 1$ states; we find the matrix elements:

$$T_z = 0: \quad \|M^0\|$$

$$T_z = \pm 1: \quad \|M^0\| \pm \frac{1}{\sqrt{2}} \|M^1\|$$

where $\|M^0\|$ and $\|M^1\|$ are the parts due to the isoscalar and isovector electromagnetic interaction respectively. Now we do not expect the isovector part to show any collective enhancement whereas the isoscalar part may. In the $A = 30$ case involving the first excited states of the $T_z = \pm 1$ nuclei and the analogue transition in $T_z = 0$ Bini *et al.* (1970) indeed find that the isoscalar part of the transition has a strength of 9.9 ± 1.9 Weisskopf units while the isovector part is only about 0.2 W.u. A similar complete analysis has been made in $A = 26$ by Schulz *et al.* (1970) who find an isoscalar strength of 12 ± 2 W.u. and an isovector strength of 0.1–1.1 W.u.

A very general relationship has recently been pointed out by M. Ram (unpublished) who shows that the *total* gamma decay widths of members of an isospin multiplet are related by

$$\Gamma_\gamma\ (T_z) = a + bT_z + cT_z^2$$

just as are the masses. It would appear difficult technically to examine this relation.

8 ISOSPIN IMPURITY

Isospin cannot be pure-except for the deuteron where it is 100% but where it has no literal meaning. It is of great interest to enquire into the degree of isospin purity under the widest range of circumstances, to find empirically the relaxation that the impurity effects in isobaric analogue symmetry and to ask theoretically for the expectation as to that purity that derives purely from the Coulomb perturbation to see whether a further, nuclear, charge dependence is demanded.

Empirically isospin purity covers a tremendous range. There are many examples where the purity is obviously very high. The low-lying $T = 3/2$ states in light nuclei of $T_z = +\frac{1}{2}$ are isospin-forbidden in their proton decay and indeed have very tiny proton widths; for example the partial widths for ground state proton emission from the lowest $T = 3/2$ states of ^{17}F and ^{25}Al are only about 50 and 17 eV respectively while the associated single-particle widths are several MeV; widths of a few tens or few hundreds of eV are everywhere typical (see e.g. Temmer (1969) for a collection of data). Extensive quantitative data are not available for the similar forbidden (proton and alpha-particle) decays of the $T_z = 0$ members of $T = 2$ multiplets but the indications are that the purity is there high also—for example the ground state proton decay of the first $T = 2$ state of ^{20}Ne has a width of about 130 eV against a single particle width of about 7 MeV (Bloch *et al.* 1967; Kuan *et al.* 1967). It is not possible, without a detailed model, to translate such numbers into isospin purities but it appears, in rather general terms, as though impurities of about 0.1% (by intensity) are typical.

But there are also cases where isospin purity is very poor. The best known example is the first $T = 1$ state of ^{8}Be which does not exist but which appears as two states at 16.63 and 16.93 MeV that are equally broad against alpha-particle decay and that behave rather like $^7\text{Li}_{\text{g.s.}} + p$ and $^7\text{Be}_{\text{g.s.}} + n$ respectively and so must (or may) both be thought of as 50–50 $T = 0$ and $T = 1$ (Marion 1965). This case may be sharply contrasted with the superficially very similar one in the neighboring ^{12}C where the first $T = 1$ state has an alpha-particle width of only 0.8 eV (Reisman *et al.* 1969) against a single-particle width of some MeV. But there are many other old-established cases of high isospin impurity (see e.g. Wilkinson 1958) in nuclei as light as ^{10}B.

It is, in fact, rather difficult to gain reliable information about the isospin purity of the low-lying states about which news is most to be desired. This is because they have no open heavy-particle isospin-forbidden channels and because tests involving their formation in nuclear reactions are obscured by the possibility that breakdown of the isospin selection rules resides in the reaction mechanism rather than in the final states. The best information comes from the gamma-ray selection rules (see e.g. Warburton and Weneser 1969) and from isospin-forbidden Fermi-decay (which, however, is very specific in its reference to mixing only with the analogue of the beta-emitter) (see e.g. Blin-Stoyle 1969b). Such information is consistent with an isospin impurity of 1% or less (by intensity).

Theoretical expectation as to isospin impurity (see e.g. Soper 1969) for low-lying states throughout the periodic table is substantially less than 1% (by intensity) and this is in accordance with experiment. It is indeed quite difficult in explicit calculations, using shell-model configurations, to produce isospin impurities of several per cent. This is true whether one uses as the charge-dependent perturbation the Coulomb force alone or supplements it by a nuclear charge-dependence of the type already indicated. It is not, on the other hand, difficult to understand the high purity that we have seen to obtain for the low-lying $T = 3/2$ states in $T_z = +\frac{1}{2}$ nuclei, for example. The reason why isospin purity remains high even when, as is the case for the $T = 3/2$ states discussed, states of the "mixing" isospin are quite abundant nearby is that the mixing matrix elements of the isospin non-conserving interaction, Coulomb or nuclear, tend to be small except with the $T_<$ states that are configurationally closely related to the $T_>$ states in question, differing from them in the extreme only by an isospin-flip when they are called the anti-analogue states. But such $T_<$ states are separated from their $T_>$ configurational companions by the symmetry term, the $t \cdot T$ term of the optical model potential, and so are several MeV away; the effect of the large matrix element is then cut down by the large perturbation denominator. Occasionally, by chance, states that are closely related configurationally may come close together in energy, as in the ^{8}Be case referred to, but this is exceptional (and also surprising). External mixing is usually ignored in these considerations of light nuclei and is felt to be relatively unimportant. It remains, however, to be investigated in more detail.

It is a rather general result that where plausible detailed calculations of isospin impurity based on specific shell model assignments can be made, including external mixing, they fail to produce adequate amounts of isospin mixing when they restrict themselves to the Coulomb perturbation, although they only fail by modest factors. This is illustrated by Barker's (1966) attempt to explain the complete isospin mixing of the 16.63 and 16.93 MeV levels of ^{8}Be already referred to where the failure is by a factor of about 2. It is also illustrated by, for example, the attempt by Bertsch (1970) to understand the apparently complete splitting into two components, separated by 70 keV, of the analogue at about 3.5 MeV in ^{56}Co of the ground state of ^{56}Fe (Dzubay *et al.* 1970) where the failure is again by a factor of about 2. One is left with the impression that the charge dependence of the nuclear force, well established as we have seen in sections 2, 3 and 4 must be brought into play as noted in section 5. (Although this would not, of course, help in the self-conjugate case of ^{8}Be). It is significant that where detailed calculations have been made using a charge-dependent nuclear force such as set out in section 4, the contribution to the isospin mixing from the nuclear charge-dependent force is indeed roughly equal to the contribution from the Coulomb interaction (see e.g. Blin-Stoyle 1969b).

Of course, when the density of $T_<$ states surrounding the $T_>$ state in question becomes sufficiently high dissolution of the analogue is inevitable; it splits into many components and we move into the regime of the isobaric analogue resonances as they are generally understood where the components are not resolved (see e.g. Lane 1969; Robson 1969). The intermediate stage, where the analogue has split into several, but still enumerable, components, is shown in the classic study of ^{41}K by Keyworth *et al.* (1966) in which, at an excitation of about 10 MeV, two analogues (corresponding to the fourth and sixth excited states of ^{41}Ar) each fragment into many (10–30) components, dissolving into the surrounding $T_<$ states which have a mean spacing of some 6–10 keV. Such major dissolution with its attendant high isospin impurity of the fragments must always be kept in mind when the density of surrounding $T_<$ states gets high and, as already emphasized, can also occur "by chance" where the $T_<$ level density is relatively small if a suitable $T_<$ mixing state presents itself close to hand.

It is also interesting to enquire into the anatomy of the isospin impurity and to ask, for example, whether the isospin impurities of analogue states within a multiplet are themselves similar. The answer, as we have it so far from very limited data, seems to be that they are definitely not. Two cases have been investigated namely the isospin-forbidden nucleon decays of the $T = 3/2$ states of ^{13}C and ^{13}N at 15.10 and 15.07 MeV respectively (Adelberger *et al.* 1969) and those of the $T = 3/2$ states of ^{17}O and ^{17}F at 11.08 and 11.20 MeV respectively (McDonald *et al.* 1970). In the first case the branching ratios in the break-up of the $T_z = +\frac{1}{2}(-\frac{1}{2})$ states by proton (neutron) emission to the ground and first excited states of ^{12}C were measured with the result:

$$\text{for } ^{13}\text{N} \quad \theta^2_{\text{g.s.}}/\theta^2_{\text{exc.}} = 1.18 \pm 0.11$$

$$^{13}\text{C} \quad \theta^2_{\text{g.s.}}/\theta^2_{\text{exc.}} = 0.17 \pm 0.03.$$

In the second case the ratio of the transitions to the ground state of ^{16}O and to one or other (or both) of the states at 6.05 and 6.13 MeV were measured. The deduced ratios of reduced widths depend on the assumption about the excited state but in either case there is as large departure from equality:

$$\left.\begin{array}{ll} \text{for } ^{17}\text{F} & \theta^2_{\text{g.s.}}/\theta^2_{\text{exc.}} = 0.16 \pm 0.05 \\ \text{for } ^{17}\text{O} & \theta^2_{\text{g.s.}}/\theta^2_{\text{exc.}} = 3.4 \pm 1.4 \end{array}\right\} E_x = 6.05 \text{ MeV}$$

$$\left.\begin{array}{l} \text{or } 0.066 \pm 0.019 \\ \text{or } 0.32 \pm 0.14 \end{array}\right\} E_x = 6.13 \text{ MeV}$$

The interpretation of these differences is not clear. We may simply be seeing the effect of differences in the separations of the contaminating levels in the two systems or differences (which will certainly exist because of the different energetics) in external mixing or coupling to the continuum. Or we may be seeing the effect of the differing Coulomb perturbation in the two T_z cases which may be expressed by writing the admixed impurity from a particular contaminating level i as proportional to:

$$\langle \alpha T_0 \| H^1 \| \alpha_i T_i \rangle \pm \langle \alpha T_0 \| H^2 \| \alpha_i T_i \rangle$$

H^1 and H^2 being the usual spherical tensor operators. In this case we understand the difference as reflecting the role of the isotensor term.

It is also interesting to note the behavior, as a function of A, of the isospin-forbidden decay of $T = 2$ states in $T_z = 0$ nuclei (McGrath *et al.* 1970). Here proton decay ($\Delta T = 1$) and alpha-particle decay ($\Delta T = 2$) compete (the neutron channel is closed). The alpha-particle percentages are:

^{20}Ne	^{24}Mg	^{28}Si	^{32}S	^{40}C
70	25	90	75	≈ 100

The apparent conclusion that $T = 0$ admixtures in the $T = 2$ states are as important as $T = 1$ admixtures is surprising but not startling since the two-body Coulomb matrix elements have isotensor components although these are normally reckoned to be relatively weak. Explicit shell model calculations using restricted configurations do not give quite enough $T = 0$ admixture in the $T = 2$ states but are not sufficiently deficient to cause alarm.

9 SYMMETRY IN FERMI BETA-DECAY

It has already been remarked that allowed Fermi transitions may take place only between members of an isospin multiplet if isospin is a perfect quantum number. The theory of the conserved vector current (Feyman and Gell-Mann 1958) further tells us that the vector beta-decay strength should not be renormalized by the strong interactions so that all nuclear Fermi matrix elements should be determined solely by the appropriate isospin Clebsch-Gordan coefficients. Furthermore if we adopt the hypothesis of universality and modify it as suggested by Cabibbo (1963) to split the vector strength between strangeness-conserving and strangeness-changing beta-decay we may use the lifetime of the muon and the Cabibbo angle to predict the strength of nuclear Fermi beta-decay in the absence of isospin mixing. In making this comparison we must attend to factors outside the weak and strong interactions that have guided us to this point namely the electromagnetic radiative corrections to the beta-decay process itself (as opposed to those that may lead to a change in the nuclear Fermi matrix element by isospin mixing and so on). These radiative corrections may be split into two parts: the "outer" radiative corrections δ^R are, to lowest order, model-independent but energy-dependent; the "inner" radiative corrections, $\Delta^R_{V,A}$ are, again to lowest order, model-dependent but energy-independent; both corrections are to order α. We may therefore write, for a Fermi transition:

$$ft = \frac{2\pi^3(\ln 2)}{G_V^2}\,[|M_V|^2(1 - \varepsilon)\,(1 + \delta^R)\,(1 + \Delta^R_V)]^{-1}$$

where the factor ε represents the modifications to the "perfect" matrix element M_V due to isospin mixing, etc. The correction δ^R has been given by Sirlin (1967). Figure 6 shows the effect of applying only the δ^R correction to the above expession for ft to the 7 best studied cases of nuclear Fermi transitions, those between $J^\pi = 0^+$ analogue states starting from the body named. We see immediately three things:

(i) The 7 cases show very nearly equal ft values; this shows that CVC is working and that ε in the above expression is either negligible or roughly constant;

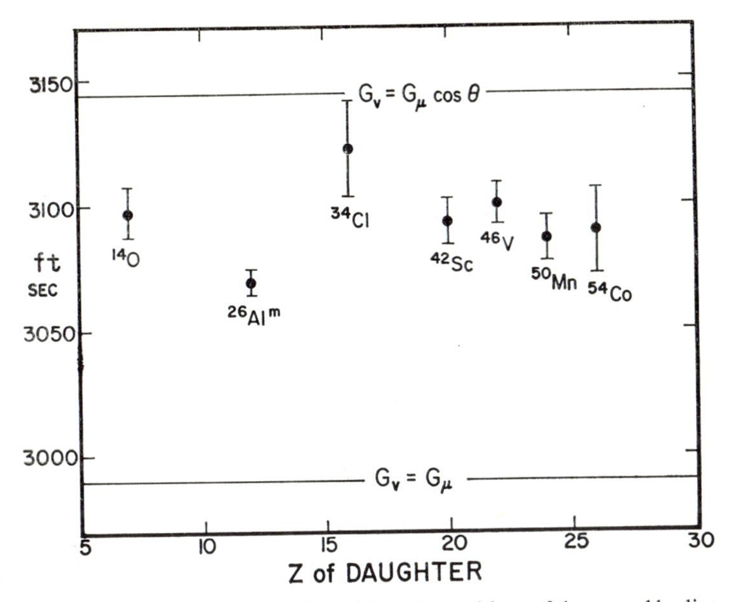

FIGURE 6 ft values for the allowed Fermi transitions of the named bodies. The experimental points have been corrected for the "outer" radiative corrections discussed in the text. The horizontal line marked $G_V = G_\mu$ shows expectation based on equality of the nuclear Fermi coupling constant and that governing muon decay; the line marked $G_V = G_\mu \cos\theta$ shows expectation based on the additional assumption of Cabibbo theory, θ being the Cabibbo angle

(ii) Nuclear Fermi transitions are significantly slower than expectation derived from muon decay alone (the line marked $G_V = G_\mu$) and simple universality;

(iii) The radiative correction δ^R has not been adequate to bring nuclear beta-decay into line with the Cabibbo expectation (marked by $G_V = G_\mu \cos\theta$). (For the Cabibbo angle we use $\sin\theta = 0.221 \pm 0.004$ from the recent analysis of Blin-Stoyle and Freeman (1970) whom we largely follow in this section.)

We may at this point pause to discuss ε. This has many components (Blin-Stoyle 1969b) but it is very sure that ε must be positive. A finite, positive value of ε, not allowed for in Fig. 6 will widen the gap that is to be made up by Δ_V^R between the experimental points and the Cabibbo expectation. Theoretical estimates of ε using both the Coulomb interaction and a charge-dependent nuclear force consistent with the indications given in sections 2, 3 and 4 give values of no more than a few tenths of a per cent, up to a per cent or so at most for the heaviest elements concerned. The especially low value of ft for ^{26}Al^m should be noted. This value appears to be significantly lower than the others and can only be understood, if correct, as reflecting a significantly lower value of ε. It is interesting that the divergence between $^{26}Al^m$ and the others (about 0.6%) is of the order of the best estimates for ε. Blin-Stoyle and Freeman (1970) have taken the view that $\varepsilon \approx 0$ for $^{26}Al^m$ and that the *ft* value of this body should be used in assessing the shortfall between nuclear Fermi-decay and the Cabibbo expectation that must be made up for by Δ_V^R. In this cas $\Delta_V^R = 2.3 \pm 0.3\%$: if we rather took the mean of the other Fermi-decays and ignored both $^{26}Al^m$ and ε we should have $\Delta_V^R \approx 1.7\%$. We repeat that these values of Δ_V^R must be increased by the factor $1 + \varepsilon$ whatever ε may be.

To go further we may either seek to assess Δ_V^R as revealed so far, in an attempt to extract information about the "inner" radiative corrections that in turn illuminate the details of the weak interaction process itself, or we may take advice as to Δ_V^R to illuminate the isospin impurity ε. It turns out that these two avenues are really the same because it is really quite difficult to generate values of Δ_V^R as high as 2.3% or even 1.7%.

Consider the model of the radiative corrections that understands the beta-decay process as mediated by a vector boson of mass M_W. We then have (Abers *et al.* 1968):

$$\Delta_V^R \approx \frac{\alpha}{2\pi}\left\{3 \ln \frac{M_W}{M} + 6\bar{Q} \ln \frac{M_W}{M_{A_1}}\right\} - 1.2 \times 10^{-3}$$

where $\bar{Q}$ is the mean charge of the underlying isodoublet (e.g. $\bar{Q} = \frac{1}{6}$ for a Gell-Mann-Zweig quark model) M_{A_1} is the mass of the A_1 axial meson, (this term stemming from the axial component). M is the nucleon mass. Figure 7 shows Δ_V^R as a function of $\bar{Q}$ and M_W and it is seen that $\Delta_V^R = 2.3 \pm 0.3\%$ is a high value, requiring $M_W \geqq 100\ M$ while $\Delta_V^R \approx 1.7$ requires $M_W \approx 50\ M$ (both for $\bar{Q} = \frac{1}{6}$). Such high values of M_W are approaching the unitarity limit which means that they cannot be taken too seriously but they are very high and higher values of Δ_V^R are not likely to be forthcoming for any M_W when the unitarity limit is avoided. We may, in passing, note that this conclusion is not firmly tied to the intermediate vector boson model since the results for Δ_V^R in that model are closely similar to those given by "no-model" models, the boson mass merely replacing the cut-off in the latter models. Reversing the argument we may say that the value of Δ_V^R is already uncomfortably large and its significant increase by a large value of ε could not be tolerated. We can probably say that ε must, in all reasonableness, be less than 1% which accords with the theoretical expectation. [It should be cautiously remarked that these present conclusions could be upset if higher-order radiative corrections, particularly those of order $Z\alpha^2$, turn out to be larger than is now thought—see for example Jaus and Rasche (1970).]

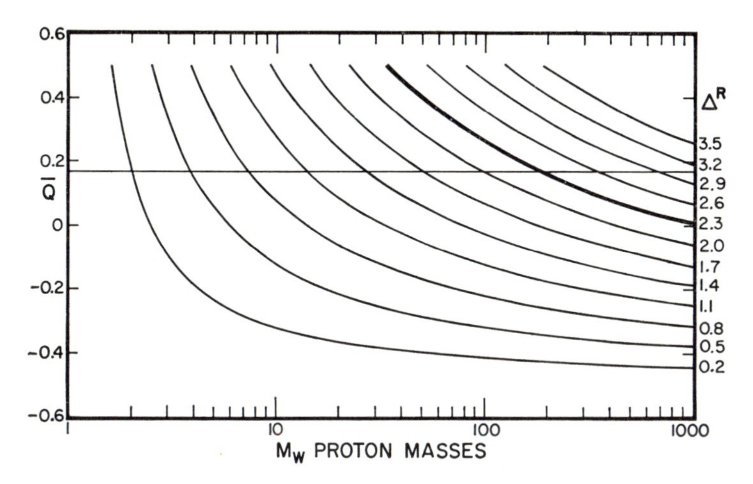

FIGURE 7 The "inner" radiative correction to allowed Fermi decay Δ^R (Δ_V^R of the text), shown as a percentage, for the intermediate vector boson model. M_W is the boson mass and $\bar{Q}$ the mean charge of the underlying isospin doublet (the horizontal line is for $\bar{Q} = \frac{1}{6}$ as on the Gell-Mann-Zweig quark model)

10 SYMMETRY IN GAMOW-TELLER BETA-DECAY

The Fermi transitions enable us to set limits on isospin impurity and simultaneously to make interesting observations on CVC and on the inner mechanism of the beta-decay process. They cannot be used unfortunately, to make a direct test of the identity of the members of isospin multiplets because two Fermi mirror transitions within a multiplet are not yet available. It seems, however, that this test is available by studying Gamow-Teller transitions because there, as for example in the mirror decays of ^{12}B, ^{12}N to the ground state of ^{12}C, the identical decays of two members of a multiplet can be compared. If the two members are identical then our initial expectation is that the *ft*-values for their decay should be the same; a departure from identity for positive and negative decay that we may write:

$$\delta = (ft)^+/(ft)^- - 1$$

would seem to indicate a lack of identity of the two decaying members of the isospin multiplet (and/or of the final states if A is odd).

The present experimental situation (Wilkinson 1970 supplemented by some later work) is shown in Fig. 8 where δ is plotted against the sum of the energy releases in the positon and negaton decay, W_0^+ and W_0^- respectively. It is seen that δ is far from zero for the larger energy releases although it con-

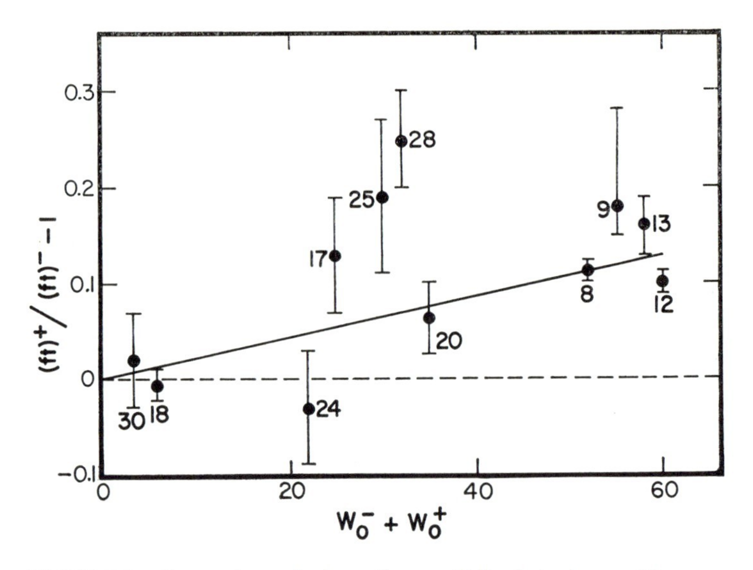

FIGURE 8 Comparison of mirror Gamow-Teller beta-decays. The numbers on the experimental points are the A-values of the decaying systems. The energies W_0^+, W_0^- of positon and negaton decay are in natural units. The solid line corresponds to $G_{IT} \approx 2 \times 10^{-3}$

forms well when the energy release is small. Must we conclude that this study, unlike all the others that we have made, is indicating substantial lack of identity within isospin multiplets? It seems as though differences of several per cent are involved; this conflicts with previous evidence and particularly with that just presented from the Fermi transitions (although there the states involved tend to be more strongly bound than those used in the present test). In some cases that show large δ-values in Fig. 8 the decaying members belong to $T = 3/2$ quartets that follow the isobaric multiplet mass parabola with great precision, namely $A = 9$ and 13, (see Fig. 1), so that it is difficult to hypothesize a gross difference between them—although in both cases, to differing degrees, decay to nucleon-unbound states in the $T_z = \pm\frac{1}{2}$ nuclei is involved so the differences may perhaps be more hopefully sought there. For the case of $A = 12$ Blin-Stoyle and Rosina (1965) have carried out a detailed investigation by the same methods as used for estimating ε in the case of Fermi decays and fail to account for differences as between the $T_z = \pm 1$ members of more than 2–3% at the most as compared with the experimental δ-value of more than 10%.

An alternative to the unwelcome suggestion that we are seeing gross and unexplained departures from isospin analogue symmetry comes from the beta-decay process itself. As Weinberg (1958) pointed out, the strong interactions can induce terms in the beta-decay Hamiltonian that have transformation properties in going from negaton to positon decay of opposite sign to those of the main terms. These are called second class currents and their existence, or otherwise, is a matter of high interest. Following Huffaker and Greuling (1963) we gain for the Hamiltonian operator in the limit of small momentum transfer:

$$H = [\gamma_\mu(1 + \lambda\gamma_5) + i\sigma_{\mu\nu}(A + B\gamma_5)\,\partial/\partial x_\nu + (C + D\gamma_5)\,\partial/\partial x_\mu]\,L_\mu.$$

Here standard notation is used and L_μ is the lepton current. In this expression unity in the first round bracket stands for the vector interaction (unrenormalized by the strong interactions as prescribed by CVC) and λ represents the renormalization of the axial vector interaction. The other terms are the induced terms due to the strong interaction. A is weak magnetism, of magnitude specified by CVC; B is the induced tensor term (with associated coupling constant G_{IT}); C is the induced scalar term (zero according to CVC); D is the induced pseudoscalar term of magnitude specified by PCAC or derivable from dispersion relations. Of these terms all but B and C are first class while those two, only B being possibly finite if CVC holds, are second class and change sign relative to the others in going from negaton to positon emission thereby giving a non-zero value to δ in the above expression.

For sufficiently large W_0 values we have:

$$\delta \approx \frac{4}{3}\,|G_V/G_A|\,G_{IT}(W_0^+ + W_0^-).$$

The straight line of Fig. 8 corresponds to $G_{IT} \approx 2 \times 10^{-3}$ which is of the gross order of the reciprocal of the nucleon mass and so is not unreasonable.

These considerations by no means establish the realtity of second class currents but if such currents do not exist we are faced with a much more systematic breakdown of isobaric analogue symmetry than has been met in the rest of our investigations and one that we are not very obviously equipped to explain.

REFERENCES

E. S. Abers, D. A. Dicus, R. E. Norton, and H. R. Quinn, *Phys. Rev.* **167** (1968) 1461

E. G. Adelberger, C. L. Cocke, C. N. Davids, and A. B. McDonald, *Phys. Rev. Lett.* **22** (1969) 352

F. C. Barker, *Nucl. Phys.* **83** (1966) 418

S. Barshay and G. M. Temmer, *Phys. Rev. Lett.* **12** (1964) 728

G. F. Bertsch, *Nucl. Phys.* **A 142** (1970) 499

M. Bini, P. G. Bizzetti, A. M. Bizzeti-Sona, M. Mando, and P. R. Maurenzig, *Nuovo Cim. Lett.* **3** (1970) 235

R. J. Blin-Stoyle, *Phys. Rev. Lett.* **23** (1969) 535 (a)

R. J. Blin-Stoyle in *Isospin in Nuclear Physics*, ed. D. H. Wilkinson (North-Holland; Amsterdam 1969) (b)

R. J. Blin-Stoyle and J. M. Freeman, *Nucl. Phys.* **A 150** (1970) 369

R. J. Blin-Stoyle and M. Rosina, *Nucl. Phys.* **70** (1965) 321

R. Bloch, R. E. Pixley, and P. Truöl, *Phys. Lett.* **25B** (1967) 215

S. D. Bloom, *Nuovo Cim.* **32** (1964) 1023

D. P. Boyd, P. F. Donovan, and J. F. Mollenauer, *Phys. Rev.* **188** (1969) 1544

P. G. Butler, N. Cohen, A. N. James, and J. P. Nicholson, *Phys. Rev. Lett.* **21** (1968) 470

N. Cabibbo, *Phys. Rev. Lett.* **10** (1963) 531

J. Cerny, *Ann. Rev. Nucl. Sci.* **18** (1968) 27

L. M. Delves and A. C. Phillips, *Rev. Mod. Phys.* **41** (1969) 497

T. G. Dzubay, R. Sherr, F. D. Becchetti, and D. Dehnhard, *Nucl. Phys.* **A 142** (1970) 488

R. P. Feynman and M. Gell-Mann, *Phys. Rev.* **109** (1958) 193

G. T. Garvey, *Ann. Rev. Nucl. Sci.* **19** (1969) 433 (a)

G. T. Garvey in *Nuclear Isospin*, eds. J. D. Anderson, S. D. Bloom, J. Cerny, and W. W. True (Academic Press, New York and London, 1969) (b)

E. E. Gross, E. V. Hungerford, J. J. Malanify, and R. Woods, *Phys. Rev.* **C 1** (1970) 1365(a)

E. E. Gross, E. Newman, W. J. Roberts, R. W. Rutkowski, and A. Zucker, *Phys. Rev. Lett.* **24** (1970) 473(b)

R. P. Haddock, R. M. Salter, M. Zeller, J. B. Czirr, and D. R. Nygren, *Phys. Rev. Lett.* **14** (1965) 318

J. C. Hardy, H. Brunnader, and J. Cerny, *Phys. Rev. Lett.* **22** (1969) 1439

D. Harting, J. C. Kluyver, A. Kusunegi, R. Rigopoulos, A. M. Sachs, G. Tibell, G. Vanderhaeger, and G. Weber, *Phys. Rev.* **119** (1960) 1716

E. M. Henley in *Isospin in Nuclear Physics*, ed. D. H. Wilkinson (North-Holland; Amsterdam 1969)

E. M. Henley and C. E. Lacy, *Phys. Rev.* **184** (1969) 1228

J. N. Huffaker and E. Greuling, *Phys. Rev.* **132** (1963) 738

J. Jähnecke in *Isospin in Nuclear Physics*, ed. D. H. Wilkinson (North-Holland; Amsterdam 1969)

W. Jaus and G. Rasche, *Nucl. Phys.* **A 143** (1970) 202

G. A. Keyworth, G. C. Kyker, E. G. Bilpuch, and H. W. Newson, *Nucl. Phys.* **89** (1966) 590

Kim Sin Nam, G. M. Osetinskii, and V. A. Sergeev, *JETP* **10** (1970) 407

H. S. Köhler, *Phys. Rev.* **118** (1960) 1345

H. M. Kuan, D. W. Heikkinen, K. A. Snover, F. Riess, and S. S. Hanna, *Phys. Lett.* **25B** (1967) 217

A. M. Lane in *Isospin in Nuclear Physics*, ed. D. H. Wilkinson (North-Holland; Amsterdam 1969)

H. T. Larson, A. D. Bacher, K. Nagatani, and T. A. Tombrello, *Nucl. Phys.* **A 149** (1970) 161

A. B. McDonald, E. G. Adelberger, H. B. Mak, D. Ashery, A. P. Shukla, C. L. Cocke, and C. N. Davids, *Phys. Lett.* **31B** (1970) 119

R. L. McGrath, J. Cerny, J. C. Hardy, G. Goth, and A. Arima, *Phys. Rev.* **C 1** (1970) 184

J. B. Marion, *Phys. Lett.* **14** (1965) 315

R. Mendelson, G. T. Wozniak, A. D. Bacher, T. M. Loiseaux, and T. Cerny *Phys. Rev. Lett.* **25** (1970) 533

M. D. Miller, M. S. Sher, P. Signell, N. R. Yoder, and D. Marker, *Phys. Lett.* **30B** (1969) 157
J. M. Nelson, N. S. Chant, and P. S. Fisher, *Phys. Lett.* **31B** (1970) 445
J. V. Noble, *Phys. Rev.* **162** (1967) 934
J. A. Nolen and J. P. Schiffer, *Ann. Rev. Nucl. Sci.* **19** (1969) 471
D. R. Nygren, Ph. D. Thesis, University of Washington (1968) (unpublished)
W. von Oertzen, J. C. Jacmart, M. Liu, F. Pougheon, J. C. Roynette, and M. Riou, *Phys. Lett.* **28B** (1969) 482
K. Okamoto and C. Lucas, *Nucl. Phys.* **82** (1967) 347
F. D. Reisman, P. I. Connors, and J. B. Marion, *Bull. Amer. Phys. Soc.* **14** (1969) 507
D. Robson in *Isospin in Nuclear Physics*, ed. D. H. Wilkinson (North-Holland; Amsterdam 1969)
N. Schulz and M. H. Shapiro, *Nucl. Phys.* **A 148** (1970) 632
A. Sirlin, *Phys. Rev.* **164** (1967) 1767
J. M. Soper in *Isospin in Nuclear Physics*, ed. D. H. Wilkinson (North-Holland; Amsterdam 1969)
G. M. Temmer in *Isospin in Nuclear Physics*, ed. D. H. Wilkinson (North-Holland; Amsterdam 1969)
G. F. Trentelman, B. M. Preedom, and E. Kashy, *Phys. Rev. Lett.* **25** (1970) 530
E. K. Warburton and J. Weneser in *Isospin in Nuclear Physics*, ed. D. H. Wilkinson (North-Holland; Amsterdam 1969)
S. Weinberg, *Phys. Rev.* **112** (1958) 1375
D. H. Wilkinson in *Proceedings of Rehovoth Conf. on Nuclear Structure*, ed. H. J. Lipkin (North-Holland; Amsterdam 1958)
D. H. Wilkinson, *Phys. Rev. Lett.* **13** (1964) 571
D. H. Wilkinson, *Phys. Lett.* **31B** (1970) 447
D. H. Wilkinson and W. D. Hay, *Phys. Lett.* **21** (1966) 80
C. T. Yap, *Nucl. Phys.* **A 100** (1967) 619
B. Zeitnitz, R. Maschuw, and P. Suhr, *Nucl. Phys.* **A 149** (1970) 449

DISCUSSION

L. Radicati What is now the experimental situation with the $\Delta T = 1$ transition $\Sigma^{\pm} \rightarrow \Lambda^{\circ} + e^{\pm} + \nu$.

D. H. Wilkinson One may look to several places in high-energy physics for evidence on second-class currents. The architypal case is the comparison of $\Sigma^{+} \rightarrow \Lambda^{\circ}$ and $\Sigma^{-} \rightarrow \Lambda^{\circ}$ beta-decay. Data are extremely hard to acquire. In the absence of second class currents we expect $\Sigma^{+}/\Sigma^{-} = 0.60$ (relative decay rates) whereas experimentally $\Sigma^{+}/\Sigma^{-} = 0.71 \pm 0.15$. There is therefore no evidence for second class currents from this process. On the other hand, the value of G_{IT} that we should need if the effect in mirror beta-decay were to be ascribed to second-class currents would lead to $\Sigma^{+}/\Sigma^{-} \simeq 0.3$ which appears to conflict with the experimental data.

Other places where one may look are to the absolute rates of muon capture and of radiative muon capture. I hope that Dr. Macq may be able to tell us the latest position.

P. Macq A limit on the B value predicted by Wilkinson may be found through a comparison of muon capture rate from the F^{+} $(2-)$ and F^{-} $(1-)$ states of muonic-^{11}B to the $1/2^{-}$ state of ^{11}Be. The transition from the F^{+} state is only due to the D wave contribution (calculated in Dubna) and the induced terms; if we take the value of weak magnetism and pseudoscalar coupling constants given by Wilkinson we arrived in our experiment (not yet completed) at an approximate limit of g_T/g_A of roughly $|10^{-3}|$ to be compared to the 2.10^{-3} value of Wilkinson.

D. H. Wilkinson I should just like to re-emphasize that muon capture is *not* specifically sensitive to the induced tensor interaction but to a combination of weak magnetism, the induced pseudoscalar interaction and the induced tensor interaction. We must confidently estimate the first two, as well as, of course, the purely nuclear structure factors, before being able to say anything about the induced tensor interaction. Radiative muon capture has a somewhat higher sensitivity to the induced tensor term, at least in certain cases, and renewed experimental and theoretical study of this process are very much to be desired.

A. Bohr Could you clarify what is the value of the quantity B in your current expression?

D. H. Wilkinson Many matrix elements enter into the evaluation of $\delta = [(ft)^+/(ft)^-] - 1$ but in the high energy limit the dominant induced tensor term is proportional to the ordinary Gamow-Teller matrix element so that δ can be written directly in terms of G_{IT} without recourse to a nuclear structure calculation.

H. P. Lipkin Wilkinson's analysis of departures from isospin symmetry of Gamow-Teller beta transitions is very interesting. However, his discussion of second class currents is somewhat misleading. It is true that the effect observed experimentally might arise if there were second class currents, as has been pointed out by Weinberg. However, the effect can also be produced if there are no second class currents, and can be absent if there are second class currents. Furthermore, Weinberg's paper is quite old (1958). The significant developments in weak interactions subsequent to Weinberg's paper include $SU(3)$ symmetry, current algebra, and the experimental discovery of CP violation, all of which may be relevant to the present discussion. It is therefore of interest to keep these points in mind in analyzing the data from a more modern point of view.

G parity has nothing directly to do with nuclear physics. It is a combination of isospin and charge conjugation invented by particle physicists to extend the application of C-invariance to include charged bosons as well as neutral bosons. Charged pions are not eigenstates of C, but they *are* eigenstates of G; this is the basis of the use of G in particle physics. In nuclear structure, C is irrelevant as long as we are only considering nuclei made of nucleons and containing no anti-nucleons. Any results obtained for nuclear systems by assuming combinations of C and G invariance *must also be obtainable from isospin invariance alone*, since G invariance automatically follows from isospin invariance and C-invariance and C-invariance is irrelevant.

In the case considered by Wilkinson, we need only consider the *isospin* transformation properties of the weak currents as given by the accepted theory of weak interactions. The weak hadron currents are assumed to transform under $SU(3)$ like members of the same octet. The components relevant to nuclear beta decay (no change in strangeness) therefore transform under isospin like *components of the same isovector*. This assumption together with the assumption of isospin symmetry for nuclear structure leads to the predictions which Wilkinson has shown to be violated by the experimental data.

There are two possible approaches to the discrepancy:

1) Isospin symmetry is broken in the nuclear wave function,

2) The accepted description of weak interactions is not valid.

In the first case, better nuclear structure calculations are necessary to explain the discrepancy and there is nothing to do except go back to the drawing boards and work. In the second case we must revise our picture of the weak currents. I should like to discuss the implications of this second case further, and assume that isospin is a good symmetry for the nuclear states (or that it is not broken by a sufficiently large amount to explain the discrepancy).

One immediate conclusion is that nuclear beta decay matrix elements cannot be given by single-nucleon operators. If the matrix element $\langle p|J|n\rangle$ of the weak current J is known between the neutron and proton states, then the matrix element $\langle n|J^{+}|p\rangle = (\langle p|J|n\rangle)^{*}$ is uniquely determined and indicates that the relevant parts of J and $J^{\dagger}$ transform under isospin rotations like two components of the same isovector. Any parts of J and $J^{\dagger}$ which transform otherwise give vanishing contributions between single nucleon states; i.e. they cannot be single nucleon operators. Thus if the nuclear states are really good isospin states, and there is a real discrepancy between the data and the isospin predictions, then the discrepancy must be due to many-body or meson-exchange contributions to the nuclear matrix elements.

That second class currents are not directly relevant is seen as follows. The weak hadron current conserves baryon number and can be divided into three parts as follows:

$$J = J_{B>0} + J_{B>0} + J_{B=0}$$

where $J_{B>0}$ acts only in the subspace of states of positive baryon number, $J_{B<0}$ acts only in the subspace of states of negative baryon number and $0_{B=0}$ acts in the subspace of states of zero baryon number.

The distinction between 1st and 2nd class currents refers to relations between $J_{B>0}$ and $J_{B<0}$ and to properties of $J_{B=0}$, but place no restrictions on $J_{B>0}$. The G-parity transformations transform the $B > 0$ subspace into the $B < 0$ subspace. The experiments discussed by Wilkinson *concern only* the properties of $J_{B>0}$, in particular the isospin transformation properties. The experimental data can be explained by modifying $J_{B>0}$ to introduce additional components, but will be unaffected by arbitrary modifications of $J_{B>0}$ and $J_{B=0}$ which can give arbitrary transformation properties under G-parity and arbitrary mixtures of first and second class currents. Thus the presence or absence of second class currents is not tested by these experiments.

The matrix elements of the current in the $B > 0$ subspace can be related to matrix elements in the $B < 0$ subspace *by making additional assumptions* regarding the transformation properties of the currents under charge conjugation. However, these assumptions are *model dependent* and the results cannot be used as a *model independent* proof of the existence of second class currents. Weinberg's model is a particular example of this type.

The most reasonable additional assumption is CPT invariance, which so far has not been contradicted by experiment. In the combination $G \cdot CPT$ the two C operations cancel, and the result is a "charge-symmetry" rotation followed by P and T. It is amusing to note the transformation properties of the single nucleon matrix element $\langle p|J|n\rangle$ under $G \cdot CPT$. The charge symmetry rotation interchanges p and n, but the time reversal interchanges initial and final states and brings the p and n back where we started from. Similarly P reverses momenta and T reverses them back again. The only transformation which is not cancelled out is the spin reversal due to T. Thus the transformation properties of the weak current under $G \cdot CPT$ *are* observable by measurements *on single nucleons*, and therefore on nuclei, even if the weak current is a one-body-operator. However, it is the *spin* dependence, rather than the *isospin* dependence of the transition rate which must be measured. Thus asymmetry in mirror transitions would not detect this effect if the nuclear matrix elements are given by single nucleon operators, but it could be detected by polarization measurements in neutrino experiments.

Since there is no unique model which gives a satisfactory description of weak interactions at present, and it is just the transformation properties under C and CP which provide the biggest puzzle, it is perhaps better to describe the nuclear data entirely in the $B > 0$ subspace, where the measurements are made, and give the model-makers of particle physics something to shoot at.

There are two possible ways to describe the discrepancy, each having different theoretical implications and experimental consequences. There can be 1) components in the weak hadron current with transformation properties of higher isospin tensors; i.e. with $\Delta T > 1$, or 2) two different and independent isovectors in the hadron current, so that J and J^+ are components of isovectors, but *not of the same isovector*. The latter possibility occurs in Weinberg's model. The former would not contribute to decays from a $T = 1$ multiplet to a $T = 0$ state, e.g. the decays $N^{12} \to C^{12}$ and $B^{12} \to C^{12}$. Thus this particular discrepancy would indicate the presence of two independent isovectors, but would not rule out the simultaneous existence of a $\Delta T = 1$ component.

The distinction between $\Delta T > 1$ and two different isovectors can be tested in decays of a quartet, to see whether the electron and positron transitions *individually* satisfy $\Delta T = 1$ relations, but do not satisfy relations between them. This would also test purity of nuclear wave functions. Such experiments are not feasible at present, since the states of a quartet with $T_z = \pm 1/2$ are not ground states but isobaric analog states where beta decay cannot be studied. One might think however of excitation of such analog states with neutrinos.

Both descriptions of the discrepancy violate the Gell-Mann current algebra which assumes octet transformation properties for the weak currents and therefore isovector transformation properties of the strangeness currents. Furthermore, J and $J^\dagger$ are assumed to belong *to the same octet* and the strangeness conserving parts *to the same isovector*. If Gell-Mann's current algebra disagrees with experiment, this will profoundly affect the theory of weak interactions, whether or not there are second class currents. It is therefore important to check the experimental data further and to look for other explanations involving isospin breaking in the relevant nuclear structure.

D. H. Wilkinson I do not think that Lipkin and I can be disagreeing about the meaning of second class currents. If I may state the situation in a different way: currents containing first together with second class parts are precisely those that are *not* generated by an infinitesimal isospin rotation and that are therefore *not* of the Feynman-Gell-Mann type. If such currents exist then we get a difference between the intrinsic rate of positon and negaton decay even though the decaying bodies are precise mirrors of each other and the nuclear matrix elements for their decay are precisely equal.

H. J. Lipkin I disagree. The presence of second class currents as defined by Weinberg is not clearly related to the behaviour of these currents under isospin rotations nor to the difference between the intrinsic rates of positon and negaton decay. The particular form for the current in the universal $V - A$ current-current interaction proposed by Marshak and Sudarshan and by Feynman and Gell-Mann is irrelevant to this discussion.

The experimental mirror asymmetry observed by Wilkinson is interesting as an indication that the currents responsible for negaton and positon decays are not members of the same isovector. This may result either from first class or second class currents, and second class currents can exist without producing this effect*. However, mirror asymmetry definitely contradicts the Gell-

* N. Cabibbo (*Particle Symmetries*, Brandeis University Summer Institute in Theoretical Physics, 1965, Volume II, Edited by M. Chretien and S. Deser, Gordon and Breach, (New York) 1966, p. 1) has given a very clear analysis of this problem and divides currents into four classes, according to whether or not they produce mirror asymmetry as observed by Wilkinson and according to whether or not they are second class by Weinberg's definition.

Mann current algebra. The current algebra approach postulates commutation relations for the currents in the same way that Prof. Heisenberg postulated the commutation relation $[p, x] = -i\hbar$ many years ago. This goes beyond the old fashioned expressions for the current used by Weinberg in 1958 and quoted by Wilkinson, which contain only terms of the form $\bar{\psi} 0 \psi$, and assume the current to be a bilinear form in nucleon field operators. These days we know that there must be other contributions to the currents which are not of this form, e.g. meson exchange currents, and it is just these many-body contributions which must be responsible for the experimental mirror asymmetry. Thus any analysis of the data which uses only currents having the form $\bar{\psi} 0 \psi$ is apt to be misleading.

L. Radicati I am not entirely in agreement with Dr. Lipkin's assumption that G parity has nothing to do with the problem. It seems to me, it has to do because we are considering the behaviour under G parity of an operator, the current, and I don't see that the fact that we are discussing nuclei and not antinuclei enters the problem. We can certainly discuss the behaviour of the electromagnetic potential under charge conjugation, and I do not see why we could not do the savue for G parity, but this is important. And it is important to know whether there are only currents with the ordinary G parity or whether there are currents with opposite G parity which is what Wilkinson discussed. Moreover to split the current operator into parts, $B > 0$, $B = 0$, $B < 0$, seems to me rather artificial. If we introduce a new operator for every nucleus we can not, for example, use the information we have gained about CVC from the pion decay: $\beta^+ \rightarrow \beta^\circ + e^+ + \nu$. We have the same operator whose matrix elements are taken between different states.

L. C. Biedenharn I want to make a very short remark. I find, that when particle physicists talk to nuclear physicists and vice versa, things are not quite as clear as they should be. One difficulty is that nuclear physics is a complicated subject whereas particle physics is in a certain sense simple, for example one generally assumes that the state vectors that one is working with are in the same representation type as the currents that one uses. But when you come to nuclear physics it is very essential to remember that whether you blame the difficulties on currents (or operators) versus states is entirely a point of view. You can have the state vectors badly distorted. In fact, you can have them as deformed states in isospin if you like, then you put all the blame on the state vectors and you can then take a simple operator. Or you can take the contrary point of view that the states are simple and the operators become *effective* operators. This is the theory of effective charges and everyone knows that one has a great deal of information about, say, e.m. transitions from the a priori point of view. I am only saying that these differences between Wilkinson and Lipkin are, to a certain extent, a matter of language, because Wilkinson is entirely right. When he says two different currents, what he means, is that they are not related to each other by an elementary isospin rotation and that is true. You may prefer to look at that as a statement that the state vectors, he is working with for these currents, are themselves deformed. But let us remember secondly that we are dealing with small effects and, therefore these points of view, these minor distortions, become the main problem.

B. R. Mottelson It seems to me that if we accept *PCT* symmetry there are significant consequences for nuclear β-decay arising from the assumption of *G*-parity—that is there are terms in the effective β-decay current for nuclei which would be eliminated if this symmetry is assumed.

D. H. Wilkinson I also fully agree that second class currents violate the Gell-Mann commutation relations: it is the business of experimental science to test such matters.

Finally I should like to reiterate that I am *not* claiming that second class currents are established in beta-decay but rather that there exists a systematic empirical phenomenon that is consistent with such currents and that, if it is due rather to nuclear effects requires that the wave-function differences between the mirror states are surprisingly large.

H. J. Lipkin The division of the current into three pieces in $B > 0$, $B < 0$ and $B = 0$ subspaces may look artificial to someone who has been brought up on field theory, is used to writing currents as bilinear products of field operators, and knows that the same term applies to both particles and antiparticles. However, I don't think that nuclear physicists should take sides in arguments between particle physicists about controversial subjects like field theory which are not directly related to the nuclear phenomena under discussion. We know that we can divide Hilbert space into three pieces according to the baryon number, and that all nuclear measurements give us information only on the $B > 0$ part of Hilbert space. We should therefore quote all nuclear results in terms of quantities defined in this space, and should not use transformation properties under *G*, *C*, *CP* or *CPT* which are not directly measured in these experiments. Every statement about nuclear results which uses *G*, *C*, *CP* or *CPT* in a non-trivial way is making some assumptions about the theory of weak interactions which at least one group of particle theorists does not believe. This point has become more significant since the discovery of *CP* violation, which still has no satisfactory explanation and may very well be related to the phenomena discussed by Wilkinson. If nuclear physicists express their results only in terms of symmetries like isospin, parity and time reversal, which are all defined within the $B > 0$ subspace, then each particle theorist can make his own additional assumptions in trying to find a model or theory to explain the data.

Rotational Motion

A. BOHR

The Niels Bohr Institute, University of Copenhagen, Denmark

THE QUEST for symmetry has always been a driving force in the attempts to understand the relationships among natural phenomena. In the development of quantal physics, the scope of symmetry concepts has been greatly extended, but also the significance of symmetry breaking has come more into focus. In the exploration of nuclear phenomena, these themes have played a prominent role, and they provide a special flavour to the study of collective motion in the nucleus.

OCCURRENCE OF ROTATIONAL SPECTRA IN QUANTAL SYSTEMS

Spectra associated with quantized rotational motion were recognized in the absorption of infrared light by molecules at a very early state of development of quantum theory. The existence of rotational motion as a well-defined degree of freedom in molecules follows immediately from the existence of a semi-rigid structure represented by the atomic nuclei in their equilibrium positions.

The possible occurrence of rotational motion in nuclei became an issue in connection with the early attempts to interpret the evidence on nuclear excitation spectra (see, for example, Teller and Wheeler, 1938). The available data, as obtained, for example, from the fine structure of α-decay, appeared to provide evidence against the occurrence of lowlying rotational excitations, but the discussion was hampered by the expectation that rotational motion would either be a property of all nuclei or generally excluded, as in atoms, and by the assumption that the moment of inertia would have the classical value as for rigid rotation.

The establishment of the nuclear shell structure, which implies that the nucleons move approximately independently in the average nuclear potential, and the recognition, which came almost simultaneously, that many nuclei have equilibrium shapes deviating from spherical symmetry, created a new basis for the discussion of rotational motion in the nucleus. It was evident that the existence of a non-spherical shape implied collective rotational degrees of freedom, but one was faced with the need for a generalized treatment of rotations applicable to quantal systems that do not have a rigid structure. Such a treatment has been gradually developed over the years and the subject has continued to exhibit new facets*.

The quantal theory of rotations may find application not only to molecules and nuclei, but also to the spectra of the hadrons, as has been much discussed in recent years, in connection with the intimately related concept of Regge trajectories. Moreover, excitations of rotational character occur in the non-classical spaces that play a prominent role in quantal physics (isospace, gauge space, etc.).

The existence of a deformation, taken in the general sense of an element of anisotropy in the structure of the system, may be recognized as the hallmark of quantal systems that exhibit rotational spectra. Indeed, such an element of anisotropy is required to make it possible to specify the orientation of the system. This definition of a deformation includes the lattice-like structure in molecules and rotating pieces of solid matter; a classical physical object that may seem perfectly spherical is in fact highly anisotropic on account of the atomic constitution of matter.

The occurrence of rotational degrees of freedom may thus be said to originate in a breaking of rotational invariance. In a similar manner, the translational degrees of freedom are based upon the existence of a localized structure. However, while the different states of translational motion of a given object are related by Lorentz invariance, there is no similar invariance applying to co-ordinate frames rotating with respect to the matter distribution of the universe. The Coriolis and centrifugal forces that act in such coordinate frames perturb the structure of a rotating object.

In a quantal system, already the frequency of the lowest rotational excitations may be so large that the Coriolis and centrifugal forces affect the structure in a major way. The condition that these perturbations be small (adiabatic condition) is intimately connected with the condition that the zero-point fluctuations in the deformation parameters be small compared with the equilibrium value of these parameters, and the adiabatic condition

---

* A more detailed presentation of the considerations in the present report, as well as references to the various steps in the development of the subject, will be contained in Vol. II of *Nuclear Structure* by A. Bohr and B. R. Mottelson, Addison-Wesley/W. A. Benjamin, Inc. Reading, Mass.

provides an alternative way of expressing the criterion for the occurrence of rotational spectra.

For sufficiently large values of the angular momentum, the rotational perturbations strongly affect the structure of the system; however, rotational sequences may still occur if the properties of the system vary smoothly with the angular momentum.

ROTATIONAL DEGREES OF FREEDOM

The occurrence of a rotational spectrum corresponds to the possibility of obtaining an approximate separation of the motion represented by a total wave function of the product form

$$\Psi = \varphi_{\text{int}}(q)\,\Phi_{\text{rot}}(\omega) \tag{1}$$

where the angular variables ω specify the orientation of the system, while the co-ordinates q characterize the intrinsic motion with respect to the body-fixed frame with orientation ω.

The specification of an orientation in three-dimensional space involves three co-ordinates, such as the Euler angles, $\omega = \theta\phi\psi$, and there are three associated quantum numbers. The overall rotational invariance for the system as a whole implies the constancy of the total angular momentum I and its component M on a fixed axis. As a third angular momentum variable, one may choose the component $I_3 = K$ of the angular momentum on one of the intrinsic axes (which are labelled $\varkappa = 1, 2, 3$), but this quantity is not, in general, a constant of the motion. The full rotational degrees of freedom imply the occurrence of $(2I + 1)^2$ states for each I and $(2I + 1)$ states for each IM; the state IKM is represented by the rotation matrix $\mathscr{D}^I_{MK}(\omega)$.

The full rotational degrees of freedom come into play only if the deformation completely breaks the rotational symmetry so that it permits a unique specification of the orientation. However, the deformation may be invariant with respect to a subgroup of rotations of the co-ordinate frame. Thus, a deformation in three-dimensional space may retain axial symmetry; the deformation may also be invariant with respect to a group of finite rotations. For example, an ellipsoidal deformation is invariant with respect to rotations $\mathscr{R}_\varkappa(\pi)$ of π about any of the principal axes ($\varkappa = 1, 2, 3$).

The rotations that leave the deformation invariant constitute a subgroup G of the total group of rotations in the space considered. The well-established nuclear deformations have axial symmetry and moreover are invariant with respect to a rotation π about an axis perpendicular to the symmetry axis ($\mathscr{R}_2(\pi)$), corresponding to the invariance group D_∞. The same invariance applies to a diatomic molecule with identical nuclei.

The elements of G are part of the intrinsic degrees of freedom, and the intrinsic states can thus be classified in terms of the representations of G. For example, if the intrinsic structure involves independent-particle motion in a deformed potential, each orbit can be classified in terms of the symmetry G; in molecules, the operations contained in G can be expressed in terms of permutations of identical nuclei, in addition to the effect on the electronic orbits.

Since the elements of G are part of the intrinsic degrees of freedom, the rotational degrees of freedom are reduced correspondingly. One can express this constraint by requiring that, for each element $\mathscr{R}$ of G, we have

$$\mathscr{R}_e = \mathscr{R}_i \qquad (\mathscr{R} \in G) \tag{2}$$

where $\mathscr{R}_i$ expresses the rotation as an operator acting on the intrinsic variables, while $\mathscr{R}_e$ accomplishes the same rotation by acting on the collective orientation angles. For example, for a deformation with axial symmetry, the rotational quantum number K is constrained to have the same value as the angular momentum component of the intrinsic motion, corresponding to the absence of collective rotations about a symmetry axis. (This constraint also implies that the total wave function (1) becomes independent of the redundant third Euler angle, ψ, that specifies the orientation about the symmetry axis).

If the intrinsic state belongs to a one-dimensional representation of G, it is an eigenstate of the operations $\mathscr{R}_i$, and the rotational spectrum (1) only contains states with the same values of $\mathscr{R}_e$. If the intrinsic state belongs to a multidimensional representation of G, the total wave function is a linear combination of products of the form (1).

Thus, for a deformation of symmetry D_∞, the representations with $K = 0$ are one-dimensional and the constraint (2) for $\mathscr{R} = \mathscr{R}_2(\pi)$ implies that the rotational spectrum consists of a sequence of states with

$$(-1)^I = r \tag{3}$$

where r is the eigenvalue of $\mathscr{R}_i$ for the intrinsic state. The representations of D_∞ with $K \neq 0$ are two-dimensional ($I_3 = \pm K$, with K taken to be positive), and the wave functions take the form

$$\Psi = \left(\frac{2I+1}{16\pi^2}\right)^{1/2} \{\varphi_K(q)\, \mathscr{D}^I_{MK}(\omega) + (-1)^{I+K}\, \varphi_{\bar{K}}(q)\, \mathscr{D}^I_{M,K}(\omega)\} \tag{4}$$

where $\varphi_{\bar{K}}$ is obtained from φ_K by the rotation $\mathscr{R}_2^{-1}(\pi)$. The band contains a state for each $I(\geqq K)$, but the phase factor in Eq. (4)

$$\sigma = (-1)^{I+K} \tag{5}$$

provides a signature dividing the band into two sequences with $\sigma = \pm 1$, each having consecutive states with $\Delta I = 2$.

The above description can be readily extended to deformations of more general type that may involve violations of other symmetries. Thus, in molecules, the deformations most frequently are unsymmetric with respect to spatial inversion; such $\mathscr{P}$ violation also occurs in nuclei, if the deformation contains components of odd multipole order. Deformations with $\mathscr{T}$ violation may occur in molecules with partial spin alignment of the electrons and also characterize the pion field produced by the nucleon spin.

A $\mathscr{P}$- or $\mathscr{T}$-violating deformation implies a two-valued collective degree of freedom; $\mathscr{P}$ violation gives rise to parity doubling and $\mathscr{T}$ violation to a doubling of states with the same values of $I\pi$. However, there may be a connection between the rotation and reflection symmetries, in the sense that the deformation may be invariant with respect to a combination of symmetries though violating the individual symmetries. For example, a molecule such as NH_3 or a nucleus with an axially symmetric octupole deformation is invariant with respect to reflection in a plane, which can be expressed as the product $\mathscr{S} = \mathscr{P}\mathscr{R}_2(\pi)$, but neither with respect to $\mathscr{P}$ nor $\mathscr{R}_2(\pi)$, separately. In such a situation, the rotational degrees of freedom are linked to that of parity so that the band contains members of both parity values ($\pi = \pm 1$) and with rotational quantum numbers satisfying the relation

$$\mathscr{S}_e = \mathscr{P} \exp\{i\pi I_2\} = \mathscr{S}_i \tag{6}$$

For example, a band with $K = 0$ contains the states

$$I = 0, 1, 2, \ldots$$

$$\pi = s(-1)^I$$

where s is the eigenvalue of $\mathscr{S}_i$ for the intrinsic state. These results are well known, but the emphasis is on a general formulation directly based on the relationship between the rotational degrees of freedom and the symmetry breaking of the deformation.

In quantal physics, attributes such as particle number and isospin that are assigned a passive role in classical physics, become dynamical operators with properties similar to those of angular momentum. In the associated angular spaces, motion of rotational type occurs when the system possesses a deformation in these dimensions.

Thus, the superfluidity in quantal many-body systems may be described in terms of deformations in the gauge space associated with particle number. These deformations are expressed in terms of densities and fields that create particles (single bosons in the case of superfluid liquid He, pairs of fermions in the case of superconductors or pair-correlated nuclei). The presence of such deformations implies the occurrence of rotation-like excitations, in which particle number takes the place of angular momentum, and for which the rotational frequency corresponds to the chemical potential.

For superfluid deformations involving a single type of particles, the orientation in gauge space is specified by a single angular variable, and the collective mode corresponds to rotations about a fixed axis. In the fermion systems, the deformation is invariant with respect to rotations of π about the gauge axis, and the condition (2) expresses the fact that the rotational excitations involve the addition or subtraction of pairs of particles.

Rotational modes of this type can be studied in superfluid nuclei and involve sequences of corresponding states, such as the ground states of even-even nuclei. A tool for directly exciting and probing the quanta of these "pair rotations" is provided by the two-nucleon transfer reaction. The corresponding mode in superconductors is involved in the Josephson junction, which can be viewed as two coupled rotors forced to rotate with respect to each other with a frequency determined by the electrostatic potential across the junction.

Pair correlations in which both neutrons and protons participate give rise to deformations in isospace as well as in the gauge space associated with total nucleon number A. The orientation thus involves four angular variables conjugate to the momentum variables ATM_TK_T. The favoured deformations in the nuclear ground states have axial symmetry in isospace and are invariant with respect to a rotation of π about an axis perpendicular to the symmetry axis in isospace followed by a rotation of $\pi/2$ in gauge space. For example, the ground states of even-even nuclei belong to a band with $K_T = 0$ that contains states with total isospin T and mass number A satisfying

$$\tfrac{1}{2}A - T = 2p \qquad (7)$$

$$p \text{ integer}$$

The empirical evidence is as yet too incomplete to test the possible occurrence of the comprehensive rotational band structure of this type.

An example of a deformation whose symmetry involves a combination of isospace with ordinary space is provided by the pion field of the nucleon. The preferred structure of the p-wave field is axially symmetric in each of the spaces and in addition is invariant with respect to arbitrary rotations performed simultaneously in space and isospace. A stable deformation of this type gives rise to a rotational band containing a sequence of states with $I = T$ (= 1/2, 3/2, ...).

These examples illustrate the great variety of symmetry patterns that may be associated with rotational motion in quantal systems.

STRUCTURE OF MATRIX ELEMENTS. EXPANSION IN POWERS OF THE ANGULAR MOMENTUM

The intimate relationship between members of a rotational band manifests itself in the regularities of the energy spectra and the intensity rules that govern transitions leading to different members of a band. The underlying deformation is expressed by the occurrence of collective transitions within a band.

For sufficiently small values of the rotational quantum numbers, the analysis of matrix elements can be based on an expansion in powers of the angular momentum. The leading-order approximation is obtained by considering the intrinsic motion for a fixed orientation. The dynamical effects of the rotation (Coriolis and centrifugal forces) give rise to perturbations of the intrinsic motion, which can be treated in terms of couplings between different bands. One can also express the rotational perturbation effects in terms of a renormalization of the effective operators acting in the unperturbed basis. Such a description corresponds to a canonical transformation that restores the coupled wave functions to the unperturbed form, now expressed in terms of new co-ordinates q, ω. The various operators, when expressed in terms of the new co-ordinates, become functions of the angular momentum. Such an approach forms a convenient basis for a phenomenological treatment exploiting the symmetry of the rotational degrees of freedom.

The matrix elements take an especially simple form for systems with axial symmetry, for which the rotational wave functions are completely specified by the quantum numbers IKM. For a spherical tensor operator $\mathscr{M}(\lambda\mu)$, the matrix elements are obtained by a transformation to the intrinsic co-ordinate system

$$\mathscr{M}(\lambda\mu) = \sum_{\nu} \mathscr{M}(\lambda\nu)\, \mathscr{D}^{\lambda}_{\mu\nu}(\omega) \tag{8}$$

In the leading-order approximation, the intrinsic moments $\mathscr{M}(\lambda\nu)$ are independent of the rotational variables (if the operator is not explicitly angular momentum-dependent), and the reduced matrix element has the form

$$\langle K'I' \| \mathscr{M}(\lambda) \| KI \rangle = (2I + 1)^{1/2} \langle IK\lambda K' - K | I'K' \rangle \langle K' | \mathscr{M}(\lambda, \nu = K' - K) | K \rangle \tag{9}$$

where the last factor is an intrinsic matrix element, which is independent of I and I'. For a system with D_∞ symmetry, the matrix element between two bands, both having $K \neq 0$, may involve two terms with $\nu = |K' - K|$ and $\nu = |K' + K|$, with a relative phase involving the signature factor $(-I)^{I+K}$. While the relation (9) for large quantum numbers has a simple classical interpretation, the interference term involving the signature is a specific quantal effect, which may occur when the operator can produce effects equivalent to a finite rotation of the entire system (exchange effects).

In higher order, the renormalized intrinsic operators $\mathcal{M}(\lambda\nu)$ depend on the rotational variables through the components $I_\varkappa$ of the angular momentum with respect to the intrinsic axes and can be expanded in powers of these operators, for sufficiently small rotational frequencies. As an example, we consider a scalar operator. For matrix elements that do not involve a change in K, the effective operator is diagonal with respect to I_3 and therefore a function of $I_1^2 + I_2^2$; hence, the matrix elements are given by the familiar expansion in powers of $I(I + 1)$. For matrix elements that involve a change in K, the effective operator is of the form $(I_1 + iI_2)^{\Delta K}$ multiplied by a function of $I_1^2 + I_2^2$. Thus, if we consider the energy of a band with $K \neq 0$, associated with the two-dimensional representations of D_∞ having $I_3 = \pm K$, we obtain terms with $\Delta K = 2K$ as well as $\Delta K = 0$,

$$E = E_K + AI(I+1) + BI^2(I+1)^2 + \cdots$$
$$+ (-1)^{I+K}\frac{(I+K)!}{(I-K)!}(A_{2K} + B_{2K}I(I+1) + \cdots) \qquad (10)$$

In a similar manner, one can derive the expansion of matrix elements of arbitrary multipole operators.

The extensive empirical evidence that has been accumulated regarding energies and matrix elements for a variety of nuclear processes shows that, in certain rather sharply defined regions in the (N, Z) plane, the spectra exhibit the pattern characteristic of deformations with D_∞ symmetry. Examples are shown in Figs. 1–3. It is found that, in favourable cases (nuclei with large stable deformations), the power series converges rather rapidly, though in some cases there are selection rules inhibiting the leading-order intrinsic matrix elements, but allowing certain higher-order terms. The spheroidal symmetry of the deformation is directly revealed by the very large $E2$ matrix elements between members of a band.

REGGE TRAJECTORIES

For systems with axial symmetry, the rotational bands can be viewed as Regge trajectories. The analytic form of the energy, as a function of the total

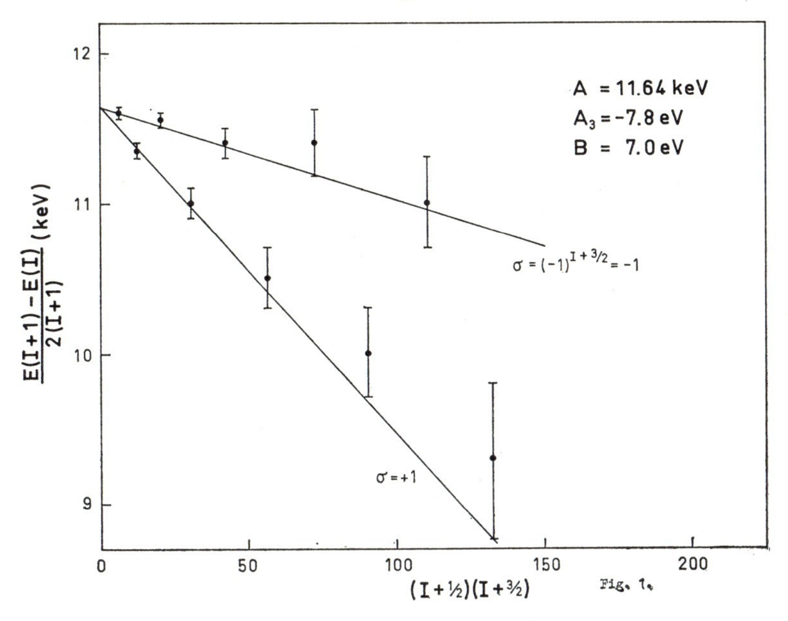

FIGURE 1 Rotational energies for ^{159}Tb. The ground-state rotational band has $K = 3/2$, and levels up to $I = 23/2$ have been populated by Coulomb excitation (Diamond *et al.*, 1963). The spectrum can be fitted by the expansion (10) with the three leading-order terms with coefficients A, B, and A_3. In the plot in the figure, such an expression corresponds to two straight lines with signature $\sigma = +1$ and -1, intersecting on the ordinate axis

angular momentum I, is given by the power series expansion (10), within the radius of convergence of this expansion. For systems with D_∞ symmetry, the presence of the signature-dependent terms in the bands with $K \neq 0$ implies that these bands involve two trajectories with $\sigma = \pm 1$.

For asymmetric systems, the bands involve a multitude of states of each I, the number increasing with I, and the problem of connecting states of different I by trajectories acquires new aspects. As an illustration, Fig. 4 shows the solution to the asymmetric rotor Hamiltonian

$$H_{\text{rot}} = \sum_{\varkappa=1}^{3} A_\varkappa I_\varkappa^2$$
$$A_\varkappa \equiv \frac{\hbar^2}{2\mathscr{I}_\varkappa} \tag{11}$$

for a particular choice of the asymmetry parameter characterizing the relative values of the inertial parameters. The Hamiltonian matrix is written in a representation in which one of the components of $I_\varkappa$ is diagonal; in this

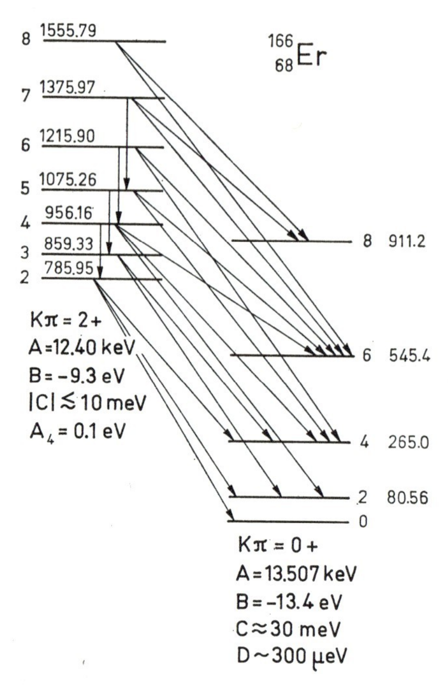

FIGURE 2 Ground-state band ($K\pi = 0+$) and excited $K\pi = 2+$ band in ^{166}Er. The energies of the ground-state band are taken from the Table of Isotopes by Lederer *et al.*, 1967, while the energies of the $K\pi = 2+$ band (γ-vibrational excitation) are from C. W. Reich and J. E. Cline, private communication. The energies of the bands can be fitted by the expansion (10) with the parameters shown in the figure. The arrows represent observed $E2$ transitions

matrix, the total angular momentum I is considered to be a parameter that can be continuously varied. For non-integer values of I, the matrix does not break off and is of infinite dimension. The trajectories in Fig. 1 show the eigenvalues of these infinite matrices as a function of I in the representation associated with the axes that have the largest and smallest moment of inertia (labelled by $\varkappa = 1$ and 3, respectively). For the intermediate axis ($\varkappa = 2$), the Hamiltonian matrix has no discrete spectrum for non-integer I. The $\varkappa = 1$ and $\varkappa = 3$ trajectories intersect at the physical values of I, but are seen to connect different sequences of states in the total two-dimensional array of levels in the band. One may attempt to use the transition probabilities to characterize dominant trajectories. For an ellipsoidal deformation, it is found that the strongest $E2$ transitions follow the $\varkappa = 3$ trajectories in the lower part of the spectrum, and the $\varkappa = 1$ trajectories for the highest states of given I.

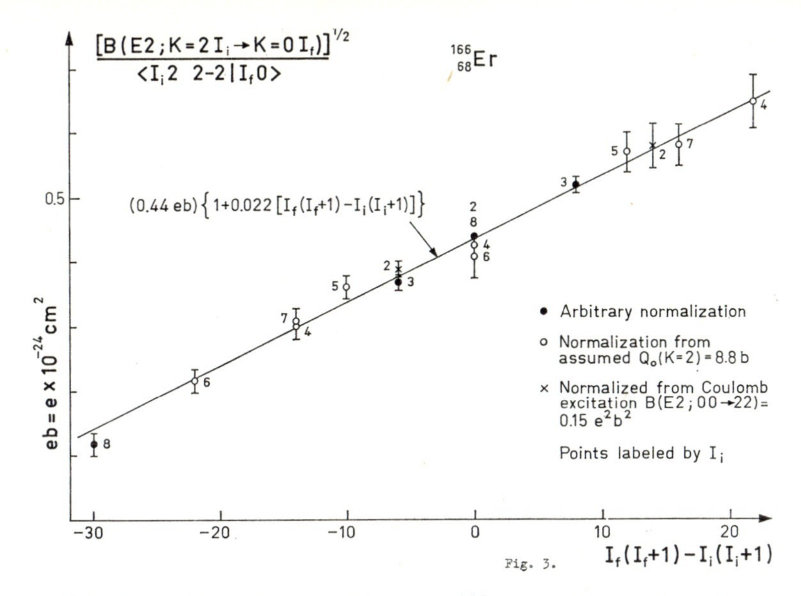

FIGURE 3 $E2$-transition amplitudes in ^{166}Er. The figure shows the measured transition probabilities for the $E2$ transitions between the $K\pi = 2+$ and $K\pi = 0+$ band shown in Fig. 2. The figure is based on the γ intensities measured by Gallagher *et al.* (1965) and by Günther and Parsignault (1967). The leading-order intensity relation (9) corresponds to a constant value for the quantity plotted in the figure, while the inclusion of the leading-order rotational coupling effects yields a straight line

The analysis of the asymmetric rotor raises questions concerning the general significance of one-dimensional Regge trajectories and the conditions under which trajectories can be defined in systems that do not allow of a separation between intrinsic and rotational motion. (For a further discussion of the analytic structure of the solutions to the asymmetric rotor Hamiltonian, see Talman, 1971).

INTERPRETATION OF PHENOMENOLOGICAL PARAMETERS

The phenomenological description of rotational spectra, as discussed in the preceding sections, provides a basis for the analysis of the experimental data in terms of the physically significant parameters that characterize the intrinsic structure of the system and its coupling to the rotational motion.

For the nuclear systems, one may attempt an interpretation of these parameters, starting from independent-particle motion in a potential of self-consistent shape. In the description of the intrinsic motion, one must include

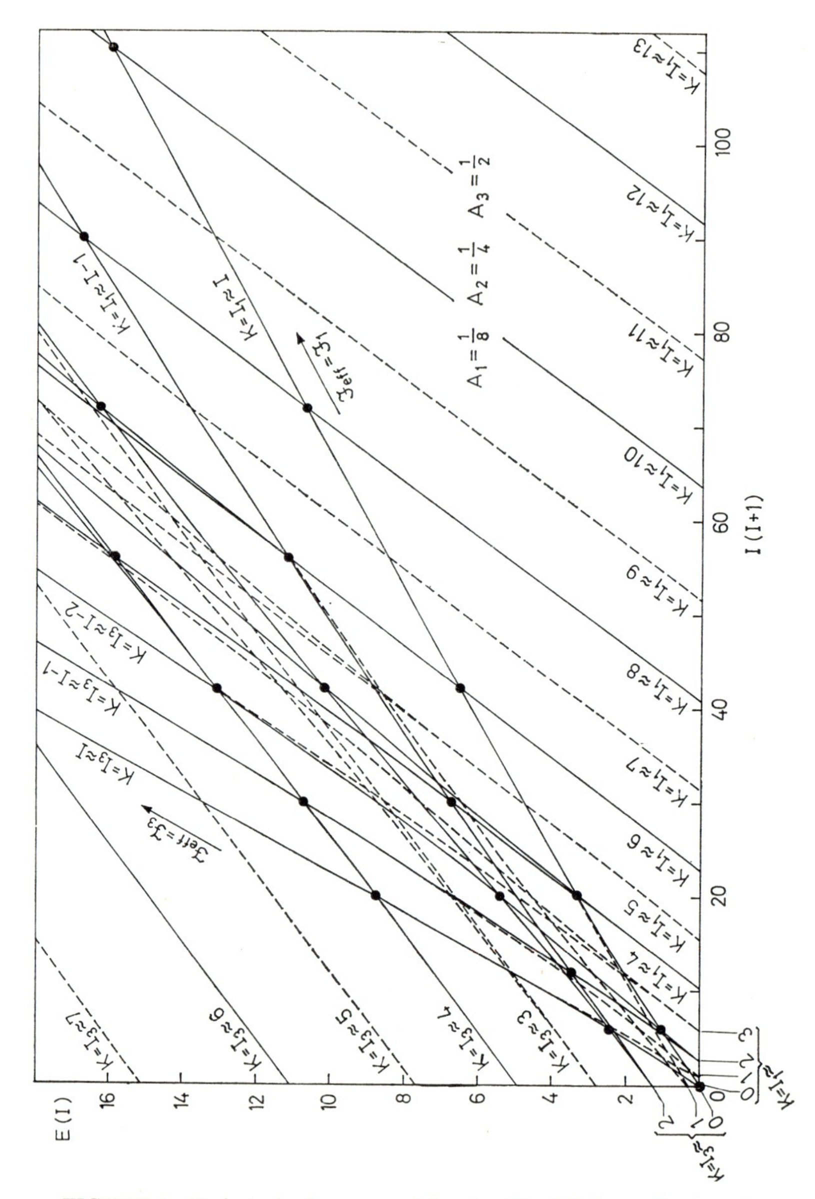

FIGURE 4 Trajectories for asymmetric rotor. The full-drawn lines are the trajectories associated with even values of K, while the trajectories with odd K are dotted. The small filled circles at the intersections of two full-drawn trajectories represent the physical states with the symmetry quantum numbers $(r_1 r_2 r_3) = (+1\ +1\ +1)$, where $r_\varkappa$ is the eigenvalue of the rotational wave function for the operation $\mathscr{R}_\varkappa(\pi) = (-1)^{I_\varkappa}$

the pair correlations as well as further interaction effects such as those that give rise to collective vibrational modes. On this basis, it is possible to account in rather great detail for the observed sequence and quantum numbers of the intrinsic excitations as well as for many features of the matrix elements associated with the I-independent terms in the effective operators.

The I-dependent terms represent the effect of the rotation on the motion of the particles, and may be analyzed in terms of the Coriolis and centrifugal forces acting in a co-ordinate frame rotating with given frequency ω_{rot}. Such an analysis gives the properties of the system as a power series in ω_{rot}, and the connection between rotational frequency and angular momentum can be obtained from the canonical relation

$$\hbar(\omega_{\mathrm{rot}})_{\varkappa} = \frac{\partial H}{\partial I_{\varkappa}} \tag{12}$$

This approach has been rather successful, in particular in accounting for the observed nuclear moments of inertia that are typically about a factor of two smaller than the values for rigid rotation.

The simple interpretation of the spectra of the deformed nuclei and the powerful tools available for the precision study of nuclear properties provide the opportunity for a very detailed exploration of the dynamics of the rotational motion in nuclei. The evidence indicates that, at least in some cases, the estimated Coriolis coupling acting on an unpaired particle is too large by about a factor of two. This evidence comes from the observed difference in moments of inertia of odd-A and even-even nuclei as well as from the effect of the Coriolis coupling on $E2$ matrix elements between bands with $\Delta K = 1$ in odd-A nuclei (see, for example, Stephens *et al.*, 1968). The finding may seem surprising in view of the adequacy of the Coriolis coupling in accounting for the total moment of inertia of even-even nuclei, and it is important to obtain more detailed evidence on these phenomena, since we are here dealing with the basic coupling between collective rotations and the motion of individual nucleons.

The interpretation of the I-dependent properties in terms of particle motion in a rotating potential (the so-called cranking model, first employed by Inglis, 1954) may appear to involve a semi-classical element. The basis of the approach has been much debated and there have been many alternative formulations that have elucidated the quantal features of the rotational motion (see, for example, Tomonaga, 1955; Peierls and Yoccoz, 1957; Bohr and Mottelson, 1958; Thouless and Valatin, 1962; Peierls and Thouless, 1962; Kerman and Klein, 1963; Villars, 1965; Marshalek and Weneser, 1969; Beliaev and Zelevinskij, 1970). A derivation that emphasizes the quantal basis of the treatment can be obtained by starting from particle motion in a deformed static potential, $V_0(r, \vartheta)$, and adding interactions that restore the rotational symmetry. These interactions can be described in terms of the coupling between the particle motion and the collective degree of freedom associated with variations in the orientation. Thus, a small change $\delta\theta$ in orientation produces an added potential

$$\delta V = -\frac{\partial V_0}{\partial \vartheta}\delta\theta \tag{13}$$

acting on the individual particles. The generation of the collective mode itself through this interaction can be treated in the same manner as the analysis of collective vibrational modes in terms of the particle-vibration coupling (see Mottelson, 1970). In the usual normal mode (or random phase) approximation, one obtains a collective mode of zero frequency with an inertial parameter equal to that given by the cranking model. Such a treatment of the rotational motion as a vibrational mode is justified for a particle-rotation interaction slightly smaller than the value (13) required by rotational invariance, in which case the nucleus oscillates with large amplitude about the orientation of the potential V_0. The slight increase in the interaction that restores the rotational invariance changes the vibrational into a rotational mode, but is not expected to significantly affect the inertial parameter.

The additional normal modes emerging from the treatment of the coupling (13) give the intrinsic spectrum of $K\pi = 1+$ excitations. Thus, the analysis at the same time leads to a removal of the spurious degrees of freedom that is present in the description of the intrinsic motion in terms of the complete spectrum of excitations for particle motion in a static deformed potential. The problem of the $K\pi = 1+$ intrinsic excitation spectrum and the possible occurrence of collective effects in this channel is closely related to the structure of the rotational motion, and empirical evidence on these modes of excitation may elucidate the problem encountered in the analysis of Coriolis coupling effects mentioned above.

The treatment of the interaction (13) also yields expressions for the collective orientation angles, in terms of the degrees of freedom of the particle motion in a potential of fixed orientation. In general, these collective variables, considered as functions of the position, momentum, and spin variables of the nucleons, have a complicated structure, corresponding to the fact that this functional relationship reflects the detailed dynamics of the rotational motion.

A very special situation arises if all the particle excitations generated by the interaction (13) have the same frequency, as for particle motion in a harmonic oscillator potential, if we consider only excitations that shift a quantum of oscillation from one direction to another and neglect the high-frequency excitations involving the creation of two quanta. From the relation

$$\frac{\partial V_0}{\partial \vartheta} = i[I_1, H_0] \tag{14}$$

where I_1 is the angular momentum conjugate to ϑ and where H_0 is the Hamiltonian for particle motion in the fixed deformed potential, it follows that, when all the excitations generated by $\partial V_0/\partial\vartheta$ have the same frequency, the operators $\partial V_0/\partial\vartheta$ and I_1 are proportional to each other. Since I_1 is the generator of infinitesimal rotations, we may conclude that the perturbations produced by $\partial V_0/\partial\vartheta$ can be described in terms of a superposition of states representing the given initial configuration oriented in different directions. The weighting of the different orientations can be obtained by exploiting the rotational invariance of the total Hamiltonian. Thus, for an axially symmetric configuration with $I_3 = K$, the total wave function has the form

$$\Psi_{IKM} = \text{const} \int \varphi_0(K; \omega)\, \mathscr{D}^I_{MK}(\omega)\, d\omega \tag{15}$$

This projection integral is equivalent to a wave function of the form (1) and includes rotational perturbations to all orders.

The special situation considered corresponds to the U_3 model of Elliott (1958), which played such an important role in clarifying the relationship between rotational and individual-particle motion. More generally, one may consider the projected state (15) as the zero'th order approximation to the wave function expressed in terms of the co-ordinates of the individual particles. However, when the intrinsic spectrum contains several frequencies, the dynamic effects of the rotation are not adequately included in the state (15).

ROTATIONAL MOTION FOR LARGE QUANTUM NUMBERS. DISCONTINUITIES IN ROTATIONAL BANDS

In the preceding, we have focussed attention on the properties of bands for values of the rotational quantum numbers that are sufficiently small to permit a description in terms of a power series expansion in the rotational frequency about the value zero. For large rotational quantum numbers, such a description becomes inadequate, even though the properties of the band vary smoothly with the rotational quantum number. Such a situation is encountered in the pair rotations for which the rotational frequency (the chemical potential) always has a significant influence on the intrinsic motion.

The familiar treatment of the pair correlations in terms of the Hamiltonian

$$H' = H - \lambda n \tag{16}$$

where λ is the chemical potential and n the particle number, is the analogue of the cranking model. The term $-\lambda n$, which is conventionally viewed as a Lagrange multiplier term associated with the constraint of fixed average particle number, acquires a dynamical interpretation in terms of the Coriolis coupling acting in the co-ordinate frame rotating in gauge space.

In recent years, there has been considerable progress in the study of the domain of convergence of the power series expansion around $I = 0$ for the rotational spectra in heavy nuclei. In many cases, it has been possible to follow the rotational bands to values of I close to 20. As illustrated by the examples in Figs. 1 and 2, the ratio of the coefficients B/A that characterize the leading-order deviations from the $I(I + 1)$ dependence of the rotational energies is in favourable cases of the order of 10^{-3}. This might indicate a radius of convergence for the power series expansion of $I \approx 30$, but the magnitude of the higher-order terms implies a considerably poorer rate of convergence. However, it has been noted (Harris, 1965; Mariscotti *et al.*,1969) that the rate of convergence is significantly improved if the energy is expressed as a power series in the rotational frequency rather than in the angular momentum. An example is illustrated in Figs. 5a and 5b, which shows the energies of the ground-state band of ^{172}Hf. Already for $I = 8$, the energy shows significant deviations from the expansion to fourth order in I, but the corresponding frequency plot, which gives the moment of inertia as a function of the square of the rotational frequency, can be represented by a straight line over a considerably wider domain. The moment of inertia is here defined as [see Eq. (12)]

$$\mathscr{I} \equiv \frac{\hbar I}{\omega_{\text{rot}}} = \frac{\hbar^2}{2}\left(\frac{\partial E}{\partial I(I + 1)}\right)^{-1} \tag{17}$$

One can also exhibit the faster convergence of the frequency expansion by noting that an energy expression containing only terms of second and fourth order in the frequency,

$$E_{\text{rot}} = \alpha\omega_{\text{rot}}^2 + \beta\omega_{\text{rot}}^4 \tag{18}$$

implies relations between the higher coefficients in the expansion in powers of $I(I + 1)$

$$\begin{aligned} \frac{C}{A} &= 4\left(\frac{B}{A}\right)^2 \\ \frac{D}{A} &= 24\left(\frac{B}{A}\right)^3 \end{aligned} \tag{19}$$

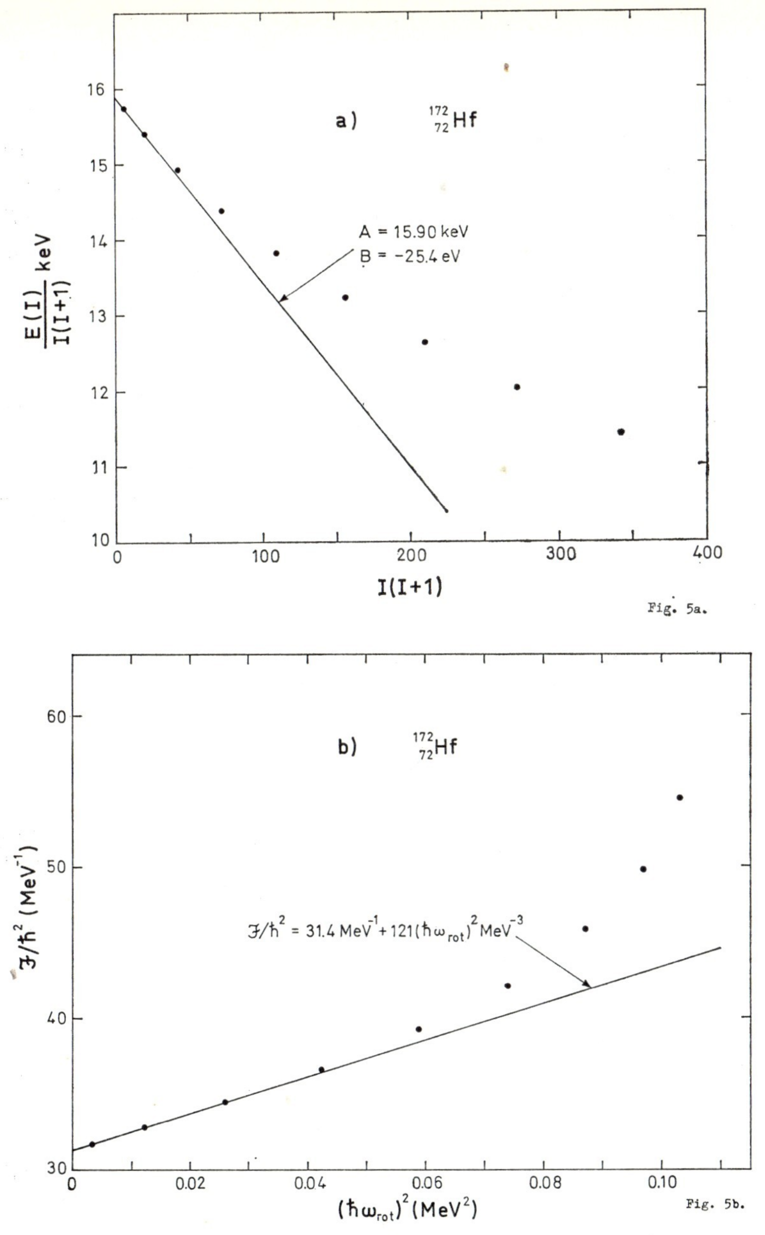

FIGURE 5 Rotational energies of the ground-state band in ^{172}Hf. The rotational levels have been studied by means of the reaction ^{165}Ho(^{11}B, 4*n*) by Stephens *et al.* (1965), and by means of the reaction ^{171}Yb(α, 3*n*) by S. Jägare, S. A. Hjorth, H. Ryde, and A. Johnson (priv. com.). In Fig. 5a, the energy is plotted as a function of $I(I+1)$, and a straight line corresponds to the expansion (10) with the leading-order terms proportional to A and B. In Fig. 5b, the moment of inertia is plotted as a function of the rotational frequency, on the basis of the relations (12) and (17)

Table 1 contains examples where precision measurements of the rotational energies have enabled the determination of the coefficients C and D; it is seen that the relations (19) are rather well fulfilled. An attempt to determine the higher-order terms in the frequency expansion (18) shows that the coefficient γ of $\omega^6{}_{rot}$ is appreciably smaller (by a factor of five or more) than β^2/α. Theoretical estimates of the coefficient γ have not so far provided an explanation of this striking feature of the expansion.

TABLE 1 Test of relations implied by linear dependence of $\mathscr{I}$ on $(\omega_{rot})^2$
The rotational energy expansion coefficients A, B, C, and D are calculated from the energies of the $I = 0, 2, 4, 6$, and 8 members of the ground-state rotational band. The data are taken from the compilations by Groshev *et al.* (1968) and (1969) and the more recent measurements on ^{156}Gd and ^{158}Gd (H. R. Koch, priv. com.)

Nucleus	A keV	$-\frac{B}{A} \times 10^3$	$\frac{C}{A} \times 10^6$	$4\left(\frac{B}{A}\right)^2 \times 10^6$	$-\frac{D}{A} \times 10^9$	$-24\left(\frac{B}{A}\right)^3 \times 10^9$
^{156}Gd	15.0332	2.362	14.3	22.3		
^{158}Gd	13.3353	1.069	4.10	4.56		
^{162}Dy	13.5174	0.934	3.74	3.49	16	20
^{164}Dy	12.2858	0.738	1.44	2.18		
^{168}Er	13.3431	0.547	1.39	1.20	4.2	4.0
^{178}Hf	15.6203	1.005	4.14	4.04	22	24

For the largest values of I that have been reached in the heavy deformed nuclei, the rotational energies show a behaviour that is not well described by an expansion about zero frequency and which suggests that the nucleus is undergoing some major modification as a result of the perturbations produced by the rotational motion (see, for example, Fig. 5b). It is possible that one is encountering a phase transition from a pair-correlated or superfluid state to a normal state without pair correlations, which is expected to rotate with the rigid-body value for thc moment of inertia. Considerable interest attaches to the further exploration of the spectrum in this domain; the nucleus offers the opportunity for a study of individual quantum states associated with such transitions, which may complement the information obtained from the study of phase transitions in macroscopic systems.

An example of a discontinuity in rotational sequences of an especially simple structure is provided by the breaks in the pair-rotational bands associated with the shell closings. Nuclei with filled major shells have no stable superfluid deformation; otherwise the shell structure would be washed out by the pair correlations. With the approach to the closed-shell configuration, the pair-rotational bands therefore become strongly coupled to the intrinsic degrees of freedom associated with the fluctuations in the pair field and, in the region of the closed-shell configurations, the excitations of the pair field acquire a vibrational character. Thus, the bands coming from one side of the discontinuity may be continued through a pair vibrational spectrum to a merging with the bands at the other side of the discontinuity. The rotational frequencies and moments of inertia at the two sides of the shell closing are shifted by amounts depending on the energy gap between the shells.

A somewhat similar discontinuity may occur in the ordinary rotational spectra in light nuclei, in situations where the energy gap between the shells prevents the particle orbits from increasing their angular momentum in response to the rotational motion. In such a situation, the band terminates when the angular momenta of the particles have been fully aligned to the extent permitted by the exclusion principle. This type of discontinuity in the band structure was first exhibited by the U_3 model (Elliott, 1958); the alignment effect can be studied more generally by an analysis, along the lines indicated above, of the rotational perturbations caused by the Coriolis interaction. In many of the light nuclei, the bands have been followed to values of I equal to the predicted maximum value I_{max}, but the evidence for a discontinuity in the band structure is still rather inconclusive, and the nature of the relationship to the states of higher I that can be obtained by shifting particles into higher shells has not so far been explored. In heavy nuclei, the values of I_{max} are very large ($\gtrsim$ 100) and are not only far beyond the domain that has been explored, but may also exceed the values of the angular momentum at which the nucleus becomes unstable with respect to fission.

The study of nuclear states with large angular momenta is rapidly advancing due, in particular, to the advent of heavy ion beams that can transfer large amounts of angular momentum into the nucleus (50 to 100 units). In such processes, the highly excited compound nucleus rapidly cools down to the neighbourhood of the "yrast" line, representing the lowest levels of given I. In the example considered in Fig. 5 (^{165}Ho(^{11}B, 4n)^{172}Hf), compound nuclei are produced with angular momenta up to $I \approx 35$, and only rather few units of angular momentum are expended before the nucleus reaches to within about an MeV of the yrast line. The further decay follows rather close to the yrast line until the main intensity flows into the ground-state band at $I \sim$ 14–16. At slightly higher values of I, it appears that the ground-state band merges with other bands.

The nuclear spectrum in the region of the yrast line, above $I \sim 20$, remains virgin territory and involves nuclear matter in a new form. Though highly excited, the nucleus is "cold", since the energy is mainly concentrated in a single or a few collective degrees of freedom that describe the modification in the nuclear structure produced by the Coriolis and centrifugal forces.

The tentative evidence provided by the heavy ion-induced reactions appears to imply the systematic occurrence of enhanced $E2$ transitions leading along paths close to the yrast line (Diamond *et al.*, 1969), but the nature of the collective motion is not yet established. One possibility may be a stabilization of the rotational motion through the development of a triaxial shape. The rotational distortion effects tend to break the axial symmetry, thereby removing the degeneracy between the moments of inertia perpendicular to the symmetry axis. The system can exploit the asymmetry by rotating about the axis with the largest moment of inertia. (The rotational coupling effects that produce the asymmetry are the same as those responsible for the coupling between the ground-state and γ-vibrational bands implied by the I-dependent term in the $E2$ matrix elements in Fig. 3).

REFERENCES

S. T. Belyaev and V. G. Zelevinsky (1970), *Yadernaja Fizika* **11**, 741 (1970)
A. Bohr and B. R. Mottelson, *Kgl. Norske Vid. Selsk. Forhandlinger*, **31**, No. 12 (1958)
R. M. Diamond, F. S. Stephens, W. H. Kelly, and D. Ward, *Phys. Rev. Letters* **22**, 546 (1969)
J. P. Elliott, *Proc. Roy. Soc.* (*London*), **A 245**, 128 and 562 (1958)
C. J. Gallagher, Jr., O. B. Nielsen, and A. W. Sunyar, *Phys. Letters* **16**, 298 (1965)
L. V. Groshev, A. M. Demidov, V. I. Pelekhov, and L. Sokolovskij, *Nuclear Data* **5**, 1 (1968) and **5**, 243 (1969)
C. Günther and D. R. Parsignault, *Phys. Rev.* **153**, 1297 (1967)
S. M. Harris, *Phys. Rev.* **138**, B 509 (1965)
D. R. Inglis, *Phys. Rev.* **96**, 1059 (1954)
A. K. Kerman and A. Klein, *Phys. Rev.* **132**, 1326 (1963)
C. M. Lederer, J. M. Hollander, and I. Perlman, *Table of Isotopes*, Sixth Ed., John Wiley and Sons, New York, 1967
M. A. J. Mariscotti, G. Scharff-Goldhaber, and B. Buck, *Phys. Rev.* **178**, 1864 (1969)
E. R. Marshalek and J. Weneser, *Ann. Phys.* (*N.Y.*) **53**, 569 (1969)
B. R. Mottelson, Report Solvay Conference 1970; this volume
R. E. Peierls and J. Yoccoz, *Proc. Phys. Soc.* (London) **A 70**, 381 (1957)
R. E. Peierls and D. J. Thouless, *Nuclear Phys.* **38**, 154 (1962)
F. S. Stephens, N. L. Lark, and R. M. Diamond, *Nuclear Phys.* **63**, 82 (1965)
F. S. Stephens, M. D. Holtz, R. M. Diamond, and J. O. Newton, *Nuclear Phys.* **A 115**, 129 (1968)
J. D. Talman, *Nuclear Phys.* **A 161**, 481 (1971)
E. Teller and J. A. Wheeler, *Phys. Rev.* **53**, 778 (1938)
D. J. Thouless and J. G. Valatin, *Nuclear Phys.* **31**, 211 (1962)
S. Tomonaga, *Prog. Theor. Phys.* **13**, 467 (1955)
F. M. H. Villars, *Nuclear Phys.* **74**, 353 (1965)

DISCUSSION

B. R. Mottelson Professor Bohr's point concerning the expansion in the frequency seems to be such an exciting clue as to the structure of the rotational motion. It somehow must be related to the nature of the changes in the internal structure as the systems rotate faster and faster but it is quite a mystery why one gets such a better description in term of the frequency. It is my personal prejudice that this feature must somehow be related to the effect of the rotational motion on the pair correlation, but it has not yet been possible to establish a quantitative relation. I wonder if anyone has an idea about this point.

E. P. Wigner Could you amplify your statement on the simpler dependence of the energy on the angular velocity ω of rotation, than on angular momentum I. Initially, that is experimentally, the energy is given as function of I. How is then ω defined?

A. Bohr From the relation (12), it follows that

$$\hbar^2(\omega_{\text{rot}})^2 = \hbar^2\left((\omega_{\text{rot}})_1^2 + (\omega_{\text{rot}})_2^2\right) = 4I(I+1)\left(\frac{\partial E}{\partial I(I+1)}\right)^2$$

Since the experimental energies vary smoothly with $I(I+1)$, the derivative can be determined with high accuracy.

E. P. Wigner You say that the power series for the energy, in terms of ω, converges better than the power series in terms of I. Could you comment how much better the convergence is? Let us say, how much smaller is the third order term in terms of ω (or ω^2) than in terms of I (or, rather $I(I+1)$)?

A. Bohr If, for example, we expand the moment of inertia $\mathscr{I} = \alpha + \beta\omega^2 + \gamma\omega^4$ the ratio γ/β is, in most cases, less than a fifth of β/α. Such a behavior of the expansion in ω^2 implies that, in the expansion in $I(I+1)$, the ratio of successive coefficients increases [see Eq. (19)].

D. R. Inglis One gets a graphical impression, graphically suggesting that there may be such a transition between a low-I regime and a high-I regime, by plotting the data on, in nuclear bands already available, up to almost $I = 20$ somewhat differently, plotting rotational energies against $I(I+1)$. In each band these approach for low I a line through the origin with a slope

A characteristic of a moment of inertia. For $I \simeq 10$ the data fall below this line and for high *I* seem to approach quite closely (but not completely convincingly) an asymptote not passing through the origin and having the same slope for all bands, suggestive of a simple rotational regime with altered internal structure.

G. Bertsch Could the poorness of convergence be due to the possibility that the spectrum is partly linear in *I* and partly proportional to $I(I + 1)$. Such a behaviour would emerge if the excitation of rotational states were described by a simple operator algebra. Any linear dependence on *I* would imply a very poor convergence as a power series in $I(I + 1)$.

A. Bohr While a term proportional to I occurs in the vibrational spectra, there is no evidence for a term of this form in the well-developed rotational spectra.

E. P. Wigner You mentioned the analogy between molecular and nuclear rotational spectra. There is a difference between the two cases. The way the intrinsic state for molecules is defined, an arbitrarily small rotation renders this state orthogonal to the non-rotated state. This is because in case of molecular wave functions, the nuclei are considered to be strictly localised in the intrinsic state. On the other hand, in the case of nuclear wave functions, in a typical case it might require a rotation by about 10° or 15° to render the scalar product of the rotated and unrotated state to become 1/2. Now, at least for large *I*, the *D* function drops essentially to zero when the angle of rotation becomes $2/I$. Hence, one would expect that if the wave function given by Dr. Bohr, and commonly used, is valid, an anomaly in the rotational spectrum occurs when $2/I$ becomes of the order of the angle at which the aforementioned scalar product drops to around 1/2. This is, as a rule, not a very large *I*. However, if the picture projected by Dr. Bohr is valid also for large *I*, one would expect the rotational spectrum to continue also beyond that anomaly, up to arbitrarily large *I*. This, of course, is not so in the usual independent particle (or shell) model in which the rotational spectrum comes to a sudden halt at a definite *I*. The questions then are:

1 Should one expect an upper limit to the *I* of nuclear rotational spectra, as indicated by the independent particle model? If not, what determines the highest *I*?
2 What should one expect at the *I* at which the argument sketched would indicate an anomaly in the energy as function of *I*? Should the rotational character of the spectrum fade away, or should the energy as function of *I* begin to increase (or possibly decrease) more rapidly at the critial *I*?

A. Bohr The quantal effects associated with the non-orthogonality of the wave-functions for intrinsic motion with different orientations are contained in the description that I attempted to outline. A sharp break in the rotational band occurs, if one can neglect the transfer of particles into higher shells as a consequence of the deformation or rotation. In such a situation, the rotation of the potential in which the particles move leads to a gradual alignment of the angular momenta of the particles. The alignment becomes complete (to the extent permitted by the exclusion principle) for a certain value I_{max} and associated frequency ω_{max}; a further increase of the frequency would lead to no increase of I, and hence the band teminates. The approach to the end point of the band is characterized by a decrease of the $E2$ transition probabilities, since the alignment effect leads towards a density distribution that is symmetric about the axis of rotation and hence does not radiate.

The above picture becomes modified when one takes into account the possibility of the particles being excited into higher shells by the deformation and by the Coriolis forces. Depending on the detailed features of the single-particle spectra, it would seem that a variety of intermediate situations may occur between an abrupt break and a smooth continuation. Further studies are needed to explore these problems.

L. C. Biedenharn I would like to make a couple of remarks. The first is that one probably takes much too literally the picture of a nucleus as made up of neutrons and protons. One should realize—this point has been emphasized by Gell-Mann, and others—that the neutron and protons of the shell model may in fact be almost as fictitious as the quarks inside the hadrons. That is to say, the neutron and proton of the shell model belong to a truncated picture in which the meson degrees of freedom have been eliminated.

This can be wrong for certain effects; for example, magnetic moments have possible contributions from exchange currents or baryonic excitations ... both basically mesonic effects. Large rotational excitations could be sensitive to (or stabilized by) mesonic degrees of freedom.

Concerning Prof. Wigner's remark on a single "rotated overlap integral", one should note that if one considers this problem as relating two coordinate frames, then it is clear (from the uncertainty principle) that one must have infinitely many states available. If one starts with a fixed shell model function only a finite number of states are available (in other words the projected states contain only a finite number of angular momenta). This is just the remark made by Wigner—that the overlap of a slightly rotated (intrinsic) wave function is not a delta function—expressed now differently.

But the different language is helpful in that we know the inter-relation of the two frames is not an inertial transformation. Hence the physics is basically approximate, and certainly change drastically for large rotational excitations—a clear warning that other degrees of freedom may be playing a significant role.

The situation is much better when one considers the rotational hadronic levels ("Regge recurrences"). If the quarks are truly fictitions—so that the hadron doesn't fly apart (rotationally) into quarks—the recurrences may extend to very large angular momenta indeed.

C. E. Brown I belive that there is an interesting argument between Professor Bohr and Wigner. From Wigner's point of view, why shouldn't the bands go in to infinite angular momentum whereas Bohr says that they cut off. The difference must come from the closed shells. They have the degrees of freedom to take the spectrum up to very high I. The more concrete result, I don't understand, is that in unrestricted Hartree-Fock calculations, which take into account the closed shells, the bands still break off.

L. C. Biedenharn I can answer that point very quickly. If you use the Hartree-Fock approach, what have you? You are definitely dealing with the system as a finite number of states: it must cut off. This is not surprising. In order to find these unlimited large bands there must be something in the calculation having a continuum or something like that. Closed shells intimately take part in that motion and so we are not really in contradiction. But the Hartree-Fock sort of picture starts by saying, I have restricted my world to a truncated space. The very fact that you truncate means that you have finiteness, so you should not be surprised that the bands terminate—, you put it in from the beginning.

A. Bohr It is not clear whether a Hartree-Fock calculation leads to a discontinuity in the band structure. To explore this point, one must consider the self consistent solutions as a function of the rotational angular momentum or frequency, i.e. a constrained Hartree-Fock problem. The constraint has the form $-\lambda I_1$, where I_1 is the component of angular momentum about the axis of rotation, and the effective self consistent potential is therefore no longer axially symmetric, even though the nucleus may have this symmetry for $I = 0$. Such analyses do not yet seem to have been undertaken.

C. E. Brown When one speaks about continuity of rotational bands, it seems to me that one should speak about continuity of properties such as transition probabilities, etc.... The energy levels may go on to very high angular momentum, and at the same time, the transition probabilities break off at the $SU(3)$ limit.

Permutation Group in Light Nuclei*

P. KRAMER

Institute of Theoretical Physics
University of Tübingen, Germany

IN THIS REPORT I shall deal with applications of the permutation group to the structure of light nuclei. Starting from general features in the supermultiplet scheme I shall proceed to the specific methods and results obtained in shell configurations, cluster configurations and nuclear reactions.

1 THE ORBITAL PARTITION

According to the Pauli principle nuclear states have very simple transformations properties under permutations. Under the application of a permutation p they are multiplied by $(-)^p$ which is equal to $+1$ if p is an even permutation and -1 if p is an odd permutation. Less trivial transformation properties arise if the permutations are applied separately to orbital, spin and isospin variables respectively.

The transformation properties of the orbital states of n particles may be characterized by a partition f of n,

$$f = \{f_1 f_2 \dots f_k\}, \quad f_1 \geqq f_2 \geqq \cdots \geqq f_k \geqq 0, \tag{1}$$

which defines a Young diagram of f_i boxes in the ith row. The partition for the spin and isospin states are related to the values of the total spin S and the total isospin T by

$$f^S = \{\tfrac{1}{2}n + S, \tfrac{1}{2}n - S\}, \quad f^T = \{\tfrac{1}{2}n + T, \tfrac{1}{2}n - T\}. \tag{2}$$

These two partitions may be coupled to the combined spin-isospin partition $\hat{f}$, the coupling being governed by the inner or Kronecker product of partitions

$$f^S \otimes f^T \to \hat{f}. \tag{3}$$

* Supported by the Bundesministerium für Wissenschaft und Bildung of the Federal Republic of Germany.

The coupling of the orbital and the spin-isospin states to antimetric states,

$$f \otimes \hat{f} \rightarrow \{1^n\}, \tag{4}$$

requires that f and $\hat{f}$ have associate Young diagrams (rows in f replaced by columns in $\hat{f}$).

Another important relation between partitions arises by starting out from product states of two different sets of n' and n'' particles characterized by partitions f' and f'' respectively. To the product states one may apply projection operators and in this way induce n-particle states with overall partition f. The corresponding outer product of partitions,

$$f' \times f'' \rightarrow f, \tag{5}$$

is governed by Littlewood's rules. These rules tell us for example that from the four spin-isospin states one may induce spin-isospin partitions $\hat{f}$ of at most four rows. Consequently the orbital partition f has at most four columns. It follows that f and $\hat{f}$ are completely determined by n and the three supermultiplet quantum numbers

$$\begin{aligned} P &= \tfrac{1}{2}(\hat{f}_1 + \hat{f}_2 - \hat{f}_3 - \hat{f}_4) \\ P' &= \tfrac{1}{2}(\hat{f}_1 - \hat{f}_2 + \hat{f}_3 - \hat{f}_4) \\ P'' &= \tfrac{1}{2}(\hat{f}_1 - \hat{f}_2 - \hat{f}_3 + \hat{f}_4) \end{aligned} \tag{6}$$

introduced by Wigner[1]. Littlewood's rules could also be used to show that for states of N neutrons and Z protons we must have

$$P \geqq \tfrac{1}{2}|N - Z|. \tag{7}$$

2 GROUND STATES OF LIGHT NUCLEI

Evidence for physical restrictions on the orbital partition f comes from the ground state properties of light nuclei. If the hamiltonian

$$\begin{aligned} H &= T + V, \\ V &= \sum_{1=s<t}^{n} V(st) \end{aligned} \tag{8}$$

with a two-body interaction

$$V(st) = -W(st) - M(st)(st) \tag{9}$$

that is independent of spin and isospin and contains the orbital transposition (st) or Majorana-operator is applied to a state of orbital symmetry, it does not change the transformation properties under permutations so that f is a good quantum number. Taking

$$M(st) = \overline{M(st)} = \text{const.} \tag{10}$$

one needs the expectation value of the Majorana interaction. This value depends only on the partition f and equals the difference between the number of symmetric and antimetric pairs,

$$\langle f| \sum_{s<t} (st)|f\rangle = n_+ - n_- = -F(nPP'P''),$$
$$F(nPP'P'') = \tfrac{1}{2}[(P)^2 + (P')^2 + (P'')^2 + 4P + 2P' + \tfrac{1}{4} n(n - 16)] \tag{11}$$

The lowest states come from orbital partitions f with the highest number of symmetric pairs, and these are the partitions with the lowest number of rows. The expectation value of the hamiltonian H written as

$$\langle f|H|f\rangle = a'(n) + b(n)\, F(nPP'P'') \tag{12}$$

gives the supermultiplet mass formula of Franzini and Radicati[2]. The lowest supermultiplet compatible with Eq. (7) characterizes the ground state and the coefficient $b(n)$ determines the position of the higher supermultiplets.

The mass formula Eq. (12) depends on the assumption of two-body forces. Corrections due to many-body forces which lead to a different dependency on $PP'P''$ have been considered by Wheeler[3] and recently by Burdet, Maguin and Partensky[4p.325].

3 CALCULATIONS IN THE SUPERMULTIPLET SCHEME.

For a more detailed study of nuclear states one introduces orbital partner states belonging to the partition f with row index r and characterized further by the total angular momentum L with component M_L and parity π,

$$|\alpha^n LM_L\pi fr\rangle. \tag{13}$$

These states are orthogonal with respect to $LM_L\pi f$ and r. The number of orthogonal partner states belonging to the partition we call $|f|$, the dimension of the representation f. To apply two-body interactions it proves convenient to choose the row label r as

$$r = f'r'f'' \tag{14}$$

with $f'r'$ and f'' being partitions and row labels that describe the transformation properties under permutations of the first $n - 2$ and the last 2 particles respectively. Similarly one writes the spin-isospin states as

$$|\gamma^n SM_S TM_T \hat{f}\hat{r}\rangle \tag{15}$$

with

$$\hat{r} = \hat{f}'\hat{r}'\hat{f}''. \tag{16}$$

The coupled antimetric states of total angular momentum J with component M are given by

$$|(\alpha\gamma)^n IM\pi\{1^n\}\, TM_T LSf\rangle$$
$$= \sum_{M_L M_S} |f|^{-1/2} \sum_r |\alpha^n LM_L \pi f r\rangle\, |\gamma^n SM_S TM_T \hat{f}\hat{r}\rangle\, \langle LM_L SM_S | IM\rangle. \tag{17}$$

The nuclear two-body interaction is then written in terms of coupled orbital tensor operators $T_q^\varkappa$ and spin tensor operators $U_q^\varkappa$ as

$$V = \sum_\varkappa V(\varkappa), \tag{18}$$

$$V(\varkappa) = \sum_{s<t} \sum_q (-)^q T_q^\varkappa(st)\, U_{-q}^\varkappa(st)$$
$$= [(n-2)!\,2!]^{-1} \sum_p p \sum_q (-)^q T_q^\varkappa(n-1n)\, U_{-q}^\varkappa(n-1n)\, p^{-1}. \tag{19}$$

The matrix elements of $V(\varkappa)$ between antimetric states are expressed as

$$\langle(\overline{\alpha\gamma})^n\, I\pi\{1^n\}\, TM_T\bar{L}\bar{S}\bar{f}\,\|\, V(\varkappa)\,\|(\alpha\gamma)^n\, I\pi\{1^n\}\, TM_T LSf\rangle$$
$$= (-)^{L+\bar{S}+I} \begin{Bmatrix} I & \bar{S} & \bar{L} \\ \varkappa & L & S \end{Bmatrix} (|\bar{f}|\,|f|)^{-1/2} \binom{n}{2}$$
$$\sum_{f'f''} \langle\bar{\alpha}^n \bar{L}\pi\bar{f}f'f''\,\|\, T^\varkappa(n-1n)\,\|\alpha^n L\pi f f'f''\rangle$$
$$\langle\gamma^n \bar{S}TM_T\hat{\bar{f}}\hat{f}'\hat{f}''\,\|\, U^\varkappa(n-1n)\,\|\gamma^n STM_T\hat{f}\hat{f}'\hat{f}''\rangle \tag{20}$$

in terms of orbital and spin-isospin matrix elements. The reduction to the contribution of the last pair results from the second line of Eq. (19) and from the application of the permutations p and p^{-1} to the antimetric states. The double bars indicate reduced matrix elements both with respect to the rotation and permutation group[5]. With the choice Eq. (14) of the row label r the orbital matrix element of $T\,(n-1n)$ becomes independent of r' and diagonal with respect to f' and f''.

The possible partitions $\bar{f}$ for given f are given by Littlewood's rules since the outer product of f' and f'' must fulfill

$$f' \times f'' \to \bar{f}, f. \tag{21}$$

This means that $\bar{f}$ and f differ at most in the position of two boxes from each other. Only a spin-dependent interaction can mix different partitions.

The expression Eq. (20) is the starting point for several calculations in the supermultiplet scheme. The spin-isospin matrix elements may be analyzed using the permutation group $S(n)$ or the unitary group $SU(4)$. This allows one to study nuclear properties that depend mainly on the spin-isospin states. Recent work along these lines covers

a) the ground states of light nuclei with $A \leqq 56$ and orbital partition $f = \{44 \dots 4k\}$ studied by Vanagas and Petrauskas[6].

These authors employ a tensor decomposition of the interaction Eq. (19) with respect to the permutation group that was proposed earlier by Mahmoud and Cooper[7] and analyze the contributions from spin-dependent forces.

TABLE 1 Allowed beta transitions between analogue states with $T = \frac{1}{2}$[8]

Transition	J^{π}	L_1	log $ft_{\text{theor.}}$	L_2	log $ft_{\text{theor.}}$	log ft_{exp}
n^1—H^1	$^1/_2{}^+$	—	—	0	3.07	3.07
H^3—He^3	$^1/_2{}^+$	—	—	0	3.07	3.06
Be^7—Li^7	$^3/_2{}^-$	—	—	1	3.27	3.25
C^{11}—B^{11}	$^3/_2{}^-$	2	3.53	1	3.27	3.60
N^{13}—C^{13}	$^1/_2{}^-$	1	3.63	—	—	3.66
O^{15}—N^{15}	$^1/_2{}^-$	1	3.63	—	—	3.64
F^{17}—O^{17}	$^5/_2{}^+$	—	—	2	3.32	3.38
Ne^{19}—F^{19}	$^1/_2{}^+$	—	—	0	3.07	3.25
Na^{21}—Ne^{21}	$^3/_2{}^+$	2	3.53	1	3.27	3.60
Mg^{23}—Na^{23}	$^3/_2{}^+$	2	3.53	1	3.27	3.70
Al^{25}—Mg^{25}	$^5/_2{}^+$	3	3.49	2	3.32	3.60
Si^{27}—Al^{27}	$^5/_2{}^+$	3	3.49	2	3.32	3.55
P^{29}—Si^{29}	$^1/_2{}^+$	1	3.63	0	3.07	3.73
S^{31}—P^{31}	$^1/_2{}^+$	1	3.63	0	3.07	3.70
Cl^{33}—S^{33}	$^3/_2{}^+$	2	3.53	1	3.27	3.70
Ar^{35}—Cl^{35}	$^3/_2{}^+$	2	3.53	1	3.27	3.80
K^{37}—Ar^{37}	$^3/_2{}^+$	2	3.53	—	—	3.66
Ca^{39}—K^{39}	$^3/_2{}^+$	2	3.53	—	—	3.60
Sc^{41}—Ca^{41}	$^7/_2{}^-$	—	—	3	3.34	3.44
Ti^{43}—Sc^{43}	$^7/_2{}^-$	4	3.43	3	3.34	3.50

b) allowed beta transitions and magnetic moments in the same region of masses[8,9]. This involves a decomposition of the transition operator in analogy to Eq. (19). With respect to the permutation group the new feature is that for one-body operators $\bar{f}$ and f are related by

$$f' \times \{1\} \rightarrow \bar{f}, f \tag{22}$$

with f' being a partition of $n - 1$.

c) the n, T-dependence of Coulomb energies studied by Hecht and Pang[10,11].

The results show clearly the usefulness of this approach for the systematics of nuclear properties. I shall turn now to the significance of the permutation group and the partition for the orbital states.

4 FRACTIONAL PARENTAGE IN SHELL CONFIGURATIONS

When building the orbital states Eq. (13) from single-particle states, it is possible to expand them in terms of states of the first $n - 2$ and the last 2 particles respectively. As is well-known, this leads to the concept of fractional parentage introduced by Racah[12] and allows the reduction of the orbital matrix elements in Eq. (20) to algebraic sums over two-body matrix elements. Similar expansions apply to one-body operators. The permutation group plays a central part in the determination of the fractional parentage coefficients (*fpc*). I shall not attempt to go into the details of this method. For single and multi-shell configurations it has been reviewed and extended by Horie[13].

An alternative and very powerful approach to shell configurations discussed recently by Moshinsky and Syamala Devi[14] is based on the unitary space spanned by the k single-particle states. The n-particle states form a basis for an irreducible representation of the unitary group $U(k)$ acting on the single-particle states. The representations of this unitary group and of the permutation group must coincide as was shown in a particularly transparent way by Moshinsky[15]. Mathematically the equivalence of these two approaches is based on relations between the unitary and permutation groups and on the concept of induced representations. For physical applications this implies that a good part of the unitary structure of shell configurations expresses properties of the permutation group[16,17,18] that are independent of the assumption of underlying single-particle states and hence may be carried over to more general states like cluster configurations to be discussed later.

The diagonalization of the energy matrix for shell configurations using *fpc* gives the eigenstates in terms of states with fixed partitions. For the p^n configuration Cohen and Kurath[19] found the best fit to the levels using an effective two-body interaction that has all features of a realistic nucleon-nucleon interaction. We have analyzed[20] the eigenstates of this interaction in terms of orbital partitions using the *fpc* given by Elliott, Hope, Jahn and van Wieringen[21,22]. For a given value of J and T, the lowest states have an overlap of about 0.98 with the state of lowest orbital partition. The calculations show also that states of different partitions may appear quite near to each other: the two states with $J^\pi = \frac{5}{2}^-$ of ^{7}Be belong mainly to the partitions $\{43\}$ and $\{421\}$ respectively but are separated by only 0.6 MeV.

TABLE 2 Eigenstates of the effective interaction[19] in the configuration p^3, $T=\frac{1}{2}$ of ^{7}Be and ^{7}Li[20]

p^3 $T=1/2$			3/2	1/2	5/2	5/2	7/2	J
300	1	1/2	0.9902	0.9978				
	3	1/2			−0.8946	0.3981	0.9893	
210	1	1/2	−0.1233	−0.0456				
		3/2	−0.0535	0.0481	0.3138	0.8657		
	2	1/2	−0.0241		0.1128	−0.0428		
		3/2	−0.0310	−0.0102	0.2975	0.3003	0.1460	
111	0	1/2		0.0006				
f	L	S						

Similar conclusions arise from the study of nuclei in the $(sd)^n$ configuration by Akiyama, Arima and Sebe[23]. These authors use a central spin-dependent interaction and a one-body spin-orbit force. The classification of the orbital states is based on the permutation group $S(n)$ and the unitary group $SU(3)$.

The main conclusion from their work is that states of nuclei in this configuration with up to eight nucleons are well described by including the two lowest supermultiplets along with the highest representations of $SU(3)$. This result may depend somewhat on the use of a one-body spin-orbit force which

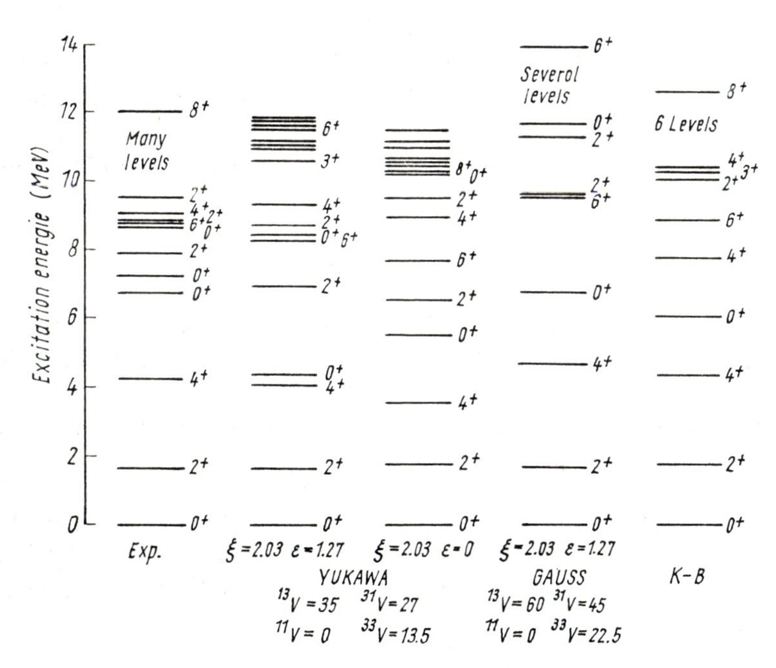

FIGURE 1 Energy levels of ^{20}Ne[23]

TABLE 3 Lowest eigenstates of ^{20}Ne in terms of orbital partitions and SU(3) representations[23]

		0_1	0_2	2_1	2_2	4_1	6_1	8_1
[4]	(80)	0.895	0.042	0.906	0.025	0.848	0.875	0.806
	(42)	0.016	0.828	0.018	0.717	0.061	0.008	
	(04)	0.025	0.053	0.004	0.105	0.001		
	(20)	0.000	0.004	0.000	0.002			
[31]	(61)	0.059	0.014	0.068	0.041	0.082	0.110	0.174
	(42)	0.000	0.000	0.000	0.019	0.002	0.001	
	(23)	0.002	0.043	0.001	0.063	0.001	0.000	
	(31)	0.001	0.009	0.001	0.014	0.002	0.000	
	(12)	0.000	0.002	0.000	0.002	0.000	0.000	
	(20)	0.000	0.000	0.000	0.000	0.000	0.000	
[22]		0.001	0.004	0.001	0.011	0.002	0.005	0.020
[211]		0.001	0.001	0.001	0.001	0.001	0.001	
[1111]		0.000	0.000	0.000	0.000	0.000	0.000	

according to Eq. (21) mixes in only the Young diagrams differing in the position of a single box.

There are shortcomings of these shell model calculations like the appearance of low-lying levels with normal and non-normal parity that belong to higher configurations. It seems that some of these levels present very simple orbital symmetries. One way of describing such states may be given by the translational-invariant oscillator model[24,25]. For the lowest configurations this model is equivalent to the oscillator shell model. Higher excitations are carried by a set of relative vectors, and therefore it appears natural to describe these states as excited cluster configurations to which I shall turn now.

5 CLUSTER PARENTAGE

I shall try to show that the description of nuclear states in terms of cluster configurations offers a possibility of finding the best orbital states for a given orbital partition f.

Cluster configurations are projected from orbital parent states of the type

$$|\alpha^n ab \ldots k) = \phi_a \phi_b \ldots \phi_k \chi_{ab \ldots k}. \tag{23}$$

Each state ϕ_j is an intrinsic orbital state of a cluster that is symmetric under internal permutations. Since for the moment we are interested in the permutation group, we identify the labels ab ... with the corresponding orbital partitions,

$$a = \{n_a\}, \quad b = \{n_b\}, \ldots \tag{24}$$

The state $\chi_{ab \ldots k}$ describes the relative motion of these clusters.

To explain the features of the projection technique I shall first discuss the more familiar angular momentum projection technique and then develop the corresponding concept for cluster configurations.

In the $SU(3)$ model[26 p. 94] or in the Hartree-Fock theory of deformed nuclei[27 p. 235] one starts out from intrinsic states

$$|\alpha^n K) \tag{25}$$

with angular momentum component K along the axis of symmetry. A state of angular momentum J is projected as

$$|\alpha^n KIM) = \frac{2I+1}{8\pi^2} \int \mathrm{d}R \langle IM\,|R|\,IK\rangle^* \, R|\alpha^n K) \tag{26}$$

with R being a rotation operator and

$$\langle IM\,|R|\,IK\rangle \tag{27}$$

its matrix element in the representation J. Clearly we must have $J \geqq K$ with the sequence of levels $J = K, K+1, \ldots$ forming a rotational band. Two projected states are orthogonal with respect to J and M while their overlap with respect to K is given by

$$(\bar{\alpha}^n \bar{K}IM\,|\,\alpha^n KIM)$$
$$= \frac{2I+1}{8\pi^2} \int \mathrm{d}R \langle I\bar{K}\,|R|\,IK\rangle^* \, (\bar{\alpha}^n \bar{K}\,|R|\,\alpha^n K) \tag{28}$$

if the product of two projection operators is contracted. Now it proves very convenient to factorize the rotation R as

$$R = R(\alpha)\, R(\beta)\, R(\gamma) \tag{29}$$

using the Euler angles $\alpha\beta\gamma$. Then the operators $R(\alpha)$ and $R(\gamma)$ may be applied to the bra and ket states of the type Eq. (25) to yield the overlap matrix elements as

$$(\bar{\alpha}^n \bar{K}IM|\alpha^n KIM) = \frac{2I+1}{2} \int_0^{\pi} \mathrm{d}\beta \sin\beta \langle I\bar{K}|R(\beta)|IK\rangle\, (\bar{\alpha}^n \bar{K}|R(\beta)|\alpha^n K) \tag{30}$$

Similar expressions hold for the matrix elements of the hamiltonian.

In a two-cluster configuration[28] the states of orbital partition f are constructed from the parent states Eq. (23) as

$$|\alpha^n abfr) = \frac{|f|}{n!} \sum_p \langle fr\,|p|\,fab\rangle\, p|\alpha^n ab) \tag{31}$$

with p running through all permutations and

$$\langle fr\,|p|\,fab\rangle$$

being the corresponding matrix elements in the representation f. The possible partitions f are given by Littlewood's rules as ($n_a \geqq n_b$)

$$a \times b = \{n_a\} \times \{n_b\} = \{n_a n_b\} + \{n_a + 1 n_b - 1\} + \cdots + \{n_a + n_b\}. \quad (32)$$

If we can find a factorization of permutations that is equivalent to Eq. (29),

$$p = \bar{h} Z_k h, \quad (33)$$

where $\bar{h}$ and h can be applied to the cluster parent states, we may write the overlap between two-cluster configurations,

$$(\bar{\alpha}^n \bar{a}\bar{b} fr\,|\alpha^n abfr) = \frac{|f|}{n!} \sum_p \langle f\bar{a}\bar{b}\,|p|\,fab\rangle\,(\bar{\alpha}^n\ \bar{a}\bar{b}\,|p|\,\alpha^n ab) \quad (34)$$

in a simplified form corresponding to Eq. (30). The factorizations Eq. (29) and Eq. (33) are known as double coset decompositions.

The operators Z_k have been constructed and may be given in terms of permutation matrices. If a permutation p maps s into $p(s)$, we put the number one into column s at the row $p(s)$ and zeros at the other rows of this column. To construct Z_k, we partition the columns of an $n \times n$ matrix as

$$n = n_{a'} + n_{a''} + n_{b'} + n_{b''} \quad (35)$$

and the rows as

$$n = n_{a'} + n_{b'} + n_{a''} + n_{b''} \quad (36)$$

subject to the restrictions

$$n_{a'} + n_{b'} = n_{\bar{a}} \qquad n_{a'} + n_{a''} = n_a$$

$$n_{a''} + n_{b''} = n_{\bar{b}} \qquad n_{b'} + n_{b''} = n_b \quad (37)$$

We put unit submatrices of dimension $n_{a'} \times n_{a'}$, $n_{b'} \times n_{b'}$, $n_{a''} \times n_{a''}$, $n_{b''} \times n_{b''}$ into the corresponding square areas of the matrix and zero submatrices elsewhere. This defines Z_k as a permutation of sets of particles:

$$Z_k = \begin{pmatrix} \mathbf{n}_{a'}\,\mathbf{n}_{a''}\,\mathbf{n}_{b'}\,\mathbf{n}_{b''} \\ \mathbf{n}_{a'}\,\mathbf{n}_{b'}\,\mathbf{n}_{a''}\,\mathbf{n}_{b''} \end{pmatrix} = \quad (38)$$

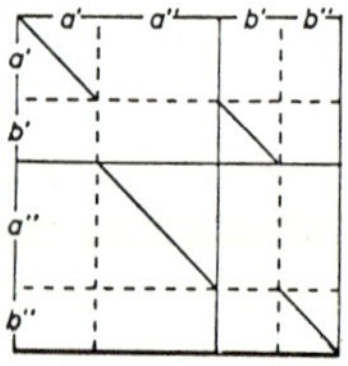

a' a'' b' b''
a'
b'
a''
b''

From the conditions Eq. (37) the number of different permutations Z_k is given by $n_b + 1$ for the choice $n_b \leqq n_{\bar{b}} \leqq n_a$. Now the expression Eq. (34) may be simplified as

$$(\bar{a}^n\bar{a}\bar{b}fr\,|\alpha^n abfr) = \frac{|f|}{n!} n_{\bar{a}}!\, n_{\bar{b}}!\, n_a!\, n_b! \sum_k m^{-1}(k) \langle f\bar{a}\bar{b}\,|Z_k|\,fab\rangle\, (\bar{\alpha}^n\bar{a}\bar{b}\,|Z_k|\,\alpha^n ab) \tag{39}$$

which differs conceptually from Eq. (30) only in the appearance of a factor $m(k)$ that counts the multiple factorization of a given permutation p and is given by

$$m(k) = n_{a'}!\, n_{a''}!\, n_{b'}!\, n_{b''}!. \tag{40}$$

To apply Eq. (39) one needs also the matrix elements of Z_k in the representation f. These matrix elements may be written more explicitly as

$$\langle f\bar{a}\bar{b}\,|Z_k|\,fab\rangle = \langle f\bar{a}(a'b')\,\bar{b}(a''b'')\,|Z_k|\,fa(a'a'')\,b(b'b'')\rangle \tag{41}$$

by indicating an appropriate subgroup decomposition. The matrix elements have been identified with special $9j$ coefficients and general Wigner coefficients of the unitary group $SU(2)$[17,18]:

$$\langle f\bar{a}(a'b')\,\bar{b}(a''b'')\,|Z_k|\,fa(a'a'')\,b(b'b'')\rangle$$

$$= [(n_{\bar{a}}+1)(n_{\bar{b}}+1)(n_a+1)(n_b+1)]^{1/2} \begin{Bmatrix} \frac{1}{2}(f_1-f_2) & \frac{1}{2}n_a & \frac{1}{2}n_b \\ \frac{1}{2}n_{\bar{a}} & \frac{1}{2}n_{a'} & \frac{1}{2}n_{b'} \\ \frac{1}{2}n_{\bar{b}} & \frac{1}{2}n_{a''} & \frac{1}{2}n_{b''} \end{Bmatrix}$$

$$= \left[\frac{(f_1+1)!\,f_2!\,n_{a'}!\,n_{a''}!\,n_{b'}!\,n_{b''}!}{n_{\bar{a}}!\,n_{\bar{b}}!\,n_a!\,n_b!]}\right]^{1/2}$$

$$(-)^{n_{a'}+n_{b''}} \begin{pmatrix} \frac{1}{2}n_a & \frac{1}{2}n_b & \frac{1}{2}(f_1-f_2) \\ \frac{1}{2}(n_{a'}-n_{a''}) & \frac{1}{2}(n_{b'}-n_{b''}) & \frac{1}{2}(n_{\bar{b}}-n_{\bar{a}}) \end{pmatrix} \tag{42}$$

For k clusters one needs special $9j$ coefficients or one-row Wigner coefficients of the unitary group $SU(k)$. An expression similar to Eq. (39) allows one to express the orbital reduced matrix elements of the hamiltonian in Eq. (20) in terms of a minimal set of basic interaction integrals between cluster parent states[28].

The concept of cluster parentage provides an intrinsic state label in complete analogy to the intrinsic label K of a rotational band. A sequence of levels projected from the same cluster parent states could be called a permutational band.

This establishes the existence of a well-defined algebraic structure of cluster configurations. The permutation group and its representations deter-

TABLE 4 Permutations Z_k, multiplicity $m(k)$ and representation matrix elements for the cluster parentage $\{4\} \times \{2\}$ and the orbital partition $\{42\}$[28]

a) $(\alpha^6\ \{4\}\ \{2\}\ \{42\} \vert \alpha^6\ \{4\}\ \{2\}\ \{42\})$			
k	0	1	2
DC symbol	$\begin{Bmatrix} 4 & 0 \\ 0 & 2 \end{Bmatrix}$	$\begin{Bmatrix} 3 & 1 \\ 1 & 1 \end{Bmatrix}$	$\begin{Bmatrix} 2 & 2 \\ 2 & 0 \end{Bmatrix}$
Z	e	$\begin{pmatrix} 4 & 5 \\ 5 & 4 \end{pmatrix}$	$\begin{pmatrix} 3 & 4 & 5 & 6 \\ 5 & 6 & 3 & 4 \end{pmatrix}$
$m(k)$	$4!2!$	$3!$	$(2!)^3$
$\begin{bmatrix} \{42\} & \{4\} & \{2\} \\ \{4\} & a'(k) & b'(k) \\ \{2\} & a''(k) & b''(k) \end{bmatrix}$	$5 \cdot 3 \begin{Bmatrix} 1 & 2 & 1 \\ 2 & 2 & 0 \\ 1 & 0 & 1 \end{Bmatrix}$	$5 \cdot 3 \begin{Bmatrix} 1 & 2 & 1 \\ 2 & \frac{3}{2} & \frac{1}{2} \\ 1 & \frac{1}{2} & \frac{1}{2} \end{Bmatrix}$	$5 \cdot 3 \begin{Bmatrix} 1 & 2 & 1 \\ 2 & 1 & 1 \\ 1 & 1 & 0 \end{Bmatrix}$

b) $(\bar{\alpha}^6\ \{3\}\ \{3\}\ \{42\} \vert \alpha^6\ \{4\}\ \{2\}\ \{42\})$			
k	0	1	2
DC symbol	$\begin{Bmatrix} 3 & 0 \\ 1 & 2 \end{Bmatrix}$	$\begin{Bmatrix} 2 & 1 \\ 2 & 1 \end{Bmatrix}$	$\begin{Bmatrix} 1 & 2 \\ 3 & 0 \end{Bmatrix}$
Z_k	e	$\begin{pmatrix} 3 & 4 & 5 \\ 5 & 3 & 4 \end{pmatrix}$	$\begin{pmatrix} 2 & 3 & 4 & 5 & 6 \\ 5 & 6 & 2 & 3 & 4 \end{pmatrix}$
$m(k)$	$3!2!$	$(2!)^2$	$3!2!$
$\begin{bmatrix} \{42\} & \{4\} & \{2\} \\ \{3\} & a'(k) & b'(k) \\ \{3\} & a''(k) & b''(k) \end{bmatrix}$	$4\sqrt{5 \cdot 3} \begin{Bmatrix} 1 & 2 & 1 \\ \frac{3}{2} & \frac{3}{2} & 0 \\ \frac{3}{2} & \frac{1}{2} & 1 \end{Bmatrix}$	$4\sqrt{5 \cdot 3} \begin{Bmatrix} 1 & 2 & 1 \\ \frac{3}{2} & 1 & \frac{1}{2} \\ \frac{3}{2} & 1 & \frac{1}{2} \end{Bmatrix}$	$4\sqrt{5 \cdot 3} \begin{Bmatrix} 1 & 2 & 1 \\ \frac{3}{2} & \frac{1}{2} & 1 \\ \frac{3}{2} & \frac{3}{2} & 0 \end{Bmatrix}$

mine both the possible types of interaction integrals and their algebraic weight factors. These weight factors in turn are identified as well-known coupling coefficients of the unitary group. It is important to note that

a) the technique can be extended to more clusters or to a shell configuration and a cluster,

b) no use is made of the details of the cluster parent states so that the technique applies equally well to nuclear reactions.

Originally the cluster parent states Eq. (23) were chosen by Wildermuth and Kanellopoulos[29,30] as oscillator states with the oscillator energy carried by the relative vectors between the clusters. Recent work[31] with these configurations has shown that the permutation group allows a complete algebraic reduction of the energy matrix to two-body matrix elements without use of *fpc*.

Before I turn to calculations with these methods I shall explain the significance of the orbital partition f for orbital states in terms of selection rules.

6 SELECTION RULES OBTAINED FROM THE ORBITAL PARTITION

The results of the last section enable us to discuss a number of selection rules resulting from the orbital partition. A good starting point are the nodes of an orbital state with fixed partition $f = \{f_1 f_2 \ldots f_k\}$. Littlewood's rules show that we can build this states as

$$\{f_1\} \times \{f_2\} \times \cdots \times \{f_k\} \rightarrow f \tag{43}$$

from subsets symmetric with respect to internal permutations of the sets of particles $1 \ldots f_1, f_1 + 1 \ldots f_1 + f_2, \ldots$ They also show that for given f we cannot include more particles into any one of these subsets. Reading Eq. (43) from right to left one sees that a state of partition f cannot be decomposed into larger symmetric sets than $\{f_1\}, \{f_2\}, \ldots$ This implies that the break-up of an excited state into clusters could at most proceed according to the rows of the partition. To get an idea of the origin of this selection rule consider an orbital state of partition f,

$$\langle x^1 x^2 \cdots x^n | \alpha^n \{f_1 f_2 \cdots f_k\} \rangle \tag{44}$$

At the points of configuration space where

$$x^1 = x^2 = \cdots = x^{f_1} = x^{f_1+1} \tag{45}$$

it follows that this state has a nodal manifold which means that one cannot combine $f_1 + 1$ particles into one cluster. The bearing of this type of selection rule on clustering in the harmonic oscillator model has been examined by Neudatchin and Smirnov[32].

A second type of nodes and corresponding selection rules applies to configurations of clusters of the same number of particles. For $n_a = n_b$ it is possible to group the operators Z_k in Eq. (39) into pairs related as

$$Z_k^* = Z_k \begin{pmatrix} n_a \, n_b \\ n_b \, n_a \end{pmatrix} \tag{46}$$

with

$$\begin{pmatrix} n_a \, n_b \\ n_b \, n_a \end{pmatrix} = (1n_a + 1)(2n_a + 2) \ldots (n_a n) \tag{47}$$

being the permutation of the two clusters.

The two representation coefficients of Z_k^* and Z_k differ only by a phase factor[17]

$$(-)^{f_{aa}} = (-)^{\{f_1 f_2\}\{n_a\}\{n_a\}} = (-)^{f_2}. \tag{48}$$

In the normalization matrix element Eq. (39) there appears then the linear combination

$$(\bar{\alpha}^n \bar{a}\bar{b}|Z_k \left\{ 1 + (-)^{f_2} \begin{pmatrix} n_a \, n_b \\ n_b \, n_a \end{pmatrix} \right\} |\alpha^n ab) \tag{49}$$

of integrals. This means that for a given partition f the parent states must have well-defined transformation properties under cluster permutations. For two clusters with the same internal orbital state this permutation affects only the state χ_{ab} and gives a factor $(-)^L$ depending on the orbital angular momentum. For example in the reaction of deuterons on deuterons with possible partitions

$$\{2\} \times \{2\} \rightarrow \{4\} + \{31\} + \{22\} \tag{50}$$

the partitions $\{4\}$ and $\{22\}$ are reached by even angular momenta while the partition $\{31\}$ comes from odd angular momenta. Similarly the reaction $^3He + {}^3He$ leads to the orbital partition $\{42\}$ and $\{33\}$ for even and odd angular momenta respectively.

An extension of this reasoning shows that orbital states of partition $\{44 \ldots 4\}$ are symmetric with respect to the interchange of α-clusters. For special positions of the relative vectors between these α-clusters in three-dimensional space the interchange of clusters may be replaced by a rotation-reflection. This leads to nodes in the orbital state for certain angular momenta and parities.

	D_{3h}	C_{2v}	$D_\infty \times I$	$C_{\infty v}$	T_d	D_{2d}	C_{3v}
0^+	●	●	●	●	●	●	●
0^-	○	○	○	○	○	○	○
1^+	○	○	○	○	○	○	○
1^-	○	●	○	●	○	○	●
2^+	●	●	●	●	○	●	●
2^-	○	●	○	○	○	●	○
3^+	○	●	○	○	○	○	●
3^-	●	●	○	●	●	●	●
4^+	●	●	●	●	●	●	●
4^-	●	●	○	○	○	●	●

TABLE 5 Nodal positions of three and four α-clusters. Corresponding point groups are indicated along with the allowed (●) and forbidden (○) combinations L^π [4p. 172]

7 CALCULATIONS WITH CLUSTER CONFIGURATIONS

In calculations with cluster configurations one assumes the general functional form Eq. (23) of the parent states, antisymmetrizes and determines the best set of states from a variational principle. This field has been thoroughly discussed by Wildermuth and McClure[33] while recent experimental and theoretical results have been reported in the Proceedings of the Conference on Clustering Phenomena in Nuclei[4].

From the point of view of the permutation group this procedure can be interpreted[28] as a natural way of constructing states of the lowest orbital partition. In most cases a cluster is associated with each row of the orbital partition. This fixes the cluster parentage and the algebra of the projection technique. In contrast to the methods used in ref.[33] one may then handle the spin-isospin states by means of standard *fpc*.

The interplay between this algebraic construction and the variational method may be illustrated by the resonating group method due to Wheeler[33,34]. If the internal states $\phi_a \phi_b$ in the cluster parent state are fixed, one obtains a Schrödinger equation for the state χ_{ab} of the relative motion. The corresponding effective hamiltonian depends on the states $\phi_a \phi_b$, on the orbital partition and is nonlocal through the appearance of the permutations Z_k. This Schrödinger equation determines a self-consistent solution that resembles the Hartree-Fock state with the difference that the function a

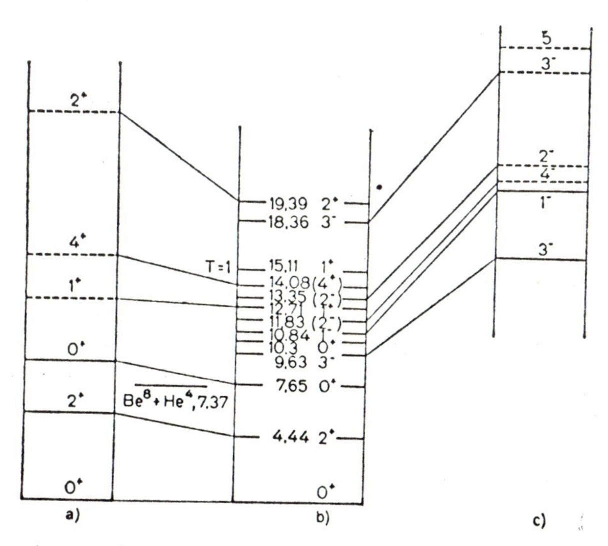

FIGURE 2 $T = 0$ states of ^{12}C with $f = \{444\}$. (a) and (c) calculated, (b) experimental[38]

form of the cluster parent states is motivated by the supermultiplet scheme rather than by the independent-particle picture.

The state description in terms of cluster parentage may be tested by the lowest states of ^{6}Li. In the oscillator model the lowest states of two quanta and partition $f = \{42\}$ are equally well projected from parent states $\{4\} \times \{2\}$ and $\{3\} \times \{3\}$. Calculations with a generalized two-cluster configuration carried out by Wodsak[35] indicate that the parentage $\{4\} \times \{2\}$ is a better starting point for these levels. In the oscillator representation this points to the need of including higher excitations. Hackenbroich[36, 37] has used superpositions of gaussian cluster parent states to describe the distortion of clusters when approaching each other. A calculation of low-lying levels of ^{12}C with the orbital partition $f = \{444\}$ using α-cluster parent states due to Hutzelmeyer and Hackenbroich[38] reproduces well the sequence of positive and negative parity states.

8 NUCLEAR REACTIONS

A new field of applications of the permutation group is opened by nuclear reaction calculations. In the reaction theory with cluster configurations as proposed by Benöhr and Wildermuth[39] and by Hackenbroich[37] one uses a superposition of quasi-bound states for the description of the internal or reaction region along with external channel states. The states used in the reaction region could be chosen according to any convenient nuclear model. The best analytic junction of the channel and reaction region is obtained from a variational principle. In a recent version[40] and [4, p. 3] of this theory the external and internal states overlap only in the nuclear surface.

The permutation group enters this reaction theory at three points:

a) the quasi-bound states may be calculated from cluster configurations as discussed in the last sections,

b) the overlap between the internal and the external region may be obtained from the projection technique,

c) the S-matrix of the reaction may be characterized by the orbital partition.

In the usual description of a channel involving two clusters one builds the overall antimetric states by inducing from the two antimetric states of the clusters ab with internal spins S^aS^b and isospins T^aT^b. To introduce the orbital partition one should first induce orbital and spin-isospin states with paritions f and $\tilde{f}$ respectively and then couple to antimetric states. The relation between these schemes may be expressed by the pattern

$$\begin{array}{ccc} a \otimes \hat{a} & \rightarrow & \{1^{n_a}\} \\ \times \quad \times & & \times \\ b \otimes \hat{b} & \rightarrow & \{1^{n_b}\} \\ \downarrow \quad \downarrow & & \downarrow \\ f \otimes \hat{f} & \rightarrow & \{1^{n}\} \end{array} \tag{51}$$

Seligman[41] has shown that the corresponding unitary transformation is determined by the spin-isospin fractional parentage coefficients

$$\langle \gamma^n ST \hat{f} \{ | \gamma^{n_a} S^a T^a \hat{a} \gamma^{n_b} S^b T^b \hat{b} \rangle . \tag{52}$$

These coefficients allow a partial symmetry analysis in reaction channels that will be useful if a resonant level corresponds mainly to a single partition.

As an example we are presently studying the reaction

$$^2\text{H} + {}^2\text{H} \rightarrow {}^3\text{H} + p \tag{53}$$

where a narrow resonance at $E_d = 100$ keV with $J = 2^+$, $T = 0$ has been found by Fick and Franz[4p.221]. The possible partitions f in the entrance channel are

$$\{2\} \times \{2\} \rightarrow \{4\} + \{31\} + \{22\} \tag{54}$$

and in the exit channel

$$\{3\} \times \{1\} \rightarrow \{4\} + \{31\} \tag{55}$$

A quasi-bound state calculation[42] by with gaussian cluster states of parentage {2} {2} shows that there is a state dominated by the partition {22} in the corresponding energy region. This is a good candidate for the resonance as it could decay into the exit channel only through admixtures of the partitions $f = \{4\}$ and $\{31\}$.

For other reactions it will be reasonable to describe the target nucleus by a shell configuration. Then it is necessary to combine the fractional parentage concept with the projection technique of clusters in order to calculate reactions between complex fragments.

CONCLUSION

I have tried to show how the permutation group underlies many different aspects of nuclei. The supermultiplet scheme provides a general framework for these considerations. Within this scheme, important properties and transitions of nuclei are governed by partitions and their Kronecker and Littlewood product. The detailed application of these concepts reveals a well-defined algebraic structure that promises to become a useful tool for the understanding of the nature and the reactions of light nuclei.

REFERENCES

1. E. P. Wigner, *Phys. Rev.* **51** (1937) 25
2. P. Franzini and L. A. Radicati, *Phys. Letters* **6** (1963) 322
3. J. A. Wheeler, *Phys. Rev.* **52** (1937) 1083
4. *Clustering Phenomena in Nuclei*, IAEA, Vienna 1969
5. P. Kramer and T. H. Seligman, *Nuclear Physics A* **123** (1969) 161
6. V. V. Vanagas and A. K. Petrauskas, *Yad. Fiz.* **5** (1967) 552; *Sov. J. Nucl. Phys.* **5** (1969) 393
7. H. M. Mahmoud and R. K. Cooper, *Ann. Phys. N.Y.* **26** (1964) 222
8. A. K. Petrauskas and V. V. Vanagas, *Yad. Fiz.* **7** (1968) 324; *Sov. J. Nucl.Phys.* **7** (1968) 216
9. A. K. Petrauskas and V. V. Vanagas, *Yad. Fiz.* **8** (1968) 463; *Sov. J. Nucl. Phys.* **8** 1969) 270
10. K. T. Hecht, *Nuclear Physics A* **114** (1968) 280
11. K. T. Hecht and S. Ch. Pang, *J. Math. Phys.* **10** (1969) 1571
12. G. Racah, *Phys. Rev.* **63** (1943) 171
13. H. Horie, *J. Phys. Soc. Japan* **19** (1964) 1783
14. M. Moshinsky and V. Syamala Devi, *J. Math. Phys.* **10** (1969) 455
15. M. Moshinsky, *J. Math. Phys.* **7** (1966) 691
16. P. Kramer, *Z. Physik* **205** (1967) 181
17. P. Kramer, *Z. Physik* **216** (1968) 68
18. P. Kramer and T. H. Seligman, *Z. Physik* **219** (1969) 105
19. S. Cohen and D. Kurath, *Nuclear Physics* **73** (1965) 1
20. G. John and P. Kramer, *Nuclear Physics*, in press
21. H. A. Jahn and H. van Wieringen, *Proc. Roy. Soc.* **A209** (1951) 502
22. J. P. Elliott, J. Hope, and H. A. Jahn, *Proc. Roy. Soc.* **A246** (1953) 241
23. Y. Akiyama, A. Arima, and T. Sebe, *Nuclear Physics* **A138** (1969) 273
24. P. Kramer and M. Moshinsky, "Group Theory of Harmonic Oscillators and Nuclear Structure", in: *Group Theory and its Applications*, ed. by E. M. Loebl, New York 1968
25. I. V. Kurdyumov, Yu. Smirnov, K. V. Shitikova, and S. Kh. El. Samarai, *Nuclear Physics* **A145** (1970) 593
26. M. Harvey, "The Nuclear SU_3 Model", in: *Advances in Nuclear Physics 1*, ed. by M. Baranger and E. Vogt, New York 1968
27. G. Ripka, "The Hartree-Fock Theory of Deformed Light Nuclei", in: *Advances in Nuclear Physics 1*, ed. by M. Baranger and E. Vogt, New York 1968
28. P. Kramer and T. H. Seligman, *Nuclear Physics* **A136** (1969) 545
29. K. Wildermuth and Th. Kanellopoulos, *Nuclear Physics* **7** (1958) 150
30. K. Wildermuth and Th. Kanellopoulos, *Nuclear Physics* **9** (1958) 449
31. P. Kramer and D. Schenzle, *Nuclear Physics*, in press
32. V. G. Neudatchin and Yu. F. Smirnov, *Atomic Energy Review* **3** (1965) 157

33. K. Wildermuth and W. McClure, *Cluster Representations of Nuclei*, New York 1966
34. J. A. Wheeler, *Phys. Rev.* **52** (1937) 1107
35. W. Wodsak, *Forschungsbericht* **K 70-02**, Bundesministerium für Bildung und Wissenschaft, ZAED, Leopoldshafen 1970
36. H. H. Hackenbroich, *Z. Physik* **231** (1970) 216
37. H. H. Hackenbroich and P. Heiss, *Z. Physik* **231** (1970) 225
38. H. Hutzelmeyer and H. H. Hackenbroich, *Z. Physik* **232** (1970) 356
39. H. C. Benöhr and K. Wildermuth, *Nuclear Physics* **A 128** (1969) 1
40. H. Federsel, E.-J. Kanellopoulos, W. Sünkel, and K. Wildermuth, *Physics Letters* **33B** (1970) 140
41. T. H. Seligman, *Forschungsbericht* **K 69-42**, Bundesministerium für Bildung und Wissenschaft, ZAED, Leopoldshafen 1969
42. H. Hutzelmeyer and G. John, to be published

DISCUSSION

M. Moshinsky You mentioned the use of the cluster model in nuclear structure calculations. I would like only to point out that many of the results can also be reproduced using a translationally invariant shell model.

P. Kramer This is certainly right, and the model you mention has been studied by several people, among them by Neudatchin and Smirnov. The algebraic problem is to obtain explicitly the fractional parentage coefficients for this model. One could also use states of the translational-invariant oscillator model within the scheme I have outlined. Both methods would not apply to reaction calculations.

I. Talmi I would like to point out that, in spite of its great elegance and mathematical beauty, the application of the super-multiplet theory, introduced by Wigner in 1937, encounters severe difficulties. I shall not speak about nuclear energies although it is important to realize that the spin-orbit interaction which determines the higher magic, numbers is also important in light nuclei. Let me just discuss allowed beta-transitions.

Gamow-Teller transitions between certain members of the same super-multiplet should be favoured. This is a very strong statement in the super-multiplet theory since the Gamow-Teller operator does not lead out of a super-multiplet. A small deviation from the supermultiplet scheme will therefore give only a small deviation in the Gamow-Teller matrix element. The situation should be contrasted with the extreme $j - j$ coupling limit. There, small admixtures to the wave-functions may give rise to large cross terms and this may change considerably the Gamow-Teller matrix elements.

The most famous case, where the supermultiplet theory predicted a favoured transition and the actual transition is highly unfavoured, is that of C^{14}. The Gamow-Teller matrix element there is reduced by a factor ~ 1000. Other, less dramatic, discrepancies are found in mirror transitions towards the end of the (s, d) shell. When subtracting the contribution of the Fermi matrix elements, we find in several cases rather small Gamow-Teller matrix elements. It seems that the actual situation in nuclei is rather complicated and cannot be given in terms of too simple a picture. There is, for example, a strong dependence on the isospin. In general, states with low isospins may be rather close to super-multiplet wave functions whereas high isospin states can be better described in the $j - j$ coupling scheme. We should thus make an attempt to understand better the regions of validity and the limitations of the various coupling schemes.

D. R. Inglis Did Cohen and Kurath obtain that attenuation by admixtures from outside the p^n configuration?

I. Talmi No, as you know it is possible to get it with a tensor force within the p-shell and their force includes a tensor force.

D. R. Inglis In your treatment of reactions it is interesting that you join a cluster treatment in the external region perhaps to some other treatment inside, emphasizing that the successful use of an α-cluster model to interpret reactions, as has been popular with some experimentalists, does not imply the existence of clusters inside.

P. Kramer I have tried to stress the connection of clustering to permutational symmetry. To associate the rows of the orbital Young diagram with different clusters does not necessarily imply spatial correlations. In fact the ^{12}C calculations by Hutzelmeyer and Hackenbroich show that the wave function contains quite extended α-clusters with considerable overlap. Therefore one does not keep the picture of a simple correlation in space, but one keeps the simple idea of finding groups of four nucleons in an internally symmetric state.

D. R. Inglis Thank you. Then these states must have some orthogonality conditions between them. It is not anymore a question of two localized clusters not having a spacial overlap but having a more complicated basis for orthogonality.

P. Kramer The orbital symmetry forbids the formation of internally symmetric groups of nucleons larger than the rows of the orbital partition. The interpretation of the wave function in space is complicated as it is expanded in a non-orthogonal basis. One should look at matrix elements of operators like for example form factors in order to test the spatial structure. I think that in Professor Brink's talk we shall hear more about the question how the clusters are arranged in space.

G. Ripka I wish to add to the comment of Prof. Talmi that at present we do not understand very well the domains of validity of various model wavefunctions. For example we do not understand yet why the isospin $T = 0$ states of the nuclei belonging to the second half of the $2s - 1d$ shell do not appear to have the rotational properties of the $SU(3)$ model wave functions which are observed in the nuclei belonging to the first half of the $2s - 1d$ shell.

P. Kramer I certainly do not imply that all problems of light nuclei can be solved easily within this approach. One open question is the relation of clustering to the rotational structure of nuclei. Besides, certain nuclei have a structure that cannot easily be interpreted in terms of cluster configurations. With increasing nucleon number, one would need clusters built from neutrons only, and it remains to be seen if they exist inside the nucleus. In general the computational difficulties of cluster calculations increase rapidly with the mass number, and we have not yet found a way of, so to say, forgetting about part of the configuration which is so easy in the shell model where you leave out closed configurations.

Oscillator Systems

M. MOSHINSKY and C. QUESNE*
Instituto de Física, Universidad de Mêxico and Comisión Nacional de Energía Nuclear. México, D.F.

INTRODUCTION AND SUMMARY

Since the advent of the nuclear shell model, harmonic oscillator states for systems of n particles have been playing an ever increasing role in nuclear structure calculations. Initially the states, restricted to one or two particles, were used simply as a convenient mathematical tool in the calculations. But soon it became clear that the symmetry properties of the harmonic oscillator would allow a drastic simplification in many calculations and, in some cases, even provide a deeper insight into the physics of the problems.

Among the important contributions in this direction we can mention the pioneering work of Talmi on the use of harmonic oscillator states in the shell model; the introduction by Elliott of the $\mathscr{SU}(3)$ symmetry associated with harmonic oscillator states and its relevance to collective phenomena in nuclei; the concept of transformation brackets for harmonic oscillator states developed by our group in Mexico and by others, and its extensive applications in the nuclear shell model, Hartree-Fock calculations, etc., etc.

Thus a considerable effort has been expended in what we could call the group theory of harmonic oscillators. But even this effort was somewhat restricted in scope, as it dealt mainly with the applications of the symmetry group of the N dimensional harmonic oscillator i.e. the unitary group $\mathscr{U}(N)$. It did not pay much attention to the deeper problem of the general role of linear canonical transformations, and their unitary representations in appropriate basis of Hilbert space, for oscillator systems. It seems though that the latter problem provides the foundations for the former applications.

Thus in the present paper we initiate our analysis by discussing in chapter I the classical canonical transformations and their unitary representations in

* Chercheur agréé, Institut Interuniversitaire des Sciences Nucléaires, on leave from Université Libre de Bruxelles.

quantum mechanics. We illustrate the applications of these ideas taking as an example the problem of a single particle in one dimensional harmonic oscillator potential. Once this example is fully understood, we proceed to discuss the symplectic group in $2N$ dimensions $Sp(2N)$ and its subgroup $\mathcal{U}(N)$, which are respectively the dynamical and symmetry groups of the N dimensional harmonic oscillator.

In chapter II we then review the applications of $\mathcal{U}(N)$ and its subgroups to problems of n particles in a three dimensional harmonic oscillator potential for which $N = 3n$. We stress the particular cases of $n = 2, 3, 4$ particles in which special techniques involving transformation brackets allow the discussion of many specific applications. The general classification of n particle states and the factorization of the fractional parentage coefficients corresponding to them are also analyzed.

Finally in chapter III we discuss the applications of the dynamical group to problems of both isotropic and anisotropic dilatation of harmonic oscillator states. We apply these results, as well as those of the previous chapters, to simple many body problems, whose solution can then be entirely given in group theoretical language. This analysis suggests a procedure that could be employed, at least in principle, for more realistic and interesting many body problems.

Each chapter is divided into sections in which the equations have the section number. To avoid confusions, when using in any of the chapters equations of the others, we add to the number of the equation the chapter number.

CHAPTER I

CANONICAL TRANSFORMATIONS AND THEIR UNITARY REPRESENTATION

IT IS WELL KNOWN that some of the powerful techniques for solving mechanics problems are based on the symmetry group of *canonical* transformations i.e. the transformations in phase space that leave the hamiltonian and the Poisson brackets of coordinates and momenta invariant. In some cases these transformations concern the coordinates alone i.e. are point transformations, as is the case when there is invariance under translations, rotations or permutations of the particles. Symmetry groups of point transformations have been discussed extensively in the literature[1, 2, 3] both in their application to classical and quantum mechanics. Groups of canonical transformations have been less extensively applied. As these are the groups that play a fundamental role in the characterization and explicit determination of states of oscillator systems, we shall start our discussion with their derivation in classical mechanics and their unitary representation in quantum mechanics.

1 CLASSICAL CANONICAL TRANSFORMATIONS AND THE SYMPLECTIC GROUP

A canonical transformation is a transformation in phase space which leaves invariant the Poisson brackets

$$\{x_i, p_j\} = \delta_{ij}, \quad \{x_i, x_j\} = \{p_i, p_j\} = 0, \quad i, j = 1, \ldots, N. \tag{1.1}$$

To understand the nature of the group with which these transformations are connected, let us introduce the notation z_α, $\alpha = 1, 2, \ldots 2N$, for a vector in phase space defined by

$$z_i \equiv x_i, \quad z_{i+N} \equiv p_i, \quad i = 1, \ldots, N. \tag{1.2}$$

The Poisson bracket of two observables f, g is then

$$\{f, g\} = \sum_{i=1}^{N} \left(\frac{\partial f}{\partial x_i} \frac{\partial g}{\partial p_i} - \frac{\partial f}{\partial p_i} \frac{\partial g}{\partial x_i} \right) = \sum_{\alpha, \beta = 1}^{2N} \frac{\partial f}{\partial z_\alpha} A_{\alpha\beta} \frac{\partial g}{\partial z_\beta}, \tag{1.3}$$

where the matrix

$$\mathbf{A} = \|A_{\alpha\beta}\| = \begin{pmatrix} 0 & I \\ -I & 0 \end{pmatrix} \tag{1.4}$$

has all its submatrices of dimension $N \times N$.

If we now pass from the vector $\{z_\alpha\}$ in phase space to a new one $\{\bar{z}_\alpha\}$ whose components are functions of the previous one, the transformation will be canonical if

$$\sum_{\gamma,\delta} \frac{\partial \bar{z}_\alpha}{\partial z_\gamma} A_{\gamma\delta} \frac{\partial \bar{z}_\beta}{\partial z_\delta} = A_{\alpha\beta}. \tag{1.5}$$

The matrix **A** is the one usually associated with the symplectic group[3,4]. We see that the canonical transformations leave this matrix invariant, and thus they can be considered as part of a general symplectic group. If in particular the $\bar{z}_\alpha$ are linear functions of the z_α i.e.

$$\bar{z}_\alpha = \sum_\beta S_{\alpha\beta} z_\beta, \tag{1.6}$$

the transformation will be canonical if

$$\mathbf{S}\mathbf{A}\tilde{\mathbf{S}} = \mathbf{A}, \tag{1.7}$$

where $\mathbf{S} = \|S_{\alpha\beta}\|$ and the $\sim$ symbol stands for the transposed.

We have thus established the group to which belong all classical canonical transformations. How does this group make its appearance in the Hilbert space of quantum mechanics? The question was answered by Dirac many decades ago and we proceed to review the main aspects of his reasoning.

2 THE UNITARY REPRESENTATION OF CANONICAL TRANSFORMATIONS IN QUANTUM MECHANICS

When we consider our original coordinates and momenta $x_i, p_i, i = 1, \dots N$, the states in which x_i is diagonal are represented in Dirac's notation by the bras and kets $\langle x'|, |x''\rangle$, where whenever something is characterized by the eigenvalues of *all* the coordinates x_i, $i = 1, \dots N$, we suppress the index i. The matrix elements of the operators x_i, p_i with respect to this basis are then[5]

$$\langle x' |x_i| x''\rangle = x'_i \delta(x' - x''), \tag{2.1a}$$

$$\langle x' |p_i| x''\rangle = -\frac{1}{i} \frac{\partial}{\partial x''_i} \delta(x' - x''), \tag{2.1b}$$

where $\hbar$ is taken as 1 and

$$\delta(x' - x'') = \prod_{i=1}^{N} \delta(x'_i - x''_i). \tag{2.1c}$$

If we now pass to another set of coordinates and momenta which are functions of the previous ones

$$\bar{x}_i(x, p), \quad \bar{p}_i(x, p) \tag{2.2}$$

and canonical, then it is clear that the matrix elements of $\bar{x}_i, \bar{p}_i$ between bras and kets $(\bar{x}'|,|\bar{x}'')$ will have the same form as (2.1). We use here a round bracket notation for the states in which $\bar{x}_i$ is diagonal to distinguish them from the states in which x_i is diagonal in case we take the same numerical value for the eigenvalues of $\bar{x}_i$ and x_i.

Clearly we could make the development

$$|\bar{x}') = \int |x'\rangle \, \mathrm{d}x' \langle x'|\bar{x}'), \tag{2.3}$$

where $\mathrm{d}x' = \mathrm{d}x'_1 \ldots \mathrm{d}x'_N$ and we have an N dimensional integration. The transformation bracket[5] then satisfies the equations

$$\bar{x}_i\left(x', \frac{1}{i}\frac{\partial}{\partial x'}\right)\langle x'|\bar{x}') = \bar{x}'_i \langle x'|\bar{x}') \quad i = 1, \ldots N. \tag{2.4}$$

It is important to notice that in general these equations do not completely determine the transformation brackets. In particular, as the equations (2.4) are linear, we could multiply the solution by some function of $\bar{x}'_i$ that normalizes the transformation bracket and even after that still keep in it a phase factor of the form

$$e^{i\phi(\bar{x}')}, \tag{2.5}$$

with arbitrary ϕ.

The normalization requirement and later the phase factor can then be used to guarantee that the matrix elements of $\bar{x}_i, \bar{p}_i$ for the bras and kets $(\bar{x}'|,|\bar{x}'')$ have the same form as those of x_i, p_i for $\langle x'|,|x''\rangle$. Thus we have that

$$\begin{aligned}(\bar{x}'|\bar{x}_i|\bar{x}'') &= \int (\bar{x}'|x'\rangle \, \bar{x}_i\left(x', \frac{1}{i}\frac{\partial}{\partial x'}\right)\langle x'|\bar{x}'') \, \mathrm{d}x' \\ &= \bar{x}'_i \int (\bar{x}'|x'\rangle \, \mathrm{d}x' \, \langle x'|\bar{x}''),\end{aligned} \tag{2.6}$$

and so the matrix element $(\bar{x}'|\bar{x}_i|\bar{x}'')$ will have the familiar form (2.1a) if the transformation brackets are normalized i.e.

$$\int (\bar{x}'|x'\rangle \, \mathrm{d}x' \, \langle x'|\bar{x}'') = \delta(\bar{x}' - \bar{x}''), \tag{2.7}$$

where[5]

$$(\bar{x}'|x'\rangle = \langle x'|\bar{x}')^{*}. \tag{2.8}$$

For the momenta we must have that

$$(\bar{x}'|\bar{p}_i|\bar{x}'') = \int (\bar{x}'|x'\rangle\, \bar{p}_i\left(x', \frac{1}{i}\frac{\partial}{\partial x'}\right)\langle x'|\bar{x}'')\, \mathrm{d}x'$$

$$= -\frac{1}{i}\frac{\partial}{\partial \bar{x}_i''}\,\delta(\bar{x}' - \bar{x}''), \tag{2.9}$$

and this can be achieved if we make a proper choice of the phase factor ϕ as will be shown explicitly in the next section for the case of linear canonical transformations in two dimensional phase space.

We now proceed to determine the unitary matrices, in the basis of the Hilbert space in which the original coordinates are diagonal, that are associated in quantum mechanics with elements of the general symplectic group of canonical transformations. For this purpose we first determine the matrix elements of $\bar{x}_i$, $\bar{p}_i$ in the basis in which the x_i are diagonal. For $\bar{x}_i$ we then get

$$\langle x'\,|\bar{x}_i|\,x''\rangle = \int \langle x'|\bar{x}')\, \mathrm{d}\bar{x}'(\bar{x}'\,|\bar{x}_i|\,\bar{x}'')\, \mathrm{d}\bar{x}''(\bar{x}''|x''\rangle$$

$$= \int \langle x'|\bar{x}')\, \mathrm{d}\bar{x}'\, \bar{x}_i'\, \delta(\bar{x}' - \bar{x}'')\, \mathrm{d}\bar{x}''(\bar{x}''|x''\rangle. \tag{2.10}$$

Now $\bar{x}_i'$, $\bar{x}_i''$ are not operators but just variables over which we carry out integrations and thus using (2.1a) we can write

$$\bar{x}_i'\,\delta(\bar{x}' - \bar{x}'') = \langle \bar{x}'\,|x_i|\,\bar{x}''\rangle, \tag{2.11}$$

where we stress the angular, rather than round, brackets of the matrix elements. To express then (2.10) as matrix multiplication, we define the matrix elements of a unitary matrix U in the basis in which the x_i are diagonal by

$$\langle x'\,|U|\,x''\rangle \equiv \langle x'|x''), \quad \text{which implies} \quad \langle x'\,|U^{-1}|\,x''\rangle = (x'|x''\rangle, \tag{2.12}$$

thus getting

$$\langle x'\,|\bar{x}_i|\,x''\rangle = \int \langle x'\,|U|\,\bar{x}'\rangle\, \mathrm{d}\bar{x}'\,\langle \bar{x}'\,|x_i|\,\bar{x}''\rangle\, \mathrm{d}\bar{x}''\,\langle \bar{x}''\,|U^{-1}|\,x''\rangle. \tag{2.13}$$

In operator language we have then that

$$\bar{x}_i = Ux_iU^{-1}, \tag{2.14a}$$

and it is clear that an entirely similar analysis gives us

$$\bar{p}_i = Up_iU^{-1}. \tag{2.14b}$$

The matrices U are representations of the general symplectic group of canonical transformations as when we carry out in succession two canonical transformations that give rise to the unitary matrices U, V, the new coordinates and momenta are affected by the transformation VU. Thus we have the quantum mechanical equivalent of the classical canonical transformations and therefore also a representation of the general symplectic group of which they form part.

We proceed to illustrate the derivation of U for the single degree of freedom problem, showing also that the linear symplectic subgroup of the general symplectic group is the dynamical group of the harmonic oscillator.

3 THE LINEAR CANONICAL TRANSFORMATIONS FOR A SINGLE PARTICLE IN ONE DIMENSION AND THEIR REPRESENTATIONS

When we have just a single coordinate x and corresponding momentum p, the most general linear transformation is

$$\begin{pmatrix} \bar{x} \\ \bar{p} \end{pmatrix} = \mathbf{S}\begin{pmatrix} x \\ p \end{pmatrix} \equiv \begin{pmatrix} a & b \\ c & d \end{pmatrix}\begin{pmatrix} x \\ p \end{pmatrix}, \tag{3.1}$$

where a, b, c, d are real numbers if we want to keep the property that $\bar{x}$, $\bar{p}$ should be hermitian operators as is the case of x, p. Furthermore, equation (1.7) requires then that

$$ad - bc = 1 \tag{3.2}$$

if the linear transformation (3.1) is canonical.

We proceed to determine the matrix

$$\langle x' |U| x'' \rangle \tag{3.3}$$

in this case. Following the analysis of the previous section we first look into the transformation bracket

$$\bar{x}\left(x', \frac{1}{i}\frac{\partial}{\partial x'}\right)\langle x'|\bar{x}'\rangle = \bar{x}'\langle x'|\bar{x}'\rangle = \left(ax' - ib\frac{\partial}{\partial x'}\right)\langle x'|\bar{x}'\rangle. \tag{3.4}$$

The solution of this equation is

$$\langle x'|\bar{x}'\rangle = f(\bar{x}')\exp\left[-\frac{i}{2ab}(\bar{x}' - ax')^2\right], \tag{3.5}$$

where $f(\bar{x}')$ is, a so far, undetermined function of $\bar{x}'$.

We first require that $\langle x'|\bar{x}'\rangle$ should be normalized in the sense of (2.7) and this implies

$$f^*(\bar{x}')f(\bar{x}'')\int_{-\infty}^{+\infty}\exp\left[\frac{i}{2ab}(\bar{x}'-ax')^2\right]\exp\left[-\frac{i}{2ab}(\bar{x}''-ax')^2\right]\mathrm{d}x'$$
$$=f^*(\bar{x}')\exp\left[\frac{i}{2ab}\bar{x}'^2\right]f(\bar{x}'')\exp\left[-\frac{i}{2ab}\bar{x}''^2\right]$$
$$\int_{-\infty}^{+\infty}\exp\left[\frac{i}{b}(\bar{x}''-\bar{x}')\,x'\right]\mathrm{d}x'=\delta(\bar{x}'-\bar{x}''), \tag{3.6}$$

which in turn leads to the relation

$$f(\bar{x}')=\frac{1}{\sqrt{2\pi\,|b|}}\exp\left[\frac{i}{2ab}\bar{x}'^2\right]\exp\,[i\,\phi(\bar{x}')], \tag{3.7}$$

where insofar $\phi(\bar{x}')$ is an arbitrary real function of $\bar{x}'$.

The transformation bracket (3.5), with $f(\bar{x}')$ given by (3.7), guarantees then that the matrix elements of $\bar{x}$ with respect to the bras and kets $(\bar{x}'|,|\bar{x}'')$ have the familiar form (2.1 a). We turn then to the momentum observable $\bar{p}$ for which the equation (2.9) must hold, i.e.

$$\int(\bar{x}'|x'\rangle\left(cx'-id\frac{\partial}{\partial x'}\right)\langle x'|\bar{x}'')\,\mathrm{d}x'=-\frac{1}{i}\frac{\partial}{\partial\bar{x}''}\delta(\bar{x}'-\bar{x}''). \tag{3.8}$$

Substituting (3.5) in (3.8) we get for the integral

$$\frac{1}{2\pi\,|b|}\int\exp\left\{-i[\phi(\bar{x}')+b^{-1}\,x'\bar{x}'-a(2b)^{-1}\,x'^2]\right\}$$
$$\times\,(-b^{-1}x'+\mathrm{d}b^{-1}\bar{x}'')\exp\left\{i[\phi(\bar{x}'')+b^{-1}x'\bar{x}''-a(2b)^{-1}x'^2]\right\}\mathrm{d}x', \tag{3.9}$$

where we made use of the relation $ad-bc=1$ imposed by the canonical nature of the transformation. If we apply now the operator $i\partial/\partial\bar{x}''$ to the last exponential in (3.9) we get the same exponential multiplied by the factor

$$-\frac{\mathrm{d}\phi(\bar{x}'')}{\mathrm{d}\bar{x}''}-\frac{x'}{b}. \tag{3.10a}$$

For this factor to coincide with the middle term in integral (3.9), we must have

$$-\frac{\mathrm{d}\phi(\bar{x}'')}{\mathrm{d}\bar{x}''}=\frac{d}{b}\bar{x}''\quad\text{or}\quad\phi(\bar{x}'')=-\frac{d}{2b}\bar{x}''^2. \tag{3.10b}$$

Introducing then this value of ϕ in (3.7), we see that with the corresponding value (3.5) of $\langle x'|\bar{x}'\rangle$, the equation (3.8) holds and thus the unitary representation of the canonical transformation (3.1) is completely determined except for a constant phase factor. We can fix the latter by requiring that the unitary representation of (3.1) tends to $\delta(x' - x'')$ when the canonical transformation tends to the identity. With this choice we get

$$\langle x' |U| x''\rangle = \langle x'|x''\rangle = \frac{e^{i\frac{\pi}{4}\,\mathrm{sign}(b)}}{\sqrt{2\pi\,|b|}} \exp\left[-\frac{i}{2b}(ax'^2 - 2x'x'' + \mathrm{d}x''^2)\right]. \tag{3.11}$$

We could now consider another canonical transformation in the linear symplectic group $Sp(2)$ characterized by the real matrix

$$\begin{pmatrix} \alpha & \beta \\ \gamma & \delta \end{pmatrix}, \quad \alpha\delta - \beta\gamma = 1. \tag{3.12}$$

The corresponding unitary representation we designate by $\langle x'|V|x''\rangle$, and it is given by (3.11) when we replace Latin by Greek letters. If we carry the transformations (3.1) and (3.12) in succession, we get an element of the linear symplectic group

$$\begin{pmatrix} \alpha & \beta \\ \gamma & \delta \end{pmatrix}\begin{pmatrix} a & b \\ c & d \end{pmatrix} = \begin{pmatrix} a\alpha + c\beta & b\alpha + d\beta \\ a\gamma + c\delta & b\gamma + d\delta \end{pmatrix}. \tag{3.13}$$

On the other hand, multiplying the corresponding representations U and V we get

$$\int \langle x' |U| x'''\rangle\, \mathrm{d}x'''\langle x''' |V| x''\rangle = \frac{e^{i\frac{\pi}{4}\{\mathrm{sign}(b) + \mathrm{sign}(\beta) - \mathrm{sign}[b\beta(b\alpha + d\beta)]\}}}{\sqrt{2\pi|b\alpha + d\beta|}}$$
$$\times \exp\left\{ -\frac{i}{2(b\alpha + d\beta)}[(a\alpha + c\beta)\, x'^2 - 2x'x'' + (b\gamma + d\delta)\, x''^2]\right\}. \tag{3.14}$$

We see therefore that the matrix (3.11) is a *ray* representation of the symplectic group, as the representation corresponding to the product of two linear symplectic transformations, differs from the product of the representations by the phase

$$e^{i\frac{\pi}{4}\{\mathrm{sign}(b) + \mathrm{sign}(\beta) - \mathrm{sign}(b\alpha + d\beta) - \mathrm{sign}[b\beta(b\alpha + d\beta)]\}}. \tag{3.15}$$

which cannot be set equal to 1 by an adjustment of the constant phase factor in (3.11). From (3.15) we see that when a, b, c, d are non negative, the representation (3.11) will become a full, rather than just a ray, representation.

Having found the unitary representations in quantum mechanics of the linear symplectic group, we shall pass to the discussion of this group in relation with the problem of the harmonic oscillator.

4 $Sp(2)$ AND $O(2)$ AS DYNAMICAL AND SYMMETRY GROUPS OF THE ONE DIMENSIONAL HARMONIC OSCILLATOR. THE DILATATION SUBGROUP.

In the previous section we discussed the group $Sp(2)$ of canonical transformations and its unitary representation in a basis in which the coordinate operator x is diagonal. We can also express the matrix U in basis in which other operators are diagonal. We shall proceed to show that the symplectic group $Sp(2)$ is actually the dynamical group of the harmonic oscillator and thus that it is interesting to find the matrix U in a basis in which the hamiltonian operator of the one dimensional harmonic oscillator is diagonal.

From the coordinate x and momentum p we can form the following quadratic monomials

$$x^2, xp, px, p^2. \tag{4.1}$$

As xp and px are identical in classical mechanics, and in the quantum case their difference is i, we see that there are only three independent second degree polynomials which we could write as[6]

$$T_1 \equiv \tfrac{1}{4}(p^2 - x^2), \quad T_2 \equiv \tfrac{1}{4}(xp + px), \quad T_3 \equiv \tfrac{1}{4}(p^2 + x^2) = \tfrac{1}{2}H, \tag{4.2}$$

where H is the hamiltonian of the one dimensional harmonic oscillator. Under the canonical transformation (3.1) the three component vector

$$\mathbf{T} = \{T_1, T_2, T_3\} \tag{4.3}$$

transforms into

$$\bar{\mathbf{T}} = \mathbf{RT}, \tag{4.4}$$

where the set of all 3×3 matrices

$$\mathbf{R} = \begin{pmatrix} \tfrac{1}{2}(a^2 - b^2 - c^2 + d^2) & cd - ab & \tfrac{1}{2}(-a^2 - b^2 + c^2 + d^2) \\ bd - ac & ad + bc & bd + ac \\ \tfrac{1}{2}(-a^2 + b^2 - c^2 + d^2) & cd + ab & \tfrac{1}{2}(a^2 + b^2 + c^2 + d^2) \end{pmatrix}$$

$$\text{with} \quad ad - bc = 1, \tag{4.5}$$

constitute a representation of the symplectic group $Sp(2)$.

From the commutation relation $[x, p] = i$ the Lie algebra of T_i's is[6]

$$[T_1, T_2] = -iT_3, \quad [T_2, T_3] = iT_1, \quad [T_3, T_1] = iT_2, \tag{4.6}$$

and, as has been extensively discussed in the literature,[6] they are the generators of the dynamical group of the one dimensional oscillator. This group is sometimes identified as the $Sp(2)$ group, as follows from the present discussion, or as the $\mathcal{SU}(1, 1)$ group that is isomorphic to it. The appearance of the latter can be seen more clearly if instead of dealing with x and p we introduce the creation (η) and annihilation (ξ) operators by the usual definition

$$\begin{pmatrix} \eta \\ \xi \end{pmatrix} = \frac{1}{\sqrt{2}} \begin{pmatrix} 1 & -i \\ 1 & i \end{pmatrix} \begin{pmatrix} x \\ p \end{pmatrix}. \tag{4.7}$$

Then the transformation (3.1) between x, p and $\bar{x}, \bar{p}$ leads to a transformation

$$\begin{pmatrix} \bar{\eta} \\ \bar{\xi} \end{pmatrix} = \begin{pmatrix} u & v \\ v^* & u^* \end{pmatrix} \begin{pmatrix} \eta \\ \xi \end{pmatrix}, \quad uu^* - vv^* = 1, \tag{4.8}$$

where

$$u = \frac{1}{2}(a + d) + \frac{i}{2}(b - c), \quad v = \frac{1}{2}(a - d) - \frac{i}{2}(b + c). \tag{4.9}$$

The transformation matrix in (4.8) is the familiar one for the $\mathcal{SU}(1, 1)$ group[7] and it can be factorized as

$$\begin{pmatrix} u & v \\ v^* & u^* \end{pmatrix} = \begin{pmatrix} e^{-i\frac{\varphi}{2}} & 0 \\ 0 & e^{i\frac{\varphi}{2}} \end{pmatrix} \begin{pmatrix} \cosh\frac{\theta}{2} & \sinh\frac{\theta}{2} \\ \sinh\frac{\theta}{2} & \cosh\frac{\theta}{2} \end{pmatrix} \begin{pmatrix} e^{-i\frac{\chi}{2}} & 0 \\ 0 & e^{i\frac{\chi}{2}} \end{pmatrix}, \tag{4.10}$$

where $0 \leqq \varphi, \chi < 2\pi$, while θ is an arbitrary real number. From (4.7) the corresponding factorization of the symplectic group $Sp(2)$ is then

$$\begin{pmatrix} a & b \\ c & d \end{pmatrix} = \begin{pmatrix} \cos\frac{\varphi}{2} & -\sin\frac{\varphi}{2} \\ \sin\frac{\varphi}{2} & \cos\frac{\varphi}{2} \end{pmatrix} \begin{pmatrix} e^{\frac{\theta}{2}} & 0 \\ 0 & e^{-\frac{\theta}{2}} \end{pmatrix} \begin{pmatrix} \cos\frac{\chi}{2} & -\sin\frac{\chi}{2} \\ \sin\frac{\chi}{2} & \cos\frac{\chi}{2} \end{pmatrix}, \tag{4.11}$$

and thus we see that it has as subgroups the orthogonal group $O(2)$ of transformations in phase space, given by the first or third matrices in (4.11), as well as the dilatation group given by the second matrix.

To the element of the group $O(2)$ associated with the angle $\varphi/2$ corresponds an $\mathbf{R}$ matrix (4.5) of the form

$$\mathbf{R}_\varphi = \begin{pmatrix} \cos\varphi & \sin\varphi & 0 \\ -\sin\varphi & \cos\varphi & 0 \\ 0 & 0 & 1 \end{pmatrix}, \tag{4.12}$$

and thus T_3, or equivalently the hamiltonian H of the harmonic oscillator, remains invariant under this group. For the vector **T** the $O(2)$ subgroup implies a rotation around the axis T_3.

For the dilatation transformation associated with $\theta/2$ the corresponding **R** matrix (4.5) takes the form

$$\mathbf{R}_\theta = \begin{pmatrix} \cosh\theta & 0 & -\sinh\theta \\ 0 & 1 & 0 \\ -\sinh\theta & 0 & \cosh\theta \end{pmatrix}, \tag{4.13}$$

and thus it implies a Lorentz transformation for the vector **T** around the axis T_2.

As the linear symplectic group $Sp(2)$ is the dynamical group of the one dimensional harmonic oscillator, while its $O(2)$ subgroup is the symmetry group of the same problem, it is of interest to consider the unitary matrix U associated with linear canonical transformations in a basis in which the hamiltonian of the harmonic oscillator is diagonal.

5 THE UNITARY REPRESENTATIONS OF $Sp(2)$ AND ITS SUBGROUPS IN THE OSCILLATOR BASIS

The unitary representation U of the elements of the $Sp(2)$ group in the basis in which the coordinate operator is diagonal was given by (3.11). If we now wish to express U in the basis in which $\frac{1}{2}(p^2 + x^2)$ is diagonal, we need to find the transformation brackets that satisfy the equation

$$\frac{1}{2}\left(-\frac{\partial^2}{\partial x'^2} + x'^2\right)\langle x'|n'\rangle = \left(n' + \frac{1}{2}\right)\langle x'|n'\rangle, \tag{5.1}$$

i.e. the familiar eigenstates of the one dimensional harmonic oscillator problem characterized by the integer n' and of eigenvalue $n' + \frac{1}{2}$ in the present units. As no confusion can occur in this case, between the kets $|n'\rangle$ with n' integer and the kets $|x'\rangle$ with x' arbitrary but real, we still keep the angular bracket notation for the eigenket $|n'\rangle$.

The normalized transformation brackets are given by the familiar function

$$\langle x'|n'\rangle = A_{n'}H_{n'}(x')\, e^{-\frac{1}{2}x'^2}, \quad A_{n'} = \pi^{-1/4}\,(2^{n'}n'!)^{-1/2}, \tag{5.2}$$

where $H_{n'}(x')$ are Hermite polynomials.

The representation of U in the oscillator basis is then given by

$$\langle n'\,|U|\,n''\rangle = \int \langle n'|x'\rangle\, dx' \langle x'\,|U|\,x''\rangle\, dx'' \langle x''|n''\rangle. \tag{5.3}$$

A convenient way to evaluate the integral is through the use of the generating function of Hermite polynomials

$$e^{-t^2+2tx} = \sum_{n=0}^{\infty} \frac{t^n}{n!} H_n(x). \tag{5.4}$$

We thus get

$$\sum_{n',n''=0}^{\infty} \frac{t'^{n'}\, t''^{n''}}{n'!\, n''!\, A_{n'} A_{n''}} \langle n' |U| n'' \rangle$$

$$= \iint_{-\infty}^{+\infty} e^{-t'^2+2t'x'-\frac{1}{2}x'^2} \langle x' |U| x'' \rangle\, e^{-t''^2+2t''x''-\frac{1}{2}x''^2}\, dx'\, dx''$$

$$= \sqrt{\frac{2\pi}{|(b-c)+i(a+d)|}} \exp\left[\frac{i}{2}\left(\frac{\pi}{2}-\varepsilon\right)\right] \exp\left\{-t'^2\right.$$

$$\left.\times \left[\frac{-(b+c)+i(a-d)}{(b-c)+i(a+d)}\right]\right\} \times \exp\left\{-t''^2\left[\frac{-(b+c)-i(a-d)}{(b-c)+i(a+d)}\right]\right\}$$

$$\exp\left\{t't''\left[\frac{4i}{(b-c)+i(a+d)}\right]\right\}. \tag{5.5}$$

where ε is the phase of the complex number $(b - c) + i(a + d)$.

Expanding then the right hand side of (5.5) in a power series of t' and t'', we obtain finally

$$\langle n' |U| n'' \rangle = \left[\frac{n'!\, n''!}{2^{n'+n''-1}\, |(b-c)+i(a+d)|}\right]^{1/2} \exp\left[\frac{i}{2}\left(\frac{\pi}{2}-\varepsilon\right)\right.$$

$$\times\, [(b-c)+i(a+d)]^{-\frac{n'+n''}{2}}\, \tfrac{1}{2}[1+(-1)^{n'-n''}]$$

$$\times \sum_{r=0}^{\min(n',n'')} \tfrac{1}{2}[1+(-1)^{n'-r}]\, \frac{(4i)^r}{\left(\frac{n'-r}{2}\right)!\left(\frac{n''-r}{2}\right)!\, r!}$$

$$\times\, [(b+c)-i(a-d)]^{\frac{n'-r}{2}}\, [(b+c)+i(a-d)]^{\frac{n''-r}{2}}. \tag{5.6}$$

We thus have the unitary representation of the $Sp(2)$ group in the basis in which the hamiltonian of the harmonic oscillator is diagonal. As the elements of $Sp(2)$ can be constructed from products of rotations, i.e. elements of $O(2)$, and dilatations, it is interesting to see to what does the general matrix element (5.6) reduce to in those two cases.

For rotations we have

$$a = \cos\frac{\varphi}{2}, \quad b = -\sin\frac{\varphi}{2}, \quad c = \sin\frac{\varphi}{2}, \quad d = \cos\frac{\varphi}{2}, \tag{5.7}$$

and substituting in (5.5) we obtain

$$\sum_{n',n''=0}^{\infty} \frac{t'^{n'} t''^{n''}}{n'!\, n''!\, A_{n'} A_{n''}} \langle n' |U| n'' \rangle = \sqrt{\pi}\, e^{-i\frac{\varphi}{4}} \exp\left[2t't''\, e^{-i\varphi/2}\right]. \tag{5.8}$$

Expanding the exponential we thus get

$$\langle n' |U| n'' \rangle = e^{-i\left(n'+\frac{1}{2}\right)\frac{\varphi}{2}} \delta_{n'n''}. \tag{5.9}$$

This last result is hardly surprising, as rotations in the two dimensional phase space leave the hamiltonian of the harmonic oscillator invariant. As the eigenstates of the one dimensional oscillator are non-degenerate, we see that the unitary representation of the symmetry group of the oscillator has to be diagonal and unitary. Thus the elements along the diagonal can only be phase factors, as we explicitly show in (5.9).

Despite the apparently trivial form of the representation (5.9) we shall see, in the next chapter, that it has far reaching applications when extended to many particles in a three dimensional harmonic oscillator potential.

We now turn to the dilatation subgroup for which

$$a = \omega^{1/2}, \quad b = c = 0, \quad d = \omega^{-1/2}, \quad \omega \equiv e^{\theta}. \tag{5.10}$$

Substituting these values in (5.6) we obtain

$$\langle n' |U| n'' \rangle = \left[\frac{n'!\, n''!}{2^{n'+n''}}\right]^{1/2} \omega^{1/4} \left(\frac{\omega+1}{2}\right)^{-1/2} \frac{1}{2}\left[1 + (-1)^{n'-n''}\right]$$
$$\times \sum_{r=0}^{\min(n',n'')} \frac{1}{2}\left[1 + (-1)^{n'-r}\right] \frac{(-1)^{\frac{n'-r}{2}} \left(\frac{\omega-1}{\omega+1}\right)^{\frac{n'+n''}{2}-r} \left(\frac{4\sqrt{\omega}}{\omega+1}\right)^{r}}{\left(\frac{n'-r}{2}\right)! \left(\frac{n''-r}{2}\right)!\, r!}. \tag{5.11}$$

As under the dilatation (5.10)

$$\bar{x} = \omega^{1/2} x, \quad \bar{p} = \omega^{-1/2} p, \quad \frac{1}{2}(\bar{p}^2 + \bar{x}^2) = \frac{1}{\omega}\frac{1}{2}(p^2 + \omega^2 x^2), \tag{5.12}$$

we see that the matrix elements $\langle n'|U|n''\rangle$ permit us to expand the eigenstates of an harmonic oscillator of frequency ω in terms of those of frequency 1.

So far we have discussed the unitary representation U of $Sp(2)$ from a general quantum mechanical standpoint. In the next section we shall rederive our results using creation and annihilation operators.

6 AN ALTERNATIVE DERIVATION, USING CREATION AND ANNIHILATION OPERATORS, OF THE UNITARY REPRESENTATION OF $Sp(2)$ AND ITS SUBGROUPS. THE QUASIPARTICLE PICTURE FOR BOSONS.

We showed in (4.11) that the most general symplectic transformation can be factorized in terms of orthogonal and dilatation transformations in the two dimensional phase space. It suffices therefore to discuss the unitary representation of elements of the $O(2)$ and the dilatation groups. We shall carry out this derivation in the present section with the help of creation and annihilation operators, rather than through the general transformation bracket analysis used previously.

We start by noting the well known fact that the eigenstate $|n\rangle$ of the one dimensional harmonic oscillator can be written as

$$|n\rangle = \frac{\eta^n}{\sqrt{n!}}|0\rangle, \tag{6.1}$$

where η is the creation operator defined by (4.7) and $|0\rangle$ is the ground state

$$|0\rangle = \pi^{-1/4}\, e^{-\frac{1}{2}x^2}. \tag{6.2}$$

Now if we restrict ourselves to the $O(2)$ subgroup of $Sp(2)$ the hamiltonian $\frac{1}{2}(p^2 + x^2)$ remains invariant and thus the ground state $|0\rangle$, which is non degenerate, can at most transform into itself, in general multiplied by a phase factor, when we apply an operation of $O(2)$. We can choose this factor as 1 and thus the ground state will be a basis for the identical representation of $O(2)$. Turning then our attention to states $|n\rangle$ of arbitrary n, we note from (4.10) that under the transformation of $O(2)$ of angle $\varphi/2$ we get

$$\bar{\eta} = e^{-i\frac{\varphi}{2}}\eta, \tag{6.3}$$

and thus

$$|n) = \frac{\bar{\eta}^n}{\sqrt{n!}}|0\rangle = e^{-in\frac{\varphi}{2}}\frac{\eta^n}{\sqrt{n!}}|0\rangle = e^{-in\frac{\varphi}{2}}|n\rangle, \tag{6.4}$$

where we use a round bracket for the state corresponding to the system $\bar{x}, \bar{p}$ to distinguish it from the state corresponding to the system x, p denoted by an angular bracket. Clearly therefore the unitary representation that takes us from $|n\rangle$ to $|n)$ is a diagonal matrix whose elements are

$$\langle n' |U| n''\rangle = e^{-in'\frac{\varphi}{2}} \delta_{n'n''}, \tag{6.5}$$

which differ from (5.9) only by the phase of the ground state transformation which in the present analysis we take as 1.

A very important point emerges already from a generalization of the analysis that we just presented. When we turn to harmonic oscillators of higher dimension or to several particles in harmonic oscillator potential of arbitrary dimension, we still have two facts to consider: a) The ground state is non-degenerate and thus can be associated with the unit representation of the symmetry group of the oscillator, b) An arbitrary excited state can be expressed in terms of a polynomial function of the creation operators acting on the ground state.[8] Thus the representation of the symmetry group of the oscillator that is carried by these excited states is fully determined by the transformation properties of the polynomial in the creation operators under the corresponding operation of the group. We shall make extensive use of this result in the next chapter.

We found the unitary representation of the $O(2)$ symmetry group of the harmonic oscillator using creation operators and we turn now to the dilatation subgroup of $Sp(2)$ for which we have

$$\bar{x} = e^{\theta/2}\, x, \quad \bar{p} = e^{-\theta/2}\, p \tag{6.6a}$$

or

$$\begin{pmatrix} \bar{\eta} \\ \bar{\xi} \end{pmatrix} = \begin{pmatrix} \cosh\theta & \sinh\theta \\ \sinh\theta & \cosh\theta \end{pmatrix} \begin{pmatrix} \eta \\ \xi \end{pmatrix}. \tag{6.6b}$$

Using (5.10) we can write

$$\cosh\theta = \tfrac{1}{2}(\omega^{1/2} + \omega^{-1/2}), \quad \sinh\theta = \tfrac{1}{2}(\omega^{1/2} - \omega^{-1/2}), \tag{6.7}$$

in which from (5.12), ω can be interpreted as the frequency of the harmonic oscillator after dilatation.

The relation (6.6b) looks remarkably as the Bogoliubov-Valatin transformation[9] in quasi-particle theory, but now applied to commuting creation and annihilation operators rather than anti-commuting ones, which reflects also in the fact that the transformation group is part of $\mathscr{SU}(1, 1)$ rather than of $\mathscr{SU}(2)$.

The determination of the unitary representation of the dilatation group presents now a problem that is quite familiar in quasiparticle theory, the ground state $|0\rangle$ no longer belongs to the unit representation of the dilatation group. In fact, under the transformation (6.6a), when we include the normalization factor we have

$$\pi^{-1/4}\, e^{-\frac{1}{2}x^2} \rightarrow \pi^{-1/4}\, \omega^{1/4}\, e^{-\frac{1}{2}\omega x^2}, \tag{6.8}$$

and the latter expression can be developed in an infinite series of oscillator functions $|n\rangle$ of frequency 1. Thus we cannot use the simple trick employed for the unitary representations of $O(2)$, in which the representation was carried only by the operator η^n of (6.1) as the ground state $|0\rangle$ corresponded to the unit representation.

We are indebted to C. Bloch for a suggestion that permits us to go around this problem. Let us apply to an arbitrary state $|n\rangle$ the operator

$$\sum_{r=0}^{\infty} \frac{(-1)^r}{r!} \eta^r \xi^r. \tag{6.9}$$

From (6.1) and the properties

$$[\xi, \eta] = 1, \quad \xi|0\rangle = 0, \tag{6.10}$$

we see that when the operator (6.9) is applied to $|n\rangle$ we obtain

$$\left[\sum_{r=0}^{n} (-1)^r \frac{n!}{r!\,(n-r)!}\right]|n\rangle = (1-1)^n|n\rangle = \delta_{n0}|0\rangle. \tag{6.11}$$

Thus clearly the operator (6.9), when applied to an arbitrary state, projects from it the ground state of the oscillator with, in general, some multiplicative factor.

The state (6.1) can also be written as

$$\frac{\eta^n}{\sqrt{n!}}\left[\sum_{r=0}^{\infty} (-1)^r \frac{\eta^r \xi^r}{r!}\right]|0\rangle. \tag{6.12}$$

When we pass from the x, p system to a new arbitrary $\bar{x}, \bar{p}$ system, we can now keep unchanged the ground state $|0\rangle$ in (6.12) and only transform the η, ξ operators into the $\bar{\eta}$, $\bar{\xi}$ of (4.8), thus getting

$$\frac{\bar{\eta}^n}{\sqrt{n!}}\left[\sum_{r=0}^{\infty} (-1)^r \frac{\bar{\eta}^r \bar{\xi}^r}{r!}\right]|0\rangle. \tag{6.13}$$

Indeed the operator in the square bracket projects out from the ground state $|0\rangle$, in our original system, the ground state $|0)$ in the system of $\bar{x}, \bar{p}$, which we designate by a round bracket. Thus (6.13) reduces to

$$C\,\frac{\bar{\eta}^n}{\sqrt{n!}}\,|0) = C|n), \tag{6.14}$$

where C is as yet an undetermined constant independent of n, equal to the transformation bracket $(0|0\rangle$ from the old to the new ground state. To develop $|n)$ in terms of $|n'\rangle$ we use (4.8) to substitute the $\bar{\eta}$, $\bar{\xi}$ in (6.13) by

$$\bar{\eta} = u\eta + v\xi, \quad \bar{\xi} = v^*\eta + u^*\xi, \tag{6.15}$$

and then make use of Wick's theorem[10] to evaluate the scalar product of (6.13) with a state $|n'\rangle$. We illustrate this procedure for

$$C\langle 0|0) = |\langle 0|0)|^2 = \langle 0| \sum_{r=0}^{\infty} (-1)^r \frac{\bar{\eta}^r \bar{\xi}^r}{r!} |0\rangle. \tag{6.16}$$

In this case we get

$$|\langle 0|0)|^2 = \sum_{r=0}^{\infty} \frac{(-1)^r}{r!} \sum_{k=0}^{\left[\frac{r}{2}\right]} a_{rk} (\overline{\bar{\eta}\bar{\eta}})^k (\overline{\bar{\xi}\bar{\xi}})^k (\overline{\bar{\eta}\bar{\xi}})^{r-2k}, \tag{6.17}$$

where $\left[\frac{r}{2}\right]$ is equal to $\frac{r}{2}$ or $\frac{r-1}{2}$ if r is even or odd respectively, and a_{rk} is a statistical factor equal to the number of ways of making the indicated contractions of the operators $\bar{\eta}$ and $\bar{\xi}$, i.e.

$$a_{rk} = \frac{(r!)^2}{[(2k)!!]^2 \, (r-2k)!}. \tag{6.18}$$

Using the result

$$\overline{\bar{\eta}\bar{\eta}} = uv, \quad \overline{\bar{\xi}\bar{\xi}} = u^*v^*, \quad \overline{\bar{\eta}\bar{\xi}} = vv^*, \tag{6.19}$$

which comes from a straightforward application of the properties (6.10) and the relation $uu^* - vv^* = 1$, (6.17) can be written in terms of v only,

$$|\langle 0|0)|^2 = \sum_{r,k,l=0}^{\infty} (-1)^r \frac{r!}{2^{2k}\, k!(r-2k)!\, l!(k-l)!} |v|^{2(r-k+l)}. \tag{6.20}$$

The summations can be carried out leading to the final simple result

$$|\langle 0|0)|^2 = [1 + |v|^2]^{-1/2} = |u|^{-1}, \tag{6.21}$$

which, when particularized to the case of dilatations with the help of (6.6), (6.7) gives

$$|\langle 0|0)|^2 = [\tfrac{1}{2}(\omega^{1/2} + \omega^{-1/2})]^{-1}. \tag{6.22}$$

A similar analysis can be carried out in general, leading to the determination of the matrix elements

$$\langle n'|n'') = \langle n' |U| n''\rangle \tag{6.23}$$

in the same form as they appear in (5.11).

Thus we are able to find the unitary representations of both the $O(2)$ and dilatation subgroups of $Sp(2)$, using the transformations of the creation and annihilation operators induced by (3.1). As we showed in (4.11) that an arbitrary element of $Sp(2)$ can be formed from products of elements of $O(2)$ and the dilatation subgroup, we can affirm that any unitary representation of $Sp(2)$ in an oscillator basis can be constructed from the transformations induced on the creation and annihilation operators.

The linear group of canonical transformations we discussed so far was homogeneous. As we could add to it translations in coordinate or momentum, we complete our discussion in the next section by considering these transformations.

7 THE UNITARY REPRESENTATION OF TRANSLATIONS IN PHASE SPACE IN THE OSCILLATOR BASIS.

The most general linear canonical transformation in the two dimensional phase space is given by a combination of the transformation (3.1) with the translation

$$\bar{x} = x + e$$
$$\bar{p} = p + f \tag{7.1}$$

where e, f are real constants. Following the analysis of section 2, we see that to obtain the unitary representation of (7.1), we must first determine the transformation brackets satisfying the equation

$$(x' + e) \langle x'|\bar{x}') = \bar{x}'\langle x'|\bar{x}'). \tag{7.2}$$

Clearly they are given by

$$\langle x'|\bar{x}') = c(\bar{x}')\, \delta(x' + e - \bar{x}'), \tag{7.3}$$

where $c(\bar{x}')$ is, so far, an arbitrary function of $\bar{x}'$. The normalization condition

$$\int (\bar{x}'|x'\rangle\, \mathrm{d}x'\langle x'|\bar{x}'') = \delta(\bar{x}' - \bar{x}'') \tag{7.4}$$

then requires that $|c(\bar{x}')|^2 = 1$ or

$$c(\bar{x}') = e^{i\phi(\bar{x}')}, \tag{7.5}$$

where $\phi(\bar{x}')$ is a real function of $\bar{x}'$. Following now the analysis of section 3 we require that

$$\int (\bar{x}'|x'\rangle \left(\frac{1}{i}\frac{\partial}{\partial \bar{x}'} + f\right) \langle x'|\bar{x}'')\, \mathrm{d}x' = -\frac{1}{i}\frac{\partial}{\partial \bar{x}''}\,\delta(\bar{x}' - \bar{x}''). \tag{7.6}$$

Substituting (7.3), with the restriction (7.5), in the integral of (7.6), we obtain for it the expression

$$\int e^{-i\phi(\bar{x}')}\,\delta(x' + e - \bar{x}')\left[\frac{1}{i}\frac{\partial}{\partial x'} + f\right][e^{i\phi(\bar{x}'')}\,\delta(x' + e - \bar{x}'')]\,\mathrm{d}x'. \quad (7.7)$$

If on the other hand we apply the operator $i\partial/\partial\bar{x}''$ to the last square bracket in (7.7), we get

$$e^{i\phi(\bar{x}'')}\left[-\frac{\mathrm{d}\phi(\bar{x}'')}{\mathrm{d}\bar{x}''} + i\frac{\partial}{\partial\bar{x}''}\right]\delta(x' + e - \bar{x}''). \quad (7.8)$$

Choosing then

$$\phi(\bar{x}') = -f\bar{x}', \quad (7.9)$$

we see that the Eq. (7.6) is satisfied as we can also write

$$\frac{\partial}{\partial\bar{x}''}\delta(x' + e - \bar{x}'') = -\frac{\partial}{\partial x'}\delta(x' + e - \bar{x}''). \quad (7.10)$$

Thus the matrix of the unitary representation in the basis in which x is diagonal, is given by

$$\langle x'|U|x''\rangle = \langle x'|x'') = e^{-ifx''}\,\delta(x' + e - x''). \quad (7.11)$$

This representation clearly reduces to $\delta(x' - x'')$ when $e = f = 0$.

We could now consider another translation characterized by

$$\bar{x} = x + \varepsilon \quad (7.12)$$
$$\bar{p} = p + \varphi,$$

for which the corresponding unitary representation designated by $\langle x'|V|x''\rangle$ is given by (7.11) when we replace e by ε and f by φ. If we carry out the transformations (7.1) and (7.12) in succession, we get

$$\bar{x} = x + e + \varepsilon \quad (7.13)$$
$$\bar{p} = p + f + \varphi,$$

while multiplying the corresponding representations U and V we get

$$\int \langle x'|U|x'''\rangle\,\mathrm{d}x'''\langle x'''|V|x''\rangle = e^{i\varepsilon f}\,e^{-i(f+\varphi)x''}\,\delta(x' + e + \varepsilon - x''). \quad (7.14)$$

As the introduction in (7.11) of a constant phase factor cannot eliminate the phase $e^{i\varepsilon f}$ in (7.14), the matrix U of (7.11) corresponds to a *ray* representation of the translation group in phase space.

As a final point in our discussion, we proceed to derive the unitary representation of the phase space translation in a basis in which the hamiltonian of the oscillator is diagonal. For this we require the evaluation of the double integral (5.3) when $\langle x'|U|x''\rangle$ is given by (7.11). Again it is convenient to make use of the generating function of the Hermite polynomials and thus we first calculate

$$\sum_{n',n''=0}^{\infty} \frac{t'^{n'}\, t''^{n''}}{n'!\, n''!\, A_{n'}\, A_{n''}} \langle n' |U| n''\rangle$$

$$= \iint_{-\infty}^{+\infty} e^{-t'^2+2t'x'-\frac{1}{2}x'^2}\, e^{-ifx''}\, \delta(x'+e-x'')\, e^{-t''^2+2t''x''-\frac{1}{2}x''^2}\, dx'\, dx''$$

$$= \sqrt{\pi}\exp\left[-\tfrac{1}{4}(e^2+2ief+f^2)\right]\exp[2t't'']\exp[-(e+if)\,t']\exp[(e-if)\,t'']. \tag{7.15}$$

Developing the right hand side in powers of t', t'', we obtain finally

$$\langle n' |U| n''\rangle = \left[\frac{n'!\, n''!}{2^{n'+n''}}\right]^{1/2} \exp\left[-\frac{1}{4}(e^2+2ief+f^2)\right]$$
$$\times \sum_{r=0}^{min(n',n'')} (-1)^{n'-r} \frac{2^r (e+if)^{n'-r} (e-if)^{n''-r}}{(n'-r)!\,(n''-r)!\,r!}. \tag{7.16}$$

In particular, when the translation is only in x, i.e. $f = 0$, the matrix U reduces to

$$\langle n' |U| n''\rangle = \left[\frac{n'!\, n''!}{2^{n'+n''}}\right]^{1/2} e^{-\frac{1}{4}e^2} \sum_{r=0}^{min(n',n'')} \frac{(-1)^{n'-r}\, 2^r\, e^{n'+n''-2r}}{(n'-r)!\,(n''-r)!\,r!}. \tag{7.17}$$

This last result allows us to develop the translated one dimensional oscillator state in terms of the original ones.

Finally, it is possible to combine the symplectic transformation (3.1) with the phase space translation (7.1) in an inhomogeneous symplectic transformation. The unitary representation of the elements of this group in a basis in which the hamiltonian of the harmonic oscillator is diagonal, is given by

$$\langle n' |U| n''\rangle = \exp\left[\frac{i}{2}\left(\frac{\pi}{2}-\varepsilon\right)\right]\left[\frac{n'!\, n''!}{2^{n'+n''-1}\,|(b-c)+i(a+d)|}\right]^{1/2}$$
$$\times \exp\left\{\frac{ce^2+2def-bf^2-i(de^2-2cef+af^2)}{2[(b-c)+i(a+d)]}\right\} \sum_{s=0}^{n'} \frac{1}{2}[1+(-1)^{n'-s}]$$
$$\times \sum_{t=0}^{n''} \frac{1}{2}[1+(-1)^{n''-t}] \frac{2^{s+t}}{\left(\frac{n'-s}{2}\right)!\left(\frac{n''-t}{2}\right)!} [(b+c)-i(a-d)]^{\frac{n'-s}{2}}$$
$$\times [(b+c)+i(a-d)]^{\frac{n''-t}{2}} \sum_{r=0}^{min(s,t)} \frac{i^r}{r!\,(s-r)!\,(t-r)!} (f-ie)^{s-r}$$
$$\times [(-ce+af)+i(de-bf)]^{t-r}\, [(b-c)+i(a+d)]^{r-\frac{n'+n''+s+t}{2}}, \tag{7.18}$$

where ε is, as in (5.6), the phase of the complex number $(b - c) + i(a + d)$. The proof of this result will be discussed elsewhere.

8 THE LINEAR SYMPLECTIC GROUP $Sp(2N)$ AND ITS UNITARY SUBGROUP $\mathscr{U}(N)$ AS DYNAMICAL AND SYMMETRY GROUPS OF THE N DIMENSIONAL HARMONIC OSCILLATOR. THE CASE $N = nq$ FOR n PARTICLES IN A q DIMENSIONAL OSCILLATOR.

When we are dealing with a $2N$ dimensional phase space, the most general linear transformation in it could be written as

$$\begin{pmatrix} \bar{\mathbf{x}} \\ \bar{\mathbf{p}} \end{pmatrix} = \begin{pmatrix} \mathbf{S}_{11} & \mathbf{S}_{12} \\ \mathbf{S}_{21} & \mathbf{S}_{22} \end{pmatrix} \begin{pmatrix} x \\ p \end{pmatrix}, \tag{8.1}$$

where $\mathbf{x} = \{x_i\}$, $\mathbf{p} = \{p_i\}$ are N dimensional vectors, and $\mathbf{S}_{\alpha\beta}$ $\alpha, \beta = 1, 2$ are $N \times N$ matrices. If the transformation is canonical, the restriction (1.7) leads to the relations

$$\begin{aligned} -\mathbf{S}_{12}\tilde{\mathbf{S}}_{11} + \mathbf{S}_{11}\tilde{\mathbf{S}}_{12} &= -\mathbf{S}_{22}\tilde{\mathbf{S}}_{21} + \mathbf{S}_{21}\tilde{\mathbf{S}}_{22} = 0, \\ \mathbf{S}_{11}\tilde{\mathbf{S}}_{22} - \mathbf{S}_{12}\tilde{\mathbf{S}}_{21} &= \mathbf{I}, \end{aligned} \tag{8.2}$$

where $\sim$ indicates the transposed. The $2N \times 2N$ matrices $\mathbf{S}$ whose components $\mathbf{S}_{\alpha\beta}$ satisfy (8.2), constitute then the elements of the real symplectic group $Sp(2N)$. We will be interested also in the subgroup of all canonical transformations for which $\mathbf{S}$ is an orthogonal matrix, i.e.

$$\mathbf{S} = \begin{pmatrix} \mathscr{O}_{11} & \mathscr{O}_{12} \\ \mathscr{O}_{21} & \mathscr{O}_{22} \end{pmatrix}, \tag{8.3}$$

where the orthogonality condition implies

$$\mathscr{O}_{11}\tilde{\mathscr{O}}_{11} + \mathscr{O}_{12}\tilde{\mathscr{O}}_{12} = \mathbf{I}, \quad \mathscr{O}_{21}\tilde{\mathscr{O}}_{21} + \mathscr{O}_{22}\tilde{\mathscr{O}}_{22} = \mathbf{I}, \quad \mathscr{O}_{11}\tilde{\mathscr{O}}_{21} + \mathscr{O}_{12}\tilde{\mathscr{O}}_{22} = 0, \tag{8.4}$$

while the restriction to canonical transformation, i.e. the relation (8.2), leads to

$$\begin{aligned} -\mathscr{O}_{12}\tilde{\mathscr{O}}_{11} + \mathscr{O}_{11}\tilde{\mathscr{O}}_{12} = 0, \ -\mathscr{O}_{22}\tilde{\mathscr{O}}_{21} + \mathscr{O}_{21}\tilde{\mathscr{O}}_{22} = 0, \\ \mathscr{O}_{11}\tilde{\mathscr{O}}_{22} - \mathscr{O}_{12}\tilde{\mathscr{O}}_{21} = \mathbf{I}. \end{aligned} \tag{8.5}$$

We can now define

$$\mathbf{U} = \mathscr{O}_{11} + i\mathscr{O}_{12}, \quad \mathbf{V} = \mathscr{O}_{22} - i\mathscr{O}_{21}, \tag{8.6}$$

which implies, as the $\mathcal{O}_{\alpha\beta}$ are real,

$$\mathbf{U}^* = \mathcal{O}_{11} - i\mathcal{O}_{12}, \quad \mathbf{V}^* = \mathcal{O}_{22} + i\mathcal{O}_{21}. \tag{8.7)(8.8}$$

The equations (8.4), (8.5) can then be expressed in terms of **U**, **V**, $\mathbf{U}^*$, $\mathbf{V}^*$ and their transposeds. As was shown trivially elsewhere,[11] the solution of these equations leads to $\mathbf{V} = \mathbf{U}$ and **U** unitary and thus, the canonical orthogonal transformations in phase space take the form

$$\mathbf{S} = \begin{bmatrix} \frac{1}{2}(\mathbf{U} + \mathbf{U}^*) & -\frac{i}{2}(\mathbf{U} - \mathbf{U}^*) \\ \frac{i}{2}(\mathbf{U} - \mathbf{U}^*) & \frac{1}{2}(\mathbf{U} + \mathbf{U}^*) \end{bmatrix} \tag{8.9}$$

and thus are a representation of the N dimensional unitary group $\mathscr{U}(N)$. We use a script $\mathscr{U}$ for this group to distinguish it from the unitary representations discussed in the previous sections.

Let us now discuss the significance of the group of canonical transformations (8.1) and its subgroup (8.9) for the N dimensional harmonic oscillator problem. For this purpose we note that from x_i, p_j we can form the independent quadratic polynomials

$$\tfrac{1}{4}(p_ip_j - x_ix_j), \quad \tfrac{1}{4}(x_ip_j + p_jx_i), \quad \tfrac{1}{4}(p_ip_j + x_ix_j). \tag{8.10}$$

$$i \leqq j = 1, \dots N \quad i, j = 1, \dots N \quad i \leqq j = 1, \dots N$$

We have

$$N^2 + 2\frac{N(N+1)}{2} = N(2N+1) \tag{8.11}$$

independent operators of this type which for $N = 1$ become just the three T_1, T_2, T_3 of the one dimensional case. It is easily seen that the commutators of the operators (8.10) lead to linear combinations of the same, so that they constitute a Lie algebra. We expect this Lie algebra to be the one associated with the symplectic group $Sp(2N)$. The simplest way of showing this is by rewriting the operators (8.10) in terms of the creation and annihilation operators

$$\begin{pmatrix} \boldsymbol{\eta} \\ \boldsymbol{\xi} \end{pmatrix} = \frac{1}{\sqrt{2}} \begin{pmatrix} \mathbf{I} & -i\mathbf{I} \\ \mathbf{I} & i\mathbf{I} \end{pmatrix} \begin{pmatrix} \mathbf{x} \\ \mathbf{p} \end{pmatrix}, \tag{8.12}$$

so that the generators of the Lie algebra can also be expressed by

$$H_i \equiv \tfrac{1}{2}(\eta_i\xi_i + \xi_i\eta_i), \quad i = 1, \dots N; \tag{8.13a}$$

$$\eta_i\eta_j, \quad i \leqq j = 1, \dots N; \quad \xi_i\xi_j, \quad i \leqq j = 1, \dots N; \quad \eta_i\xi_j, \quad i \neq j. \tag{8.13b}$$

From the commutation relations

$$[\xi_i, \xi_j] = [\eta_i, \eta_j] = 0, \quad [\xi_i, \eta_j] = \delta_{ij}, \tag{8.14}$$

we see that the operators H_i in (8.13a) commute among themselves, and that their commutator with any of the operators of (8.13b) gives the latter multiplied by a constant, i.e.

$$[H_i, E_\alpha] = a_{\alpha i} E_\alpha, \tag{8.15}$$

where E_α is a general notation for the operators (8.13b). The N dimensional vectors $\{a_{\alpha i}\}$, where α is fixed and $i = 1, \dots N$ give the root vectors of our Lie algebra.[12, 13] These root vectors can be expressed as linear combinations of the vectors $\mathbf{e}_i$ whose components are zero everywhere except at the position i, where they take the value 1. We immediately see from (8.14), that the different elements of (8.13b) correspond to the following root vectors

$$\begin{aligned} &\eta_i\eta_j \quad i < j, && \mathbf{e}_i + \mathbf{e}_j, \\ &\xi_i\xi_j \quad i < j, && -\mathbf{e}_i - \mathbf{e}_j, \\ &\eta_i\xi_j \quad i \neq j, && \mathbf{e}_i - \mathbf{e}_j, \\ &\eta_i^2, && 2\mathbf{e}_i, \\ &\xi_i^2, && -2\mathbf{e}_i. \end{aligned} \tag{8.16}$$

As is well known,[13] these root vectors correspond to the Lie algebra of a symplectic group in $2N$ dimensions, and thus the operators (8.13) are the generators of this group.

When we pass to the orthogonal subgroup (8.3) of $Sp(2N)$, we note that with respect to the creation and annihilation operators defined by (8.12), the transformation becomes

$$\begin{pmatrix} \bar{\boldsymbol{\eta}} \\ \bar{\boldsymbol{\xi}} \end{pmatrix} = \begin{pmatrix} \mathscr{U} & \mathbf{0} \\ \mathbf{0} & \mathscr{U}^* \end{pmatrix} \begin{pmatrix} \boldsymbol{\eta} \\ \boldsymbol{\xi} \end{pmatrix}. \tag{8.17}$$

The generators of this unitary subgroup must be a subset of (8.13) and in fact, they are given by

$$H_i = \tfrac{1}{2}(\eta_i\xi_i + \xi_i\eta_i) \equiv \mathscr{C}_{ii} + \tfrac{1}{2}, \quad i = 1, \dots N, \quad \mathscr{C}_{ij} \equiv \eta_i\xi_j, \quad i \neq j, \tag{8.18}$$

which close under commutation, i.e.

$$[\mathscr{C}_{ij}, \mathscr{C}_{i'j'}] = \mathscr{C}_{ij'}\delta_{i'j} - \mathscr{C}_{i'j}\delta_{ij'}, \tag{8.19}$$

and for which the root vectors[13] are

$$\eta_i\xi_j \quad i \neq j, \quad \mathbf{e}_i - \mathbf{e}_j. \tag{8.20}$$

Now we note that H_i of (8.13a) are just

$$H_i = \tfrac{1}{2}(\eta_i\xi_i + \xi_i\eta_i) = \tfrac{1}{2}(p_i^2 + x_i^2), \tag{8.21}$$

and thus the hamiltonian of the N dimensional harmonic oscillator becomes

$$H = \sum_{i=1}^{N} H_i. \tag{8.22}$$

The group $Sp(2N)$, whose generators are (8.13), is then the dynamical group of the N dimensional harmonic oscillator. The group $\mathscr{U}(N)$ whose generators are (8.18), is the symmetry group of the harmonic oscillator, as can be seen directly, because

$$[\mathscr{C}_{ij}, H] = 0, \tag{8.23}$$

and also from the fact that the orthogonal group of canonical transformations clearly leaves H invariant.

Finally, we would like to remark that if we have n particles in a q dimensional harmonic oscillator, the hamiltonian is identical to that of a single particle in an $N = nq$ dimensional harmonic oscillator. Thus we also have the dynamical and invariance groups for this problem.

CHAPTER II

THE SYMMETRY GROUP OF THE HARMONIC OSCILLATOR AND ITS APPLICATIONS.

WE SAW IN SECTION 8 of the previous chapter, that the symmetry group of the N dimensional harmonic oscillator is the unitary group $\mathscr{U}(N)$. We mentioned also in section 6, that the unitary representation of this group, in a basis in which the hamiltonian of the harmonic oscillator is diagonal, is carried out entirely by the polynomial in the creation operators that we apply to the ground state to get any other state of the problem.

In this chapter we shall make use of these facts to construct states of one, two, three, four and then n particles in an harmonic oscillator potential, that are characterized by irreducible representations (IR) of chains of subgroups of $\mathscr{U}(3n)$, as the dimension of our space is 3 and we are dealing, in general, with n particles.

1 THE SINGLE PARTICLE IN A THREE DIMENSIONAL OSCILLATOR. CHARACTERIZATION OF THE STATES BY THE IR OF THE CHAIN OF GROUPS $\mathscr{U}(3) \supset O(3) \supset O(2)$.

The single particle states in a three dimensional harmonic oscillator[14] can be written as

$$\psi_{nlm}(\mathbf{r}) = P_{nlm}(\boldsymbol{\eta})\,|0\rangle, \tag{1.1}$$

where $|0\rangle$ is the ground state wave function given by

$$|0\rangle = \pi^{-3/4}\, e^{-\frac{1}{2}r^2}, \tag{1.2}$$

and $P_{nlm}(\boldsymbol{\eta})$ is a polynomial in the creation operator that satisfies the equations

$$(H - \tfrac{3}{2})\, P_{nlm}(\boldsymbol{\eta}) = \boldsymbol{\eta} \cdot \boldsymbol{\xi}\, P_{nlm}(\boldsymbol{\eta}) = (2n + l)\, P_{nlm}(\boldsymbol{\eta}), \tag{1.3}$$

$$L^2 P_{nlm}(\boldsymbol{\eta}) = -(\boldsymbol{\eta} \times \boldsymbol{\xi})^2\, P_{nlm}(\boldsymbol{\eta}) = l(l + 1)\, P_{nlm}(\boldsymbol{\eta}), \tag{1.4}$$

$$L_z P_{nlm}(\boldsymbol{\eta}) = \frac{1}{i}(\boldsymbol{\eta} \times \boldsymbol{\xi})_z\, P_{nlm}(\boldsymbol{\eta}) = m P_{nlm}(\boldsymbol{\eta}), \tag{1.5}$$

where, because of the commutation relations (I.8.14), we can interpret ξ_i as

$$\xi_i = \frac{\partial}{\partial \eta_i} \qquad i = 1, 2, 3. \tag{1.6}$$

The solution of equations (1.3–1.5) is the polynomial[14]

$$P_{nlm}(\boldsymbol{\eta}) = A_{nl}(\boldsymbol{\eta} \cdot \boldsymbol{\eta})^n \mathscr{Y}_{lm}(\boldsymbol{\eta}), \tag{1.7}$$

where

$$\mathscr{Y}_{lm}(\mathbf{r}) = r^l \, Y_{lm}(\theta, \varphi) \tag{1.8}$$

is the solid spherical harmonic and

$$A_{nl} = (-1)^n \left[\frac{4\pi}{(2n + 2l + 1)!! \, (2n)!!} \right]^{1/2}. \tag{1.9}$$

For a fixed value of $(2n + l)$, the set of polynomials (1.7) clearly form a basis[11] for an IR of the $\mathscr{U}(3)$ group characterized by $2n + l$. The states are further characterized by the IR l of the subgroup $O(3)$ of $\mathscr{U}(3)$ whose generators are

$$\mathscr{C}_{ij} - \mathscr{C}_{ji} \quad i, j = 1, 2, 3, \tag{1.10}$$

with $\mathscr{C}_{ij}$ defined by (I.8.18), and by the IR m of the subgroup $O(2)$ of $O(3)$. Thus the state (1.1) is fully characterized by the IR of the chain of groups

$$\mathscr{U}(3) \supset O(3) \supset O(2). \tag{1.11}$$

2 TWO PARTICLES IN THE HARMONIC OSCILLATOR POTENTIAL. THE CHAINS OF GROUPS $\mathscr{U}(6) \supset \begin{pmatrix} \mathscr{U}'(3) & 0 \\ 0 & \mathscr{U}''(3) \end{pmatrix}$ AND $\mathscr{U}(6) \supset \mathscr{U}(3) \times \mathscr{U}(2)$. TRANSFORMATION BRACKETS BETWEEN THE TWO CHAINS.

When dealing with n particles in a q dimensional oscillator we shall use the notation

$$\eta_{is} = \frac{1}{\sqrt{2}}(x_{is} - ip_{is}), \quad \xi_{is} = \frac{1}{\sqrt{2}}(x_{is} + ip_{is}) \quad i = 1, \dots q, \quad s = 1, \dots n \tag{2.1}$$

for the creation and annihilation operators. If in particular $q = 3$, we designate the creation operators by the vectors

$$\boldsymbol{\eta}_s \quad s = 1, \dots n. \tag{2.2}$$

For the present problem $n = 2$ and the corresponding set of states can be written as

$$P_{n_1 l_1 m_1}(\boldsymbol{\eta}_1)\, P_{n_2 l_2 m_2}(\boldsymbol{\eta}_2)\, |0\rangle \tag{2.3}$$

where now

$$|0\rangle = \pi^{-\frac{3}{4}n} \exp\left[-\tfrac{1}{2}\sum_{s=1}^{n} r_s^2\right] \tag{2.4}$$

with $n = 2$.

Clearly the states (2.3) belong to the IR of the following chain of groups

$$\mathscr{U}(6) \supset \begin{pmatrix} \mathscr{U}'(3) & 0 \\ 0 & \mathscr{U}''(3) \end{pmatrix} \supset \begin{pmatrix} O'(3) & 0 \\ 0 & O''(3) \end{pmatrix} \supset \left(\begin{array}{cc|cc} O'(2) & 0 & & \\ 0 & 1 & \multicolumn{2}{c}{O} \\ \hline & & O''(2) & 0 \\ \multicolumn{2}{c|}{O} & 0 & 1 \end{array}\right), \tag{2.5}$$

$[2n_1 + l_1 + 2n_2 + l_2]$; $\{2n_1 + l_1\}$, $\{2n_2 + l_2\}$; l_1, l_2; m_1, m_2

where below each group we give the IR by which the state is characterized. While each single particle state belongs to an IR of $\mathscr{U}(3)$, their product is no longer irreducible, as is quite familiar from the corresponding problem in angular momentum theory, i.e. for the $O(3)$ group. We can then use the Clebsch-Gordan coefficients of $\mathscr{U}(3)$ (or more exactly[11] $\mathscr{S}\mathscr{U}(3)$), to construct linear combinations of the states (2.3) that are characterized by IR of $\mathscr{U}(3)$. These Clebsch-Gordan coefficients factorize into a standard Clebsch-Gordan coefficient of $O(3)$ and an isoscalar factor of $\mathscr{U}(3)$ in the $\mathscr{U}(3) \supset O(3)$ chain, which can be written as[15]

$$\left\langle \begin{array}{cc|c} \{2n_1 + l_1\}, & \{2n_2 + l_2\} & \{h_1 h_2\} \\ l_1 & l_2 & \Omega L \end{array} \right\rangle. \tag{2.6}$$

In (2.6), $\{2n_1 + l_1\}$, $\{2n_2 + l_2\}$ are the IR of $\mathscr{U}(3)$ and l_1, l_2 the IR of $O(3)$ of each particle. The IR of $\mathscr{U}(3)$ for the product is denoted by $\{h_1 h_2 0\} \equiv \{h_1 h_2\}$. It has, at most, two rows as it is built from direct product of two single row IR of $\mathscr{U}(3)$. The IR of $O(3)$ for the product is denoted by L. As for a two rowed representation of $\mathscr{U}(3)$, the IR of the subgroup $O(3)$ may be repeated, we introduce an extra index Ω to distinguish between these repeated representations. Possible operators whose eigenvalues could play the role of Ω, have been discussed elsewhere.[16]

The state characterized by the IR of $\mathscr{U}(3)$, $O(3)$ can then be written as

$$\sum_{l_1, l_2} \left\langle \begin{array}{cc|c} \{2n_1 + l_1\}, & \{2n_2 + l_2\} & \{h_1 h_2\} \\ l_1 & l_2 & \Omega L \end{array} \right\rangle [P_{n_1 l_1}(\boldsymbol{\eta}_1)\, P_{n_2 l_2}(\boldsymbol{\eta}_2)]_{LM}\, |0\rangle \tag{2.7}$$

where the square bracket stands for vector coupling of the angular momenta to a total value L and projection M.

The states (2.7) are also characterized by the IR of a $\mathscr{U}(2)$ group as we proceed to show. First we note that in the notation (2.1) the generators of $\mathscr{U}(6)$ are

$$\mathbf{C}_{is,jt} = \eta_{is}\xi_{jt} \quad i,j = 1,2,3, \quad s = 1,2. \tag{2.8}$$

Contracting with respect to s or i, we get the generators of a $\mathscr{U}(3)$ and a $\mathscr{U}(2)$ subgroups respectively given by

$$\mathscr{C}_{ij} = \sum_s \mathbf{C}_{is,js}, \quad C_{st} = \sum_i \mathbf{C}_{is,it}. \tag{2.9a, b}$$

The state (2.7) clearly belongs to the IR $\{h_1 h_2\}$ of the $\mathscr{U}(3)$ group whose generators are $\mathscr{C}_{ij}$. It also belongs to the single row IR of $\mathscr{U}(6)$ characterized by $[2n_1 + l_1 + 2n_2 + l_2]$. Thus it is related as well[15, 16] to the IR $\{h_1 h_2\}$ of the $\mathscr{U}(2)$ group. The IR of the $\mathscr{SU}(2)$ subgroup of $\mathscr{U}(2)$ is characterized by

$$f = \tfrac{1}{2}(h_1 - h_2), \tag{2.10}$$

to which the name pseudo spin was given.[16] Furthermore, the third component of the pseudo spin is, from (2.9b), the operator

$$\tfrac{1}{2}(C_{11} - C_{22}), \tag{2.11}$$

which, when applied to the state (2.7), gives the eigenvalue

$$\nu \equiv \tfrac{1}{2}[(2n_1 + l_1) - (2n_2 + l_2)]. \tag{2.12}$$

As from (2.6) we have

$$(2n_1 + l_1) + (2n_2 + l_2) = h_1 + h_2 \equiv \varrho, \tag{2.13}$$

we see that we could also characterize the state (2.7) by the ket

$$\begin{matrix} \mathscr{U}(6) & \mathscr{SU}(2) & O(2) & O(3) & O(2) & \\ |\varrho & f & \nu & \Omega & L & M\rangle, \end{matrix} \tag{2.14}$$

where over each quantum number, when relevant, we put the group for which it characterizes the IR. Note that there are two $O(2)$ subgroups present, the first with eigenvalue ν is the $O(2)$ subgroup of $\mathscr{SU}(2)$, while the second with eigenvalue M is the subgroup $O(2)$ of $O(3)$ which in turn is a subgroup of $\mathscr{U}(3)$.

We shall make use of the states (2.14) in a group theoretical derivation of the transformation brackets for harmonic oscillator states, brackets that have played an important role in the Nuclear Physics of the past decade.

3 TRANSFORMATION BRACKETS FOR THE TWO PARTICLE SYSTEM AND THEIR DEVELOPMENT IN TERMS OF IR OF $\mathcal{SU}(2)$ AND ISOSCALAR FACTORS OF $\mathcal{U}(3)$.

The two particle states (2.3) can be coupled to a total angular momentum L and projection M giving the ket

$$|n_1 l_1 n_2 l_2 LM\rangle \equiv [P_{n_1 l_1}(\boldsymbol{\eta}_1)\, P_{n_2 l_2}(\boldsymbol{\eta}_2)]_{LM}\, |0\rangle. \tag{3.1}$$

We can now carry out a unitary unimodular transformation $\mathcal{SU}(2)$ on the particle index $s = 1, 2$ of the creation operators, i.e.

$$\bar{\eta}_{is} = \sum_{s'=1}^{2} \mathcal{U}_{ss'} \eta_{is'} \quad s = 1, 2, \tag{3.2}$$

and ask what will be the unitary representation of this transformation for the states (3.1). Using the round bracket notation for the kets in which $\bar{\eta}_{is}$ replaces η_{ts}, i.e.

$$|n_1 l_1 n_2 l_2 LM) \equiv [P_{n_1 l_1}(\bar{\boldsymbol{\eta}}_1)\, P_{n_2 l_2}(\bar{\boldsymbol{\eta}}_2)]_{LM}\, |0\rangle, \tag{3.3}$$

where the ground state $|0\rangle$ still has the form (2.4), we see that the unitary transformation we wish to determine is related to the expansion of the states (3.3) in terms of the states (3.1), i.e. we must evaluate

$$\langle n'_1 l'_1 n'_2 l'_2 L | n''_1 l''_1 n''_2 l''_2 L) \equiv \langle n'_1 l'_1 n'_2 l'_2 L\, |U|\, n''_1 l''_1 n''_2 l''_2 L\rangle. \tag{3.4}$$

As both (3.1) and (3.3) are characterized by the IR of the same $O(3)$ and $O(2)$ groups, the transformation bracket (3.4) is diagonal in L, M and independent[17] of M.

We can now expand both bra and ket in (3.4) in terms of the states (2.14) characterized by the IR of the chain of groups

$$\mathcal{U}(6) \supset \mathcal{U}(3) \times \mathcal{U}(2), \quad \mathcal{U}(3) \supset O(3) \supset O(2), \quad \mathcal{U}(2) \supset O(2), \tag{3.5}$$

thus getting

$$\sum_{f'v'\Omega'} \sum_{f''v''\Omega''} \langle n'_1 l'_1 n'_2 l'_2 L | \varrho' f' v' \Omega' L\rangle$$
$$\times \langle \varrho' f' v' \Omega' L | \varrho'' f'' v'' \Omega'' L)\, (\varrho'' f'' v'' \Omega'' L | n''_1 l''_1 n''_2 l''_2 L), \tag{3.6}$$

where no summation over ϱ', ϱ'' appears as they are given by (2.13). The first and third brackets in (3.6) are nothing more than the isoscalar factors (2.6) (conjugate in the third bracket) of the $\mathcal{U}(3)$ group in the $\mathcal{U}(3) \supset O(3)$ chain, in a notation in which ϱ, f of (2.13), (2.10) replace h_1, h_2. The second bracket is the scalar product of a round ket in the chain (3.5) with an angular bra in the same chain. As the transformation (3.2) affects only the subgroup $\mathcal{SU}(2)$, clearly the second bracket is diagonal in ϱ, Ω and independent of them, while it is also diagonal in f, but depending on it and on v', v'' through the familiar[18] IR of the $\mathscr{SU}(2)$ group

$$\mathscr{D}^{f}_{\nu'\nu''}(\mathscr{U}^{-1}), \tag{3.7}$$

where $\mathscr{U}$ is the 2×2 matrix in (3.2). Thus we can write finally

$$\begin{aligned}\langle n_1'l_1'n_2'l_2'L|n_1''l_1''n_2''l_2''L\rangle &= \langle n_1'l_1'n_2'l_2'L|\mathscr{U}|n_1''l_1''n_2''l_2''L\rangle \\ &= \sum_{\Omega, f}\sum_{\nu', \nu''} \{\langle n_1'l_1'n_2'l_2'L|\varrho f\nu'\Omega L\rangle\, \mathscr{D}^{f}_{\nu'\nu''}(\mathscr{U}^{-1}) \\ &\quad \times \langle n_1''l_1''n_2''l_2''L|\varrho f\nu''\Omega L\rangle^*\},\end{aligned} \tag{3.8}$$

where

$$\varrho = 2n_1' + l_1' + 2n_2' + l_2' = 2n_1'' + l_1'' + 2n_2'' + l_2''. \tag{3.9}$$

If we restrict the unitary transformation (3.2) to a real orthogonal one of the form

$$\mathscr{U}_\beta = \begin{pmatrix} \cos\frac{1}{2}\beta & -\sin\frac{1}{2}\beta \\ \sin\frac{1}{2}\beta & \cos\frac{1}{2}\beta \end{pmatrix}, \tag{3.10}$$

this implies that the same transformation holds also for the coordinates $\mathbf{r}_s$ $s = 1, 2$, i.e.

$$\begin{pmatrix} \bar{\mathbf{r}}_1 \\ \bar{\mathbf{r}}_2 \end{pmatrix} = \mathscr{U}_\beta \begin{pmatrix} \mathbf{r}_1 \\ \mathbf{r}_2 \end{pmatrix}, \tag{3.11}$$

and thus (3.8) gives the unitary representation for the point transformation (3.11) if we replace the $\mathscr{U}^{-1}$ in the Wigner $\mathscr{D}$ function by $\mathscr{U}_\beta^{-1}$. If in particular $\beta = \pi/2$ the transformation (3.11) becomes

$$\bar{\mathbf{r}}_1 = \frac{1}{\sqrt{2}}(\mathbf{r}_1 - \mathbf{r}_2), \quad \bar{\mathbf{r}}_2 = \frac{1}{\sqrt{2}}(\mathbf{r}_1 + \mathbf{r}_2), \tag{3.12}$$

and it essentially takes us from the original coordinates to the relative and center of mass coordinates.

It is customary in the literature[19] to suppress $\mathscr{U}_\beta$ in (3.8) when $\beta = \pi/2$ and to denote the transformation bracket as just

$$\langle n_1''l_1''n_2''l_2''L|n_1'l_1'n_2'l_2'L\rangle = (-1)^{l_2'+l_2''}\langle n_1'l_1'n_2'l_2'L|n_1''l_1''n_2''l_2''L\rangle. \tag{3.13}$$

The quantum numbers are interchanged in the bracket (3.13) as compared to (3.8), because in the tabulated brackets one develops a two particle state in the coordinates $\mathbf{r}_1, \mathbf{r}_2$ in terms of that in $\bar{\mathbf{r}}_1, \bar{\mathbf{r}}_2$ of (3.11) and not inversely. If we wish to keep the single primed quantum numbers in the bra and the double primed in the ket, we need[19] the phase factor $(-1)^{l_2'+l_2''}$. Explicit expressions of (3.13) were obtained for $n_1' = n_2' = 0$ long ago[20] together with a recursion relation for all the other possible values of n_1', n_2'. They were used by Brody and Moshinsky[19] to tabulate the transformation brackets in all the range of interest in Nuclear Physics of the quantum numbers appearing in (3.13).

Later closed formulas for the transformation brackets for all cases have been derived by several authors,[21, 22] some of which can also be used effectively for computations. We wish to stress, in particular, that the matrix element

$$\langle n_1' l_1' n_2' l_2' L \,|U_\beta|\, n_1'' l_1'' n_2'' l_2'' L\rangle \tag{3.14}$$

for arbitrary β can be developed[21] in terms of the transformation brackets (3.13), that have been tabulated. For this purpose we only need to note that we can write $\mathscr{U}_\beta^{-1}$ of (3.10) as

$$\begin{pmatrix} \cos\frac{1}{2}\beta & \sin\frac{1}{2}\beta \\ -\sin\frac{1}{2}\beta & \cos\frac{1}{2}\beta \end{pmatrix} = \begin{pmatrix} 1 & 0 \\ 0 & -i \end{pmatrix} \begin{pmatrix} 0 & 1 \\ 1 & 0 \end{pmatrix} \begin{pmatrix} \frac{1}{\sqrt{2}} & \frac{1}{\sqrt{2}} \\ -\frac{1}{\sqrt{2}} & \frac{1}{\sqrt{2}} \end{pmatrix}$$

$$\times \begin{pmatrix} 0 & 1 \\ 1 & 0 \end{pmatrix} \begin{pmatrix} e^{-\frac{1}{2}i\beta} & 0 \\ 0 & e^{\frac{1}{2}i\beta} \end{pmatrix} \begin{pmatrix} \frac{1}{\sqrt{2}} & \frac{1}{\sqrt{2}} \\ -\frac{1}{\sqrt{2}} & \frac{1}{\sqrt{2}} \end{pmatrix} \begin{pmatrix} 1 & 0 \\ 0 & i \end{pmatrix}$$

$$= \mathbf{ABCDEFG}, \tag{3.15}$$

where the capital letters correspond to the matrices in the order indicated. We see at once that **C**, **F** lead to the transformation brackets of the type (3.13). The **A**, **E**, **G** matrices introduce only phase factors,[14] while $\mathbf{B} = \mathbf{D}$ interchange the operators $\boldsymbol{\eta}_1, \boldsymbol{\eta}_2$, which is equivalent essentially to exchanging the corresponding quantum numbers. Thus our development has been proved.

We proceed to discuss in the next section some of the most important applications of the transformation brackets (3.13) in Nuclear Physics.

4 APPLICATIONS OF TRANSFORMATION BRACKETS IN NUCLEAR PHYSICS: SHELL MODEL THEORY, HARTREE-FOCK CALCULATIONS, ETC.

In the shell model approach to nuclear structure calculations we need to determine the matrix elements of the model hamiltonian with respect to n nucleon states in a definite configuration. The usual procedure for these problems[12, 23] employs fractional parentage coefficients for the configurations involved to reduce the evaluations to one and two body matrix elements. With the help of $9j$ coefficients, the latter can be transformed into two body matrix elements in configuration space of a function $\mathscr{V}(\bar{\mathbf{r}}_1)$ of the relative coordinate $\bar{\mathbf{r}}_1$ defined in (3.12) i.e.

$$\langle \bar{n}_1' \bar{l}_1' \bar{n}_2' \bar{l}_2' \bar{L} | \mathscr{V}(\bar{\mathbf{r}}_1) | n_1' l_1' n_2' l_2' L \rangle . \tag{4.1}$$

The possibilities for the $\mathscr{V}(\bar{\mathbf{r}}_1)$ are limited to the following:[19] For central forces it is a function of the magnitude of $\bar{\mathbf{r}}_1$ only

$$\mathscr{V}(\bar{\mathbf{r}}_1) = V(\bar{r}_1). \tag{4.2a}$$

For two particle spin orbit interactions it takes the form

$$\mathscr{V}(\bar{\mathbf{r}}_1) = V(\bar{r}_1)\, l_{1q} = V(\bar{r}_1)\, (\bar{\mathbf{r}}_1 \times \bar{\mathbf{p}}_1)_q \quad q = 1, 0, -1, \tag{4.2b}$$

where $\mathbf{l}_1$ is the relative angular momentum of the two particles. For tensor forces

$$\mathscr{V}(\bar{\mathbf{r}}_1) = V(\bar{r}_1)\, Y_{2m}(\bar{\theta}_1, \bar{\varphi}_1) \quad m = 2, \ldots -2, \tag{4.2c}$$

where Y_{2m} is a spherical harmonic in the angles of the relative coordinate.

Using the transformation brackets (3.13), we can reduce the matrix elements to single particle ones and then the latter to Talmi integrals.[24] As this has been extensively discussed in the literature,[19] we only illustrate it for central forces for which

$$\begin{aligned}
&\langle \bar{n}_1' \bar{l}_1' \bar{n}_2' \bar{l}_2' L \,| V(\bar{r}_1) |\, n_1' l_1' n_2' l_2' L \rangle \\
&\quad = \sum_{\bar{n}_1'' n_1'' l_1'' n_2'' l_2''} \{ \langle \bar{n}_1'' l_1'' n_2'' l_2'' L | \bar{n}_1' \bar{l}_1' \bar{n}_2' \bar{l}_2' L \rangle \langle \bar{n}_1'' l_1'' \,\| V(\bar{r}_1) \|\, n_1'' l_1'' \rangle \\
&\qquad \times \langle n_1'' l_1'' n_2'' l_2'' L | n_1' l_1' n_2' l_2' L \rangle \} .
\end{aligned} \tag{4.3}$$

The single particle matrix element in (4.3) is given by

$$\begin{aligned}
\langle \bar{n} l \,\| V(r) \|\, n l \rangle &= \int_0^\infty R_{\bar{n}l}(r)\, V(r)\, R_{nl}(r)\, r^2 \, \mathrm{d}r \\
&= \sum_p B(\bar{n}l, nl, p)\, I_p ,
\end{aligned} \tag{4.4a}$$

with

$$I_p = \frac{2}{\Gamma(p + \frac{3}{2})} \int_0^\infty r^{2p+2}\, V(r)\, e^{-r^2} \, \mathrm{d}r . \tag{4.4b}$$

In (4.4) $R_{nl}(r)$ are the radial wave functions of the harmonic oscillator[14] that can be written as polynomials in r multiplied by $e^{-\frac{1}{2}r^2}$. Thus the integral in (4.4a) can be expressed as a linear combination of the Talmi integrals (4.4b), with coefficients that depend only on the properties of the radial functions and not on $V(r)$. These coefficients have also been tabulated by Brody and Moshinsky,[19] and thus all calculations of two body matrix elements are reduced to the evaluation of Talmi integrals. The same can be said of one body matrix elements when we take in consideration the development (4.4a).

Another important field of application of the transformation brackets concerns Hartree-Fock (HF) calculations. As is well known,[14] if the single particle states ψ_i we use in a HF approach can be expanded in terms of a given set of kets $|\alpha\rangle$

$$\psi_i = \sum_{\alpha} c_i^{\alpha} \, |\alpha\rangle, \tag{4.5}$$

the coefficients c_i^{α} satisfy the equations

$$\sum_{\gamma} \langle\alpha\, |H_0|\, \gamma\rangle\, c_i^{\gamma} + \sum_{\gamma}\sum_{j}\sum_{\beta,\delta} [c_j^{\beta *} \langle\alpha\beta\, |V|\, \gamma\delta\rangle\, c_j^{\delta}]\, c_i^{\gamma} = \varepsilon_i c_i^{\alpha}. \tag{4.6}$$

In (4.6) H_0 and V correspond respectively to the one and two body components of the hamiltonian. If the set of states $|\alpha\rangle$ is of the harmonic oscillator type for the configuration space part, the one body matrix elements can be reduced to Talmi integrals through (4.4a), while for the two body part we can couple the two particle states in the bra and ket to definite total orbital angular momentum, and then use (4.3) and (4.4) to again reduce the calculation to the evaluation of Talmi integrals. Extensive use of this approach has been made in many calculations.[25]

Many other applications of the transformation brackets and the expansion (4.4) have been given in the hundreds of papers that have made use of them. A summary of some of these applications can be found in ref.[11] and [14].

5 THE TRANSLATIONALLY INVARIANT THREE PARTICLE PROBLEM.

In Nuclear Physics the states should be translationally invariant, as we have only interactions between the nucleons themselves and not, as in atomic or molecular physics, interactions with heavier particles that can be assumed in a first approximation to be of infinite mass. Thus it will be useful if, at least for light nuclei, our trial wave functions are translationally invariant from the beginning. For the three nucleon problem this can be achieved if we introduce from the start, the Jacobi coordinates

$$\mathbf{r}_1' = \frac{1}{\sqrt{2}}(\mathbf{r}_1 - \mathbf{r}_2)$$

$$\mathbf{r}_2' = \frac{1}{\sqrt{6}}(\mathbf{r}_1 + \mathbf{r}_2 - 2\mathbf{r}_3) \tag{5.1}$$

$$\mathbf{r}_3' = \frac{1}{\sqrt{3}}(\mathbf{r}_1 + \mathbf{r}_2 + \mathbf{r}_3),$$

and make our trial wave function dependent only on $\mathbf{r}_1', \mathbf{r}_2'$.

We can now consider harmonic oscillator states depending on $\mathbf{r}_1', \mathbf{r}_2'$ of the form

$$|n_1'l_1'n_2'l_2'LM\rangle \equiv [\psi_{n_1'l_1'}(\mathbf{r}_1')\,\psi_{n_2'l_2'}(\mathbf{r}_2')]_{LM}$$
$$= [P_{n_1'l_1'}(\boldsymbol{\eta}_1')\,P_{n_2'l_2'}(\boldsymbol{\eta}_2')]_{LM}\,|0\rangle, \tag{5.2}$$

where $\psi_{nlm}(\mathbf{r})$ are the single particle oscillator wave functions (1.1) with $P_{nlm}(\boldsymbol{\eta})$ the corresponding polynomial in the creation operator. Note that $\boldsymbol{\eta}_1', \boldsymbol{\eta}_2', \boldsymbol{\eta}_3'$ are related with $\boldsymbol{\eta}_1, \boldsymbol{\eta}_2, \boldsymbol{\eta}_3$ by the same transformation (5.1) that we have between $\mathbf{r}_1', \mathbf{r}_2', \mathbf{r}_3'$ and $\mathbf{r}_1, \mathbf{r}_2, \mathbf{r}_3$. Furthermore, the ground state $|0\rangle$ is now

$$|0\rangle = \pi^{-3/2}\, e^{-\frac{1}{2}(r_1'^2 + r_2'^2)} = \pi^{-3/2}\, e^{-\frac{1}{2}[(r_1^2+r_2^2+r_3^2) - \frac{1}{3}(\mathbf{r}_1+\mathbf{r}_2+\mathbf{r}_3)^2]}, \tag{5.3}$$

and thus is symmetric under any permutation of the vectors $\mathbf{r}_s$ $s = 1, 2, 3$.

The first problem that we have to face in connection with the state (5.2) is that it does not correspond to a basis[14] for an IR of the symmetric group $S(3)$. This requirement is important if we want to combine (5.2) with appropriate spin-isospin wave functions, so that the resulting state satisfies the Pauli principle.

While we could use familiar projection techniques[14] to get from (5.2) the states corresponding to the IR $\{3\}$, $\{21\}$, $\{111\}$ of $S(3)$, a simpler procedure is to introduce new creation operators by the definition[14]

$$\begin{pmatrix} \boldsymbol{\eta}_1'' \\ \boldsymbol{\eta}_2'' \end{pmatrix} = \begin{pmatrix} -\frac{i}{\sqrt{2}} & \frac{1}{\sqrt{2}} \\ \frac{i}{\sqrt{2}} & \frac{1}{\sqrt{2}} \end{pmatrix} \begin{pmatrix} \boldsymbol{\eta}_1' \\ \boldsymbol{\eta}_2' \end{pmatrix}, \tag{5.4}$$

and then consider the ket

$$|n_1''l_1''n_2''l_2''LM) \equiv [P_{n_1''l_1''}(\boldsymbol{\eta}_1'')\,P_{n_2''l_2''}(\boldsymbol{\eta}_2'')]_{LM}\,|0\rangle, \tag{5.5}$$

where $|0\rangle$ continues to be given by (5.3).

The states (5.5) can be expanded in terms of the states (5.2) with coefficients given by

$$\langle n_1'l_1'n_2'l_2'L|n_1''l_1''n_2''l_2''L) \equiv \langle n_1'l_1'n_2'l_2'L\,|\mathscr{U}|\,n_1''l_1''n_2''l_2''L\rangle, \tag{5.6}$$

where $\mathscr{U}$ is the unitary representation associated with the canonical transformation (5.4), i.e.

$$\begin{pmatrix} -\dfrac{i}{\sqrt{2}} & \dfrac{1}{\sqrt{2}} \\ \dfrac{i}{\sqrt{2}} & \dfrac{1}{\sqrt{2}} \end{pmatrix} \equiv \begin{pmatrix} \dfrac{1}{\sqrt{2}} & \dfrac{1}{\sqrt{2}} \\ -\dfrac{1}{\sqrt{2}} & \dfrac{1}{\sqrt{2}} \end{pmatrix} \begin{pmatrix} -i & 0 \\ 0 & 1 \end{pmatrix}. \tag{5.7}$$

The effect of the last transformation in (5.7) is just to multiply the ket (5.2) by a factor $(-i)^{2n_1'+l_1'}$, while the first gives us the transformation bracket (3.13), but now from $\mathbf{r}_1, \mathbf{r}_2$ to $\bar{\mathbf{r}}_1, \bar{\mathbf{r}}_2$, thus obtaining

$$\langle n_1'l_1'n_2'l_2'L\,|\mathscr{U}|\,n_1''l_1''n_2''l_2''L\rangle = (-i)^{2n_1'+l_1'}\,\langle n_1'l_1'n_2'l_2'L|n'_1l'_1n'_2l'_2L\rangle. \tag{5.8}$$

The advantage of the kets (5.5) is that states corresponding to definite IR of $S(3)$ can be projected very simply from them.[14] In fact, we only need to discuss the congruence relation

$$2n_1'' + l_1'' - 2n_2'' - l_2'' \equiv \mu (\mathrm{mod}\ 3). \tag{5.9}$$

If $\mu = 0$ the states

$$\frac{1}{\sqrt{2}}\left[|n_1''l_1''n_2''l_2''LM\rangle \pm (-1)^{l_1''+l_2''-L}\,|n_2''l_2''n_1''l_1''LM\rangle\right] \tag{5.10}$$

will correspond to the partitions $\{3\}$ (symmetric) or $\{111\}$ (antisymmetric) depending on whether we use + or − in (5.10). If $\mu = 1$ or 2, the states (5.10) belong to the IR $\{21\}$ of $S(3)$ and to the Yamanouchi symbol (211), if we use the + sign or (121) (when multiplied by $(-1)^\mu$), if we use the − sign.[26]

Once we have translationally invariant states of definite partition $f = \{f_1f_2f_3\}$ and Yamanouchi symbol (r_3r_21) in configuration space, we can combine them with the corresponding states in spin isospin space to obtain wave functions that satisfy the Pauli principle and have definite total angular momentum J and isotopic spin T. It is also easy to obtain[26] the transformation brackets that relate these wave functions with states in which we separate the part depending on the relative coordinate, the spin and isospin of particles 1 and 2 from the rest. In this way we can reduce the evaluation of matrix elements of the hamiltonian to this part which, through (4.4) reduces again to the evaluation of Talmi integrals. Applications to three nucleon problems and to the form factor of the proton as a system of three quarks, have been given elsewhere.[14]

6 THE TRANSLATIONALLY INVARIANT FOUR PARTICLE PROBLEM. APPLICATIONS TO THE BINDING ENERGY, CORRELATION EFFECTS AND FORM FACTOR OF THE α PARTICLE.

As in the case of three particles discussed in the previous section, our trial wave functions will be translationally invariant if they depend only on the relative Jacobi vectors which for four particles take the form

$$\begin{aligned} \mathbf{r}'_1 &= \frac{1}{\sqrt{2}}(\mathbf{r}_1 - \mathbf{r}_2) \\ \mathbf{r}'_2 &= \frac{1}{\sqrt{6}}(\mathbf{r}_1 + \mathbf{r}_2 - 2\mathbf{r}_3) \\ \mathbf{r}'_3 &= \frac{1}{\sqrt{12}}(\mathbf{r}_1 + \mathbf{r}_2 + \mathbf{r}_3 - 3\mathbf{r}_4). \end{aligned} \tag{6.1 a}$$

The transformation from the original to the Jacobi coordinates is completed when we add to (6.1 a) the center of mass coordinate in the form

$$\mathbf{r}'_4 = \tfrac{1}{2}(\mathbf{r}_1 + \mathbf{r}_2 + \mathbf{r}_3 + \mathbf{r}_4). \tag{6.1 b}$$

We can now consider harmonic oscillator states depending on $\mathbf{r}'_1, \mathbf{r}'_2, \mathbf{r}'_3$ of the type

$$|n'_1 l'_1 n'_2 l'_2 (L') \, n'_3 l'_3 \Lambda M\rangle \equiv [[\varphi_{n_1' l_1'}(\mathbf{r}'_1)\, \varphi_{n_2' l_2'}(\mathbf{r}'_2)]_{L'}\, \varphi_{n_3' l_3'}(\mathbf{r}'_3)]_{\Lambda M}. \tag{6.2}$$

The states (6.2) form a complete set of translationally invariant four particle wave functions, but as in the case of the three particle problem, it is difficult to project from them states that correspond to a definite IR of $S(4)$, i.e. characterized by the partition $f = \{f_1 f_2 f_3 f_4\}$ and Yamanouchi symbol $(r_4 r_3 r_2 1)$. For the latter purpose, it is more convenient to introduce the symmetric translationally invariant coordinates

$$\begin{aligned} \mathbf{r}''_1 &= \tfrac{1}{2}(\mathbf{r}_1 + \mathbf{r}_4 - \mathbf{r}_2 - \mathbf{r}_3) \\ \mathbf{r}''_2 &= \tfrac{1}{2}(\mathbf{r}_2 + \mathbf{r}_4 - \mathbf{r}_1 - \mathbf{r}_3) \\ \mathbf{r}''_3 &= \tfrac{1}{2}(\mathbf{r}_3 + \mathbf{r}_4 - \mathbf{r}_1 - \mathbf{r}_2) \end{aligned} \tag{6.3}$$

and their corresponding harmonic oscillator states

$$|n_1''l_1''n_2''l_2''(L'')\,n_3''l_3''\Lambda M) \equiv [[\psi_{n_1''l_1''}(\mathbf{r}_1'')\,\psi_{n_2''l_2''}(\mathbf{r}_2'')]_{L''}\,\psi_{n_3''l_3''}(\mathbf{r}_3'')]_{\Lambda M}. \tag{6.4}$$

Now the $S(4)$ group can be expressed as a semidirect product[11, 27]

$$S(4) = D(2) \wedge \mathscr{S}(3), \tag{6.5}$$

where $D(2)$ has the elements

$$e, \quad d_1 \equiv (2,3)\,(1,4), \quad d_2 \equiv (1,3)\,(2,4), \quad d_3 \equiv (1,2)\,(3,4), \tag{6.6}$$

while $S(3)$, whose elements we designate by p, is the permutation group of the coordinates $\mathbf{r}_1, \mathbf{r}_2, \mathbf{r}_3$. It can be seen immediately that when $\mathscr{D}(2)$ acts on a vector whose three components are $\mathbf{r}_s''$ $s = 1, 2, 3$, its representation is given by

$$e = \begin{pmatrix} 1 & 0 & 0 \\ 0 & 1 & 0 \\ 0 & 0 & 1 \end{pmatrix}, \quad d_1 = \begin{pmatrix} 1 & 0 & 0 \\ 0 & -1 & 0 \\ 0 & 0 & -1 \end{pmatrix}, \quad d_2 = \begin{pmatrix} -1 & 0 & 0 \\ 0 & 1 & 0 \\ 0 & 0 & -1 \end{pmatrix},$$

$$d_3 = \begin{pmatrix} -1 & 0 & 0 \\ 0 & -1 & 0 \\ 0 & 0 & 1 \end{pmatrix}, \tag{6.7}$$

while the effect of the elements of $S(3)$ on the same vector is the same as they have on the one whose components are $\mathbf{r}_s$ $s = 1, 2, 3$. Thus the projection of states of definite f and r out of (6.4) reduces essentially to the same problem as the one we have to face for a general three particle case.[11, 27]

We shall illustrate our procedure for the important case of symmetric states, i.e. those characterized by the partition $\{4\}$, that are of even parity and total orbital angular momentum $\Lambda = 0$. We note that the latter restriction implies that $L'' = l_3''$ in (6.4) and so we can use the notation

$$|n_1''l_1'', n_2''l_2''; n_3''l_3'') \tag{6.8}$$

for these states. In a similar fashion we can define the states

$$|n_1''l_1''; n_2''l_2'', n_3''l_3'') \tag{6.9}$$

in which the last two wave functions are coupled to angular momentum $L'' = l_1''$ with the resulting total orbital angular momentum again being $\Lambda = 0$. The states (6.9) can be expanded in terms of (6.8) or viceversa with the help of Racah coefficients. It is easily seen though that because $\Lambda = 0$, there is only one term in the expansion, of coefficient unity, so that (6.8) and (6.9) are identical and we can use for them the generic notation

$$|n_1''l_1'', n_2''l_2'', n_3''l_3''). \tag{6.10}$$

Furthermore, if the states (6.10) have even parity, we see from (6.4) that this implies

$$(-1)^{l_1''+l_2''+l_3''} = 1 \tag{6.11}$$

and thus two l'' must be odd and one even, or all must be even.

We wish now to project out of (6.10) a symmetric state under the exchange of the coordinates of the four particles. This implies the application of the projection operator

$$\sum_{d,p} dp = (e + d_1 + d_2 + d_3) \sum_p p \tag{6.12}$$

to (6.10). We remember[29] that the application of the operator dp to a state implies applying $(dp)^{-1} = p^{-1}d^{-1}$ to the coordinates. Thus, as the inverses of the elements of $D(2)$ are the same elements,

$$\begin{aligned}(e + d_1 + d_2 + d_3)\,|n_1''l_1'', n_2''l_2'', n_3''l_3'') \\ = [1 + (-1)^{l_2''+l_3''} + (-1)^{l_1''+l_3''} + (-1)^{l_1''+l_2''}]\,|n_1''l_1'', n_2''l_2'', n_3''l_3''),\end{aligned} \tag{6.13}$$

and therefore the projection will give zero for even parity states (6.10), unless all three l_1'', l_2'', l_3'' are even. The application of the operators p to the states permutes the coordinates by p^{-1} or, equivalently,[27] the quantum numbers and thus the symmetric state of $\Lambda = 0$ and even parity is given by

$$\begin{aligned}|n_1''l_1'', n_2''l_2'', n_3''l_3'')_S \equiv \mathcal{N}[&|n_1''l_1'', n_2''l_2'', n_3''l_3'') + |n_1''l_1'', n_3''l_3'', n_2''l_2'') \\ &+ |n_2''l_2'', n_3''l_3'', n_1''l_1'') + |n_2''l_2'', n_1''l_1'', n_3''l_3'') + |n_3''l_3'', n_1''l_1'', n_2''l_2'') \\ &+ |n_3''l_3'', n_2''l_2'', n_1''l_1'')]\end{aligned} \tag{6.14}$$

with $\mathcal{N}$ being $1/\sqrt{6}$ if all three pairs $n_s''l_s''$ are distinct, $1/\sqrt{12}$ if any two pairs are equal, and $1/6$ if all pairs are equal.

So far we have only discussed how to construct states belonging to definite f and r out of the kets (6.4). If we want to calculate matrix elements of the hamiltonian with respect to these states, it is very convenient to express them in terms of Jacobi coordinates (6.1a), rather than of the symmetric coordinates (6.3), as the former include the relative coordinate $\frac{1}{\sqrt{2}}(\mathbf{r}_1 - \mathbf{r}_2)$ with respect to which we can calculate the matrix elements of two body operators using essentially (4.4). Thus we require the expansion of the states (6.4) in terms of (6.2). Restricting ourselves again to states for which $\Lambda = 0$, we need then to determine the transformation brackets

$$\langle n_1'l_1', n_2'l_2', n_3'l_3'|n_1''l_1'', n_2''l_2'', n_3''l_3''). \tag{6.15}$$

We note from (6.3) and (6.1a) that the Jacobi and symmetric vectors are related by

$$\begin{pmatrix} \mathbf{r}_1'' \\ \mathbf{r}_2'' \\ \mathbf{r}_3'' \end{pmatrix} = \mathbf{M} \begin{pmatrix} \mathbf{r}_1' \\ \mathbf{r}_2' \\ \mathbf{r}_3' \end{pmatrix}, \tag{6.16}$$

where $\mathbf{M}$ is the matrix

$$\mathbf{M} = \begin{pmatrix} \frac{1}{\sqrt{2}} & \frac{1}{\sqrt{6}} & -\frac{1}{\sqrt{3}} \\ -\frac{1}{\sqrt{2}} & \frac{1}{\sqrt{6}} & -\frac{1}{\sqrt{3}} \\ 0 & -\sqrt{\frac{2}{3}} & -\frac{1}{\sqrt{3}} \end{pmatrix} = \begin{pmatrix} \frac{1}{\sqrt{2}} & \frac{1}{\sqrt{2}} & 0 \\ -\frac{1}{\sqrt{2}} & \frac{1}{\sqrt{2}} & 0 \\ 0 & 0 & 1 \end{pmatrix}$$

$$\times \begin{pmatrix} 1 & 0 & 0 \\ 0 & \frac{1}{\sqrt{3}} & \sqrt{\frac{2}{3}} \\ 0 & -\sqrt{\frac{2}{3}} & \frac{1}{\sqrt{3}} \end{pmatrix} \begin{pmatrix} 1 & 0 & 0 \\ 0 & 1 & 0 \\ 0 & 0 & -1 \end{pmatrix} = \mathbf{M}_1 \mathbf{M}_2 \mathbf{M}_3, \tag{6.17}$$

and $\mathbf{M}_\alpha\ \alpha = 1, 2, 3$ are the three matrices in (6.17) in the order indicated. Now the effect of $\mathbf{M}_3$ on the ket $|n_1'l_1',\ n_2'l_2', n_3'l_3'\rangle$ is to multiply it by a phase factor $(-1)^{l_3'}$. For $\mathbf{M}_1$ we just have to think of the same ket in the form (6.8) to see that it gives rise to a transformation bracket (3.13). For $\mathbf{M}_2$ we have to consider that $|n_1''l_1'', n_2''l_2'', n_3''l_3'')$ can also be written in the form (6.9), to realize that it gives rise to the matrix element (3.8) of the $\mathscr{U}_\beta^{-1}$ of (3.10), when $\beta = 2 \text{ arcos } \frac{1}{\sqrt{3}} = 109.47°$. As we showed in turn at the end of section 3 that this matrix element can be expanded in terms of the transformation brackets (3.13), we see that the bracket (6.15), for the four particle problem, can be developed in terms of two particle transformation brackets that have been tabulated. Clearly a similar result holds for the general four particle transformation bracket for which $\Lambda \neq 0$, though in this case the recoupling of the angular momenta introduces also Racah coefficients in the development.

We shall now proceed to apply the above results to the discussion of some properties of the ground state of the α particle.

We begin with the determination of the binding energy and the ground state wave function[30] for the α particle when the interaction is given by the soft core potential of Tang,[31] whose parameters are presented in Table 1.

Our trial wave function is a linear combination of the states (6.14), that are symmetric, i.e. $f = \{4\}$, and of total orbital angular momentum $\Lambda = 0$ and positive parity. The spin-isospin part of the wave function then corresponds to the partition $\{1111\}$ and has $S = 0$, $T = 0$. Thus our state has $J = 0$. The states we consider here are all those of the type (6.14) for up to ten quanta, i.e.

$$2n_1'' + l_1'' + 2n_2'' + l_2'' + 2n_3'' + l_3'' \leqq 10. \tag{6.18}$$

There are altogether 37 states of this type.

Table of the parameters of the effective Tang potential

$$V(r_{12}) = \tfrac{1}{2} \sum_{s=1}^{6} A_s \, e^{-(r_{12}/\beta_s)^2}, \; A_s \text{ in MeV}, \; \beta_s \text{ in fm.}$$

SINGLET			TRIPLET		
A_1	A_2	A_3	A_4	A_5	A_6
880	−67.1	−21	1000	−143.4	−43
β_1	β_2	β_3	β_4	β_5	β_6
0.4385	1.27	1.6222	0.4303	1.1043	1.291

We calculated the matrix elements of the intrinsic hamiltonian

$$\begin{aligned}\mathscr{H}_I &= \sum_{s=1}^{A} \frac{p_s^2}{2m} + \sum_{s<t}^{A} V_{st} - \frac{1}{2mA}\left(\sum_{s=1}^{A} \mathbf{p}_s\right)^2 \\ &= \sum_{s<t}^{A} \left[\frac{1}{2mA}(\mathbf{p}_s - \mathbf{p}_t)^2 + V_{st}\right], \quad A = 4\end{aligned} \tag{6.19}$$

with respect to our states, noting that because the spin of the states is $S = 0$, the equivalent potential $\mathscr{V}_{st}$ for our problem is one of a Wigner type equal to the semi sum of the triplet and singlet parts of the original potential. These matrix elements depend on the frequency $\hbar\omega$ of the oscillator. The diagonalization of these matrices was carried for all states up to $\mathscr{N} = 0, 2, 4, \dots 10$ quanta. In Fig. 1 we draw first the ratio of the calculated to the experimental binding energy of the α particle as function of the parameter $\hbar\omega$. The curve marked $\mathscr{N} = 0$ indicates the result for the single gaussian wave function. That marked $\mathscr{N} = 10$ corresponds to the lowest eigenvalue of the 37×37 matrix of the 10 quanta approximation.

For the Tang potential the best fit at 0 quanta is achieved around $\hbar\omega \simeq$ 20 MeV. On the other hand, for $\mathscr{N} = 10$ the best fit comes around $\hbar\omega \simeq$ 65 MeV. Taking the optimum $\hbar\omega$ for $\mathscr{N} = 0$ and 10, we indicate in Fig. 2 how the approach to the experimental binding energy improves as we increase the number of quanta in the approximation. As we expect, the bind-

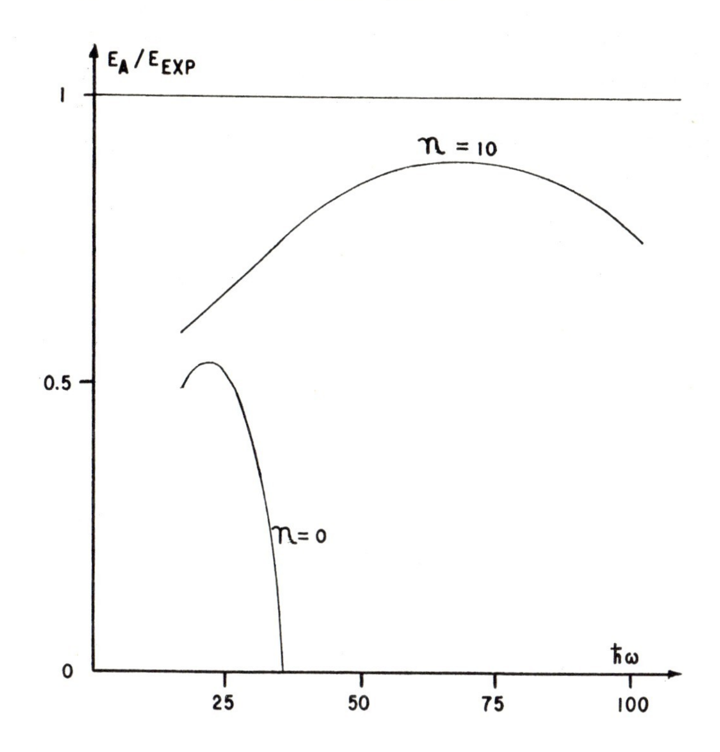

FIGURE 1 Ratio of the calculated (E_A) to the experimental (E_{exp}) binding energies of the α particle as a function of the parameter $\hbar\omega$. The curves for $\mathcal{N} = 0$ and $\mathcal{N} = 10$ quanta are presented

ing energy for $\hbar\omega = 20$ MeV improves very slowly while for $\hbar\omega = 65$ MeV (in which case there is no binding at all at 0 quanta) it reaches 90% of the experimental value at $\mathcal{N} = 10$.

Together with the binding energy we have, of course, the ground state as a linear combination of the harmonic oscillator states mentioned above. We now proceed to use this state to discuss the correlation effects in the α particle.

For this purpose we consider the expectation value, with respect to our ground state, of the operator

$$P = \frac{2}{A(A-1)} \sum_{s<t}^{A} \exp[-(\mathbf{r}_s - \mathbf{r}_t)^2/\beta^2], \quad A = 4. \tag{6.20}$$

Clearly this operator measures the probability that all pairs are at relative distances smaller than β, as otherwise the exponentials approach zero. The operator (6.20) takes the limiting value 1 when $\beta \to \infty$.

We now take for β the average (.4344 fm) of the range of the repulsive gaussians that correspond to the triplet and singlet soft cores in the Tang potential. The expectation value of (6.20) measures then the probability that the pairs of particles are at relative distance closer than the radius of the

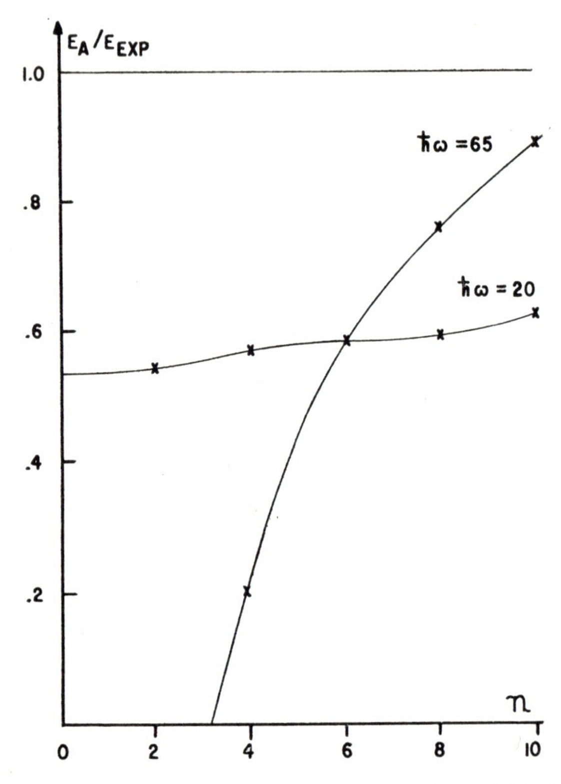

FIGURE 2 Ratio of the calculated (E_A) to the experimental (E_{exp}) binding energies of the α particle as a function of the maximum number of quanta $\mathscr{N}$ in the trial wave function. The values of $\hbar\omega = 20$ and 65 MeV are those that give the optimum binding energy for $\mathscr{N} = 0$ and 10 quanta respectively. Only the points marked with a cross on the curve are significant

core. The results are plotted in Fig. 3 for the ground states corresponding to $\mathscr{N} = 0, 2, 4 \ldots 10$ quanta and for the value $\hbar\omega = 65$ MeV. We see how the correlation between the nucleons increases as we increase the number of quanta, diminishing the probability that the particles be at relative distance smaller than the radius of the core. This probability becomes less than 1% at $\mathscr{N} = 10$. Similar results hold for other soft core potentials such as the Eikemeier-Hackenbroich one.[32]

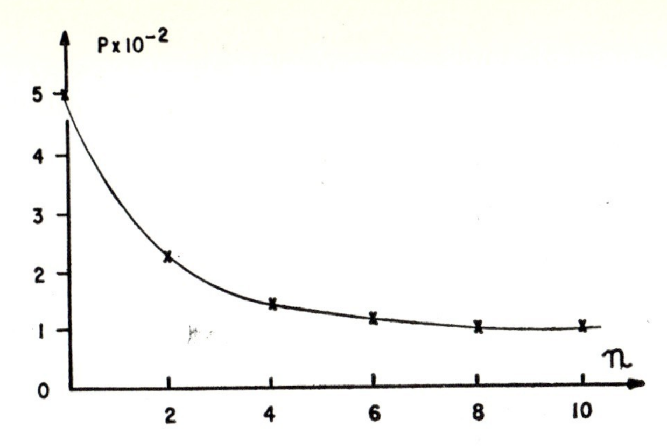

FIGURE 3 Behaviour of the expectation value of the operator P as a function of the number of quanta $\mathcal{N}$ in the trial wave function with $\hbar\omega =$ 65 MeV. Only the points marked with a cross on the curve are significant

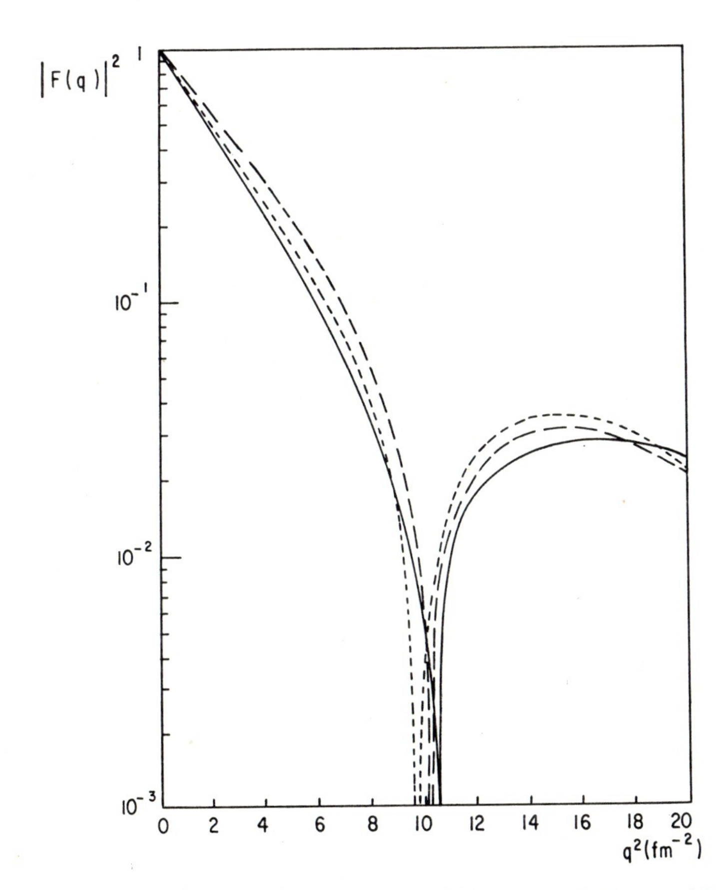

FIGURE 4 Comparison of the experimental form factor of the α particle (full line) with two theoretical calculations in which the wave functions are given by different combinations of the 0 and 4 quanta states

The ground state wave function can also be used to calculate the form factor of the α particle on which experimental information is available through the scattering of high energy electrons.[33] Unfortunately the 10 quanta ground state is rather complicated to apply to it the form factor operator.[14] We have therefore only discussed the effect on the form factor of adding to the zero quantum wave function (a gaussian for which the form factor in a logarithmic scale, as a function of the square of the momentum transfer q^2, appears as a straight line), states of either two or four quanta. The addition of two quanta states gives too low diffraction effects, while the addition of a particular four quanta state with an appropriate coefficient, can fit the observed diffraction effect very nicely, as seen in Fig. 4.

More elaborate calculations on the α particle, including its excited states, are in progress.

7 THE n PARTICLE PROBLEM IN THE THREE DIMENSIONAL HARMONIC OSCILLATOR.

a The Symmetry Group $\mathscr{U}(3n)$ and its Subgroups $\mathscr{U}(3) \times \mathscr{U}(n)$, $\mathscr{U}(3) \supset O(3) \supset O(2)$.

The discussions carried out in the previous sections were particular cases of the general problem to which we now turn, that of n particles in an harmonic oscillator potential. As we indicated in section 8 of chapter I, the symmetry group in this case is $\mathscr{U}(3n)$. Furthermore, an extension of the analysis presented in section 2 of this chapter indicates that our states can be characterized by the chains of groups

$$\mathscr{U}(3n) \supset \mathscr{U}(3) \times \mathscr{U}(n), \quad \mathscr{U}(3) \supset O(3) \supset O(2). \qquad (7.1\text{a, b})$$

The IR of $\mathscr{U}(3n)$ is characterized[15] by one non vanishing partition number $[\mathscr{N}\, 0^{3n-1}] \equiv [\mathscr{N}]$, where $\mathscr{N}$ is now the total number of quanta for the system. The IR of $\mathscr{U}(3)$ is $\{h_1 h_2 h_3\}$ with $h_1 \geqq h_2 \geqq h_3 \geqq 0$ while that of $O(3)$ is L and $O(2)$ is M. In this chain an extra quantum number Ω has to be introduced, as in the two particle case, to distinguish between repeated IR of $O(3)$ in a given IR of $\mathscr{U}(3)$. The IR of $\mathscr{U}(n)$ can be denoted by the partition

$$\{k_{ln}\} = \{k_{1n} \ldots k_{nn}\}, \qquad (7.2\text{a})$$

and because of the chain (7.1a) and the IR of $\mathscr{U}(3n)$ we have[15]

$$k_{ln} = h_l \quad l = 1, 2, 3, \quad k_{ln} = 0 \quad l > 3. \qquad (7.2\text{b})$$

The states of n particles are not completely characterized by the above IR, but require also the IR of subgroups of $\mathscr{U}(n)$. The mathematically simplest chain of subgroups of $\mathscr{U}(n)$ is the canonical one of the type

$$\mathscr{U}(n) \supset \mathscr{U}(n-1) \supset \cdots \supset \mathscr{U}(m) \supset \cdots \supset \mathscr{U}(1), \tag{7.3}$$

whose IR are $\{k_{lm}\}$ $1 \leqq l \leqq m \leqq n$. If we designate by $[k_{lm}]$ the full set of these partitions for $m = 1, 2, \ldots n$, our n particle state can be written as

$$|\{h_1 h_2 h_3\}\, \Omega LM, [k_{lm}]\rangle . \tag{7.4}$$

Explicit procedures exist for constructing the states (7.4) either from the application of the lowering operators to the highest weight state,[15] or from coupling[34] with Wigner coefficients of $\mathscr{S}\mathscr{U}(3)$. The states (7.4) have though the drawback that, insofar as $\mathscr{U}(n)$ and its subgroups are concerned, they are not physically relevant, as they do not include the permutation group $S(n)$ in the chain, the importance of which we discussed in the two previous sections for $n = 3$ and 4.

We then proceed to analyze other chains of subgroups in which this drawback is eliminated.

b Physically Relevant Subgroups of $\mathscr{U}(n)$: The Permutation Group $S(n)$, the Group of Diagonal Unitary Matrices $A(n)$, the Semi-Direct Product $K(n) \equiv A(n) \wedge S(n)$.

The group of all unitary matrices of dimension n has certainly as subgroup the permutation matrices of the same number of dimensions. It also has as a subgroup the one formed by all diagonal n dimensional unitary matrices. We shall designate by $\mathscr{S}(n)$ the former and by $A(n)$ the latter, and denote their corresponding elements by p and a.

We note now that

$$pap^{-1} \equiv \bar{a} \tag{7.5}$$

gives another diagonal unitary matrix whose elements are the same as those of "a", but permuted by p. Let us now define

$$\varkappa \equiv ap, \quad \varkappa' \equiv a'p', \tag{7.6}$$

where a, a' and p, p' are arbitrary elements of the corresponding subgroups. We have then that

$$\varkappa\varkappa' = apa'p' = apa'p^{-1}pp' = a\bar{a}'pp' \tag{7.7}$$

and thus $\varkappa\varkappa'$ is also of the form (7.6), so that then the set of matrices $\varkappa$ form a group we designate by $K(n)$ in honor of P. Kramer, who first proposed it. Furthermore, from (7.5) $A(n)$ is an invariant subgroup of $K(n)$, and as $A(n)$, $S(n)$ have only the unit element in common, we conclude that $K(n)$ is a semi-direct product of $A(n)$ and $S(n)$, i.e.

$$K(n) = A(n) \wedge S(n). \tag{7.8}$$

We proceed now to see the physical significance of the group $K(n)$ for a system of n particles in an harmonic oscillator potential.

c The chain of groups $\mathscr{U}(n) \supset K(n) \supset S(n)$ and the shell structure in the harmonic oscillator potential.

The states (7.4), for fixed $\{k_{ln}\} = \{h_1h_2h_3\}, \Omega, L, M$, form a basis for an IR of $\mathscr{U}(n)$. Thus they are also a basis for a, in general, reducible representation of its subgroups $K(n)$ and $S(n)$. We shall use this fact to get an idea of how the IR of $K(n)$ are characterized, as well as in what way we can get basis for IR of $K(n)$ and $S(n)$ from linear combinations of the states (7.4) with the above restrictions.

The states (7.4) have also a definite weight, i.e. they are characterized by the set of eigenvalues

$$(w_1 w_2 \ldots w_m \ldots w_n) \tag{7.9}$$

of the commuting operators

$$C_{11}, C_{22}, \ldots C_{mm}, \ldots C_{nn}, \tag{7.10}$$

defined in (2.9b). In fact, it can be easily seen[15] that

$$w_m = \sum_{l=1}^{m} k_{lm} - \sum_{l=1}^{m-1} k_{lm-1}. \tag{7.11}$$

We note now that the operators (7.10) transform into themselves under the action of any permutation p of $S(n)$. Furthermore, the operators C_{mm} are not changed under any of the operations of $A(n)$ as

$$\bar{C}_{mm} = \sum_i \bar{\eta}_{im} \bar{\xi}_{im} = \sum_i \eta_{im}\, e^{i\alpha_m} (e^{i\alpha_m})^* \,\xi_{im} = C_{mm}. \tag{7.12}$$

Therefore when applying a group element $\varkappa$ of $K(n)$ to the operators (7.10), we get the same operators in different order. This clearly indicates the convenience of classifying the set of all Gelfand states corresponding to a given IR $\{k_{ln}\}$ of $\mathscr{U}(n)$ into subsets characterized by their weight, where all weights that differ only by the permutation of the weight elements, will be considered equivalent. We could then introduce the concept of reference weight as one in which the weight elements are ordered by increasing magnitude. Thus the states characterized by IR of the groups $K(n) \supset S(n)$ can be formed from subsets of the sets of states (7.4), associated with a definite $\{k_{ln}\}$ where each subset is characterized by the reference weight

$$\mathbf{w} = (w_1^{n_1} w_2^{n_2} \dots w_p^{n_p}) \quad w_1 < w_2 < \cdots < w_p, \tag{7.13}$$

where n_i, $i = 1, 2, \dots p$, is the number of weight components equal to w_i and, of course,

$$n_1 + n_2 + \cdots + n_p = n. \tag{7.14}$$

We can now define the group of the weight or little group W, as the subgroup of $S(n)$ that leaves the reference weight invariant. From (7.13) this group is

$$W = S(n_1) \oplus S(n_2) \oplus \cdots \oplus S(n_p), \tag{7.15}$$

where $S(n_1)$, $S(n_2)$, ... are symmetric groups for particles $1, \dots n_1$ of weight component w_1, particles $n_1 + 1, \dots n_1 + n_2$ of weight component w_2, etc. The IR of $S(n_i)$ $i = 1, \dots p$ are characterized by the partitions

$$f^i \equiv \{f_1^i f_2^i \dots f_{n_i}^i\}, \tag{7.16}$$

so the IR f_w of the group W is given by

$$f_w = f^1 f^2 \dots f^p. \tag{7.17}$$

It was shown elsewhere[15] that the IR of $K(n)$ are fully characterized by both the reference weight and the IR of the group W, i.e. by

$$(\mathbf{w}, f_w). \tag{7.18}$$

In turn, the IR of the chain of groups $S(n) \supset S(n-1) \supset \cdots \supset \mathscr{S}(1)$ is given by the partition

$$f = \{f_1 f_2 \dots f_n\} \tag{7.19}$$

and the Yamanouchi symbol

$$r = (r_n r_{n-1} \dots r_2\, 1). \tag{7.20}$$

Thus the states characterized by the IR of the chain of groups (7.1 a, b) and

$$\mathscr{U}(n) \supset K(n) \supset S(n) \supset S(n-1) \supset \cdots \supset S(1) \tag{7.21}$$

are given by the kets

$$|\{h_1 h_2 h_3\}\, \Omega L M, \{k_{ln}\}\, \chi(\mathbf{w}, f_w)\, \varphi f r\rangle \tag{7.22}$$

where χ represents the set of quantum numbers that distinguish between the repeated IR of $K(n)$ in a given IR of $\mathscr{U}(n)$, while φ plays the same role for the repeated IR of $S(n)$ in a given IR of $K(n)$.

The states (7.22) will be the physically significant ones for the system of n particles in an harmonic oscillator potential, as we can combine them with the corresponding spin isospin part of the associate partition $\tilde{f}$, so as to get states that satisfy the Pauli principle.

We note also that the exponents $n_1, \ldots n_p$ in the reference weight give the number of particles in the levels of the harmonic oscillator at number of quanta $w_1, \ldots w_p$, and thus indicate the shell structure of the states. The calculation of one and two body matrix elements with respect to these states requires the determination of the corresponding fractional parentage coefficients. We shall briefly discuss this problem in the next section.

8 FACTORIZATION OF THE FRACTIONAL PARENTAGE COEFFICIENTS FOR A SYSTEM OF n PARTICLES IN A SINGLE LEVEL OF THE HARMONIC OSCILLATOR POTENTIAL.

If we consider a system of n particles in a single level of w quanta of the harmonic oscillator potential, the reference weight becomes

$$\mathbf{w} = (w^n), \tag{8.1}$$

and the group of the weight W coincides with $S(n)$. Thus the state (7.22) can be written as

$$|\{h_1h_2h_3\}\,\Omega LM, \chi(w^n, f)\, r\rangle \tag{8.2}$$

where the IR of W is now just the partition

$$f = \{f_1 \ldots f_n\}, \quad f_1 \geqq f_2 \geqq \cdots \geqq f_n \geqq 0, \quad f_1 + f_2 + \cdots + f_n = n, \tag{8.3a}$$

associated with $S(n)$, so that φ, f in (7.22) become redundant and therefore were suppressed in (8.2). We also suppress $\{k_{ln}\}$ due to the relation (7.2b). We note[35] that in the state (8.2)

$$h_1 + h_2 + h_3 = nw. \tag{8.3b}$$

The fractional parentage coefficient (*fpc*) appears when we develop the state (8.2) as

$$\begin{aligned}
&|\{h_1h_2h_3\}\,\Omega LM, \chi(w^n, f)\, r\rangle \\
&\quad = \sum_{\{h_1'h_2'h_3'\}\Omega'L'l} [|\{h_1'h_2'h_3'\}\,\Omega'L', \chi'(w^{n-1}, f')\, r'\rangle\, |\{w\}\, l\rangle]_{LM} \\
&\quad\quad \times \langle\{h_1'h_2'h_3'\}\,\Omega'L'\chi'(w^{n-1}, f');\, \{w\}\, l\, |\}\, \{h_1h_2h_3\}\,\Omega L\chi(w^n, f)\rangle,
\end{aligned} \tag{8.4}$$

where the first ket on the right hand side represents the state of $n - 1$ particles with the quantum numbers having the same meaning as in (8.2); the second ket represents the state associated with the particle n that corresponds to the IR $\{w\}$ of $\mathscr{U}(3)$ and angular momentum l. The f', r' in the first ket are obtained from f and r when we suppress the box marked n in the Young tableaux. The two kets on the right hand side are coupled to total angular momentum L and projection M. The last factor in (8.4) is then the *fpc* for the n particle states in a single level of the harmonic oscillator potential. It is independent of r', r, as the IR of $S(n-2) \supset \cdots \supset S(1)$ are common to both sides of (8.4), and r can be obtained from the knowledge of r' and f.

The dependence on Ω', L', l and Ω, L of the *fpc* is clearly given by an isoscalar factor of $\mathscr{U}(3)$, as could be seen immediately if we had coupled the two kets on the right hand side of (8.4) to a total IR $\{h_1h_2h_3\}\ \Omega L$ of $\mathscr{U}(3) \supset O(3)$. Thus the *fpc* can be factorized as

$$\langle\{h'_1h'_2h'_3\}\ \Omega'L'\chi'(w^{n-1},f');\ \{w\}\ l|\}\ \{h_1h_2h_3\}\ \Omega L\chi(w^n,f)\rangle$$

$$= \langle\{h'_1h'_2h'_3\}\ \Omega'L',\ \{w\}\ l|\}\ \{h_1h_2h_3\}\ \Omega L\rangle$$

$$\times \langle\{f'_1 \ldots f'_{n-1}\}\ \chi'\{h'_1h'_2h'_3\},\ \{1\}\ \{w\}|\{f_1 \ldots f_n\}\ \chi\{h_1h_2h_3\}\rangle. \quad (8.5)$$

The first term on the right hand side is the isoscalar factor of $\mathscr{U}(3)$. The second is the remaining term which has been rewritten in a form that is familiar from the second quantized picture. To understand it better let us recall that for the level of w quanta in the harmonic oscillator, the number of single particle states is

$$d = \tfrac{1}{2}(w+1)(w+2). \quad (8.6)$$

In the second quantized picture we start with the unitary group $\mathscr{U}(d)$ and consider its subgroup

$$\mathscr{U}(d) \supset \mathscr{D}^{\{w\}}(\mathscr{U}(3)). \quad (8.7)$$

The IR of $\mathscr{U}(d)$ are characterized by the same partition[35] $f = \{f_1 \ldots f_n\}$ that gives the properties of the state (8.2) with respect to the permutation group $S(n)$. The last factor in (8.5) is then the isoscalar factor for the direct product of IR $\{f'_1 \ldots f'_{n-1}\} \otimes \{1\} \to \{f_1 \ldots f_n\}$ of the chain (8.7). This factor has some interesting properties, discussed in detail in reference 34, which we quote here. If $\{h_1h_2h_3\}$ is just the single row IR $\{h\}$, then $\{h'_1h'_2h'_3\}$ is also a single row IR $\{h'\}$ and this implies[34] $\{f'\} = \{n-1\}$, $\{f\} = \{n\}$. The last term in (8.5) reduces then to unity. If the IR of $\mathscr{U}(3)$ has only two rows $\{h_1h_2\}$, this implies that $h'_3 = 0$ and thus from (8.3b)

$$h_1 + h_2 = nw, \quad h'_1 + h'_2 = (n-1)\,w. \quad (8.8)$$

Defining now

$$t_n \equiv \tfrac{1}{2}(h_1 - h_2), \quad t_{n-1} \equiv \tfrac{1}{2}(h_1' - h_2'), \quad t \equiv \tfrac{1}{2}w, \tag{8.9}$$

we can, in the two rowed case, rewrite the last term in (8.5) as

$$\langle\{f_1' \dots f_{n-1}'\}\,\chi' t_{n-1}, \{1\}\, t | \{f_1 \dots f_n\}\,\chi t_n\rangle. \tag{8.10}$$

This is more than a change in notation, as it was proved in reference 34 that if we have a two rowed representation of $\mathscr{U}(3)$, the chain (8.7) can be interpreted as

$$\mathscr{U}(2t + 1) \supset \mathscr{D}^t(\mathscr{U}(2)), \tag{8.11}$$

and thus the isoscalar factor (8.10) is nothing more than the familiar[12] *fpc* in a shell of angular momentum t. For $w = 2$, i.e. the $2s - 1d$ shell of the harmonic oscillator, $t = 1$ and (8.10) are the *fpc* for the p shell. The IR of $S(n)$ are restricted for the two rowed IR of $\mathscr{U}(3)$ to, at most, $2t + 1$ rows. When the IR of $\mathscr{U}(3)$ has three rows, there is no simplification in the isoscalar factors of the chain (8.7), but many techniques have become available for evaluating them.[36]

The *fpc* (8.5), and the corresponding ones in which two particles are separated, have been used in calculations of the matrix elements of one and two body operators in the $2s - 1d$ shell for realistic interactions.[37]

9 THE FORM FACTOR FOR A SYSTEM OF PARTICLES IN AN HARMONIC OSCILLATOR POTENTIAL INTERACTING THROUGH A REPULSIVE HARMONIC OSCILLATOR FORCE. IMPLICATIONS FOR THE HARD CORE EFFECT IN THE FORM FACTORS OF NUCLEI.

If one assumes a shell model description for nuclei it is quite easy to calculate their form factors. The electron scattering experiments permit us to measure these form factors and they show significant differences[38] from the values predicted by the shell theory. An effort has been made to explain these discrepancies through the correlation effect introduced by the hard core, which cannot be incorporated into the shell model common potential.

It seems interesting to explore, within the context of an exactly solvable problem, what will be the effect of a repulsive interaction on the form factor of a system of particles in a common attractive potential. We could then expect this model problem to provide us with insight on the effect of the repulsive core on the form factor of nuclei.

The model is one that was proposed recently under the name of pseudo-atom.[39] It consists in replacing in the atomic hamiltonian the Coulomb interactions, both attractive and repulsive, with harmonic oscillator potentials with the appropriate signs, i.e.

$$-\frac{1}{r} \to \frac{1}{2} r^2, \qquad \frac{1}{|\mathbf{r}_1 - \mathbf{r}_2|} \to -\frac{1}{2} (\mathbf{r}_1 - \mathbf{r}_2)^2. \tag{9.1}$$

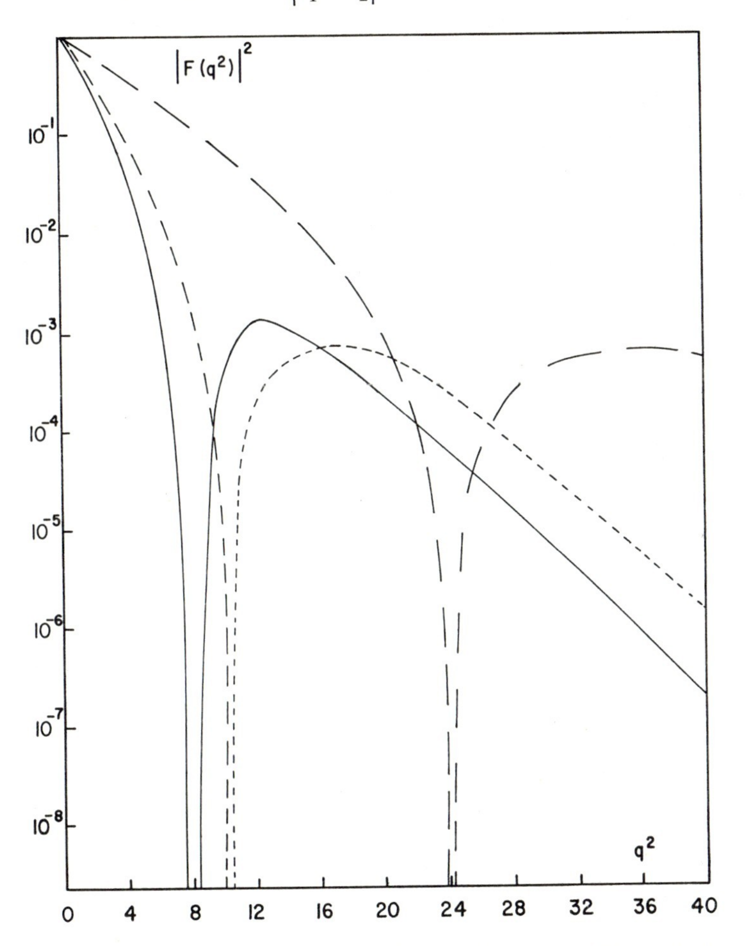

FIGURE 5 Form factor for the ground state of a system of 8 particles in an oscillator potential interacting through a repulsive oscillator force. The full line corresponds to the exact solution. The short dashed line to the Hartree-Fock approximation. The long dashed line to the particles without interaction in the configuration $(1s)^2 (1p)^6$. The shift to the lower momentum transfer when we introduce the interaction is also to be expected for the effect of the repulsive hard core on the shell model configuration of a given nucleus

Using some of the symmetry properties of harmonic oscillator states discussed above, we showed that the problem is exactly soluble, as well as in a Hartree Fock approximation (HFA) and in the case when we suppress the repulsive interaction. We have in particular discussed a problem of 8 particles of spin $\frac{1}{2}$ that fill the shell $(1s)^2\,(1p)^6$ of the harmonic oscillator. Using the exact and approximate states, we have calculated the form factor as a function of the square of the momentum transfer. The square of this form factor, in a logarithmic scale, is plotted in Fig. 5. The full line corresponds to the exact form factor. The dashed line is the form factor for the state in the Hartree-Fock approximation (HFA), while the dotted line corresponds to the state in which we suppressed the interaction. We see that the corlations due to the interaction displace the curve of $|F(q^2)|^2$ to the left of the value it has in the HFA, and even more so if we compare it to the solution where there was no interaction. This is to be expected as the interaction is repulsive and so the charge density should spread out when it is introduced, which in turn implies that the form factor (which is the Fourier transform of the charge density), should contract in, as observed.

The effect of the repulsive core on the form factor of nuclei predicted by the shell model should also be in the direction of displacing it to small values of the square of the momentum transfer. This has been observed in some of the calculations that have tried to incorporate this effect.[38]

CHAPTER III

THE DYNAMICAL GROUP OF THE HARMONIC OSCILLATOR AND ITS APPLICATIONS.

IN SECTION 8 OF CHAPTER I we saw that the $2N$ dimensional real symplectic group $Sp(2N)$ is the dynamical, and its N dimensional unitary subgroup $\mathscr{U}(N)$ is the symmetry group, of the N dimensional harmonic oscillator.

In chapter II we made extensive use of the symmetry group in the characterization, transformation and application of harmonic oscillator states with arbitrary number of particles. In the present chapter we wish to make similar use of the dynamical group itself. When discussing the single particle in one dimension, we saw that the elements of $Sp(2)$ could be expressed as products of rotations in phase space and dilatations. As the former become unitary transformations when we deal with creation and annihilation operators, we see that the elements of $Sp(2)$ are products of those of $\mathscr{U}(1)$ with dilatations. A similar result will hold for the examples we wish to discuss in $2N$ dimensional phase space, i.e. the elements of $Sp(2N)$ whose unitary representation we wish to consider will be products of elements of $\mathscr{U}(N)$ and generalized dilatations. Thus we will be mainly concerned in this chapter with the effects of dilatations as superimposed on the unitary transformations already analyzed in the previous chapter.

We shall start our discussion with the isotropic dilatation effect in the three dimensional harmonic oscillator and then extend it to the anisotropic case. Afterwards we shall discuss the solution of a problem of two particles in an oscillator potential interacting through an oscillator force, through the use of unitary and dilatation canonical transformations. Thus we will try to increase our understanding of the group theory involved in the solution of many body problems.

1 ISOTROPIC DILATATION OF THE THREE DIMENSIONAL HARMONIC OSCILLATOR.

In an isotropic dilatation the canonical transformation, which is actually a point transformation, is given by

$$\bar{x}_i = \sqrt{\omega}\, x_i, \quad \bar{p}_i = \frac{1}{\sqrt{\omega}} p_i, \quad i = 1, 2, 3. \tag{1.1}$$

The analysis of section 2 of chapter I indicates that we need to determine the normalized transformation bracket that satisfies the equations

$$\sqrt{\omega}\, x_i' \langle \mathbf{x}' | \bar{\mathbf{x}}') = \bar{x}_i' \langle \mathbf{x}' | \bar{\mathbf{x}}'), \tag{1.2}$$

where, whenever something is characterized by the eigenvalues of *all* coordinates, we suppress the index i as in $\langle \mathbf{x}' | \bar{\mathbf{x}}')$ but replace the letters by bold face ones. The solution of (1.2) is then given by

$$\langle \mathbf{x}' | \bar{\mathbf{x}}') = \omega^{3/4} \prod_{i=1}^{3} \delta(\bar{x}_i' - \sqrt{\omega}\, x_i') = \omega^{3/4}\, \delta(\bar{\mathbf{x}}' - \sqrt{\omega}\, \mathbf{x}'), \tag{1.3}$$

and in the notation of (I.2.12) it can be written as

$$\langle \mathbf{x}' | U | \mathbf{x}'' \rangle = \omega^{3/4}\, \delta(\mathbf{x}'' - \sqrt{\omega}\, \mathbf{x}'). \tag{1.4}$$

Thus we have the matrix elements for the unitary representation of the canonical transformation (1.1) in a basis in which the coordinates x_i are diagonal.

We can consider also basis in which the hamiltonian H of the harmonic oscillator, the angular momentum L^2 and its projection L_z are diagonal. This gives us the transformation bracket

$$\langle \mathbf{x}' | n'l'm' \rangle \equiv \frac{1}{r'} u_{n'l'}(r')\, Y_{l'm'}(\theta', \varphi'), \tag{1.5}$$

in which Y_{lm} is a spherical harmonic and the radial function is given by

$$u_{nl}(r) = A_{nl}\, r^{l+1}\, e^{-\frac{1}{2}r^2}\, L_n^{l+\frac{1}{2}}(r^2), \tag{1.6}$$

where $L_n^{l+\frac{1}{2}}$ is the associated Laguerre[40] polynomial and A_{nl} the normalization factor

$$A_{nl} = \left[\frac{2n!}{[\Gamma(n + l + \frac{3}{2})]^3} \right]^{\frac{1}{2}}. \tag{1.7}$$

The matrix elements of U in this new basis are given by

$$\langle n'l'm' | U | n''l''m'' \rangle$$
$$= \iint \langle n'l'm' | \mathbf{x}' \rangle\, d\mathbf{x}' \langle \mathbf{x}' | U | \mathbf{x}'' \rangle\, d\mathbf{x}'' \langle \mathbf{x}'' | n''l''m'' \rangle. \tag{1.8}$$

As from (1.4) we can also write

$$\langle \mathbf{x}' | U | \mathbf{x}'' \rangle = \omega^{1/4} \frac{1}{r'r'' \sin\theta'} \delta(r'' - \sqrt{\omega} r')\, \delta(\theta'' - \theta')\, \delta(\varphi'' - \varphi'), \tag{1.9}$$

we see that the matrix elements (1.8) are diagonal in l, m and independent of the latter, and are given by the radial double integral

$$\langle n'l|U|n''l\rangle = \int_0^\infty \int_0^\infty u_{n'l}(r')\,\omega^{1/4}\,\delta(r'' - \sqrt{\omega}r')\,u_{n''l}(r'')\,\mathrm{d}r'\,\mathrm{d}r''. \quad (1.10)$$

To evaluate this integral, a convenient procedure is to use the generating function of the Laguerre polynomials

$$\frac{\exp\left[-\dfrac{tz}{1-t}\right]}{(1-t)^{a+1}} = \sum_{n=0}^{\infty} \frac{t^n}{\Gamma(n+a+1)} L_n^a(z), \quad (1.11)$$

to write

$$\sum_{n',n''=0}^{\infty} \frac{t'^{n'}\,t''^{n''}\,\langle n'l|U|n''l\rangle}{\Gamma(n'+l+\frac{3}{2})\,\Gamma(n''+l+\frac{3}{2})\,A_{n'l}A_{n''l}}$$

$$= \int_0^\infty \int_0^\infty (r')^{l+1}\,e^{-\frac{1}{2}r'^2}\,\frac{\exp\left[-\dfrac{t'}{1-t'}r'^2\right]}{(1-t')^{l+\frac{3}{2}}}\,\omega^{1/4}\,\delta(r'' - \sqrt{\omega}\,r')$$

$$\times (r'')^{l+1}\,e^{-\frac{1}{2}r''^2}\,\frac{\exp\left[-\dfrac{t''}{1-t''}r''^2\right]}{(1-t'')^{l+\frac{3}{2}}}\,\mathrm{d}r'\,\mathrm{d}r''$$

$$= \omega^{\frac{2l+3}{4}}\,[(1-t')(1-t'')]^{-\left(l+\frac{3}{2}\right)}\left[\frac{1+\omega}{2} + \frac{t'}{1-t'} + \frac{\omega t''}{1-t''}\right]^{-\left(l+\frac{3}{2}\right)}$$

$$\times \frac{(2l+1)!!}{2^{l+2}}\sqrt{\pi}. \quad (1.12)$$

If we now expand the right hand side in powers of t', t'', and compare with the left hand side, we obtain finally

$$\langle n'l|U|n''l\rangle$$

$$= \left[\frac{2\pi\,n'!\,n''!}{\Gamma(n'+l+\frac{3}{2})\,\Gamma(n''+l+\frac{3}{2})}\right]^{\frac{1}{2}} \sum_{s=0}^{n'} \sum_{t=0}^{n''} (-1)^{n'+n''-s-t}$$

$$\times \frac{(2n'+2n''+2l-2s-2t+1)!!\,(2n'+2l+1)!!\,(2n''+2l+1)!!}{(n'-s)!\,(n''-t)!\,(2s)!!\,(2t)!!\,(2n'+2l-2s+1)!!\,(2n''+2l-2t+1)!!}$$

$$\times\,\omega^{\frac{2n''+l+3/2}{2}-t}\,(1+\omega)^{-\frac{(2n'+l+3/2)+(2n''+l+3/2)}{2}+s+t}. \quad (1.13)$$

We have thus determined the transformation matrix for the expansion of dilatated three dimensional harmonic oscillator states in terms of the original states of the same l and m.

2 ANISOTROPIC DILATATION OF THE THREE DIMENSIONAL HARMONIC OSCILLATOR.

In an anisotropic dilatation the canonical transformation is given by

$$\bar{x}_i = \sqrt{\omega_i}\, x_i, \quad \bar{p}_i = \frac{1}{\sqrt{\omega_i}} p_i \quad i = 1, 2, 3. \tag{2.1}$$

The unitary representation of this transformation in a basis in which the coordinates are diagonal, is then

$$\langle \mathbf{x}' | U | \mathbf{x}'' \rangle = \prod_{i=1}^{3} \omega_i^{1/4}\, \delta(x_i'' - \sqrt{\omega_i}\, x_i'). \tag{2.2}$$

For any other basis, we require, as in (1.8), the transformation brackets between it and the coordinate basis.

If we now wish to have the unitary representation in a basis in which the harmonic oscillator hamiltonians $H_i = \frac{1}{2}(p_i^2 + x_i^2)$, $i = 1, 2, 3$, are diagonal, we see immediately that it becomes a product of three matrix elements of the form (I.5.11). More frequently we wish to expand the dilatated states in terms of oscillator states of frequency 1, but in spherical coordinates, i.e. characterized by the eigenvalues of $H = \frac{1}{2}(p^2 + r^2)$, L^2, L_z. This we could achieve in two steps: First we express U in a basis in which the H_i are diagonal, for which we have the explicit expression we just indicated. Then we make use of the transformation brackets

$$\langle nlm | n_1 n_2 n_3 \rangle, \tag{2.3}$$

applied to the bra only of the U matrix to pass from the H_1, H_2, H_3 to the H, L^2, L_z basis.

The simplest way to obtain the brackets (2.3) is by realizing that they are from (I.6.1), (II.1.7), the scalar product of the states

$$\frac{\eta_1^{n_1}\, \eta_2^{n_2}\, \eta_3^{n_3}}{\sqrt{n_1!\, n_2!\, n_3!}} |0\rangle, \tag{2.4 a}$$

$$A_{nl}\, (\boldsymbol{\eta}|\boldsymbol{\eta})^n\, Y_{lm}\, (\boldsymbol{\eta}) |0\rangle. \tag{2.4 b}$$

As the ground state $|0\rangle$ is the same in both cases, it is sufficient to expand the polynomial in the creation operators in (2.4b) in terms of the monomials

(2.4a). This was done by Chacón and de Llano,[41] who gave explicit expressions for the transformation brackets (2.3).

Thus one can obtain the coefficients of the expansion of an anisotropically dilatated oscillator in terms of the eigenstates of H, L^2, L_z for unit frequency. A similar reasoning also applies if we have an axially symmetrical dilatation i.e.

$$\omega_1 = \omega_2 \neq \omega_3, \tag{2.5}$$

in which case, as for example in the Nilsson model,[42] it is convenient to consider the dilatated states in cylindrical coordinates, i.e. given by the kets

$$|n_\varrho n_3 \Lambda), \tag{2.6}$$

associated with

$$H_\varrho = \tfrac{1}{2}[p_1^2 + p_2^2 + \omega_1^2(x_1^2 + x_2^2)], \quad H_3 = \tfrac{1}{2}(p_3^2 + \omega_3^2 x_3^2),$$
$$L_z = x_1 p_2 - x_2 p_1. \tag{2.7}$$

The development of the states (2.6) in terms of $|nlm\rangle$ can be achieved by a reasoning similar to the above.

Explicit expressions of the unitary transformations for all cases of anisotropic dilatation will be published elsewhere.

3 SOLUTION OF A SIMPLE MANY BODY PROBLEM THROUGH THE UNITARY REPRESENTATION OF ITS DYNAMICAL GROUP.

So far in this chapter we have used the dynamical group only in connection with dilatation effects. Possibly its most interesting application is related though with the insight it provides for the solution of many body problems. We proceed to present a simple many body problem which we then solve with the help of unitary representations of the dynamical group.

a The One Dimensional Two Particle Pseudo-Atom.

If we disregard relativistic effects an atom, or more generally an ion of charge Z, is a system of n fermions of spin $\frac{1}{2}$ moving in an attractive Coulomb potential of strength Z and interacting through a repulsive Coulomb potential of strength 1. We define a pseudo-atom[39] as the corresponding problem in which harmonic oscillator potentials of the appropriate signs replace the Coulomb ones. Thus a two particle one dimensional pseudo-atom has the hamiltonian

$$H = \tfrac{1}{2}(p_1^2 + Zx_1^2) + \tfrac{1}{2}(p_2^2 + Zx_2^2) - \tfrac{1}{2}(x_1 - x_2)^2 \tag{3.1}$$
$$\equiv H_0' - \tfrac{1}{2}(x_1 - x_2)^2.$$

We can carry out a scale transformation for the coordinates in (3.1) of the type

$$x_i \to Z^{-1/4} x_i \quad i = 1, 2, \tag{3.2}$$

so that the hamiltonian can be written as

$$H' \equiv \frac{H}{\sqrt{Z}} = H_0 - \delta' \tfrac{1}{2}(x_1 - x_2)^2, \tag{3.3}$$

where

$$H_0 = \tfrac{1}{2}(p_1^2 + x_1^2) + \tfrac{1}{2}(p_2^2 + x_2^2), \quad \delta' = \frac{1}{Z}. \tag{3.4a, b}$$

Before proceeding with the discussion of the solution of (3.3) with the help of the unitary representation of the dynamical group, we analyze the corresponding problem when the starting point is given by the Hartree-Fock hamiltonian.

b The Hartree-Fock (HF) Approximation for the Pseudo-Atom.

As the ground state of the exact solution[39] of (3.1) is symmetric in the coordinates, and thus antisymmetric in the spins, we can propose for the configuration space part of the HF solution two particles in the same orbital, i.e.

$$\Psi_{HF}(x_1, x_2) = \psi(x_1)\,\psi(x_2). \tag{3.5}$$

The standard analysis[14] gives then, for the single particle state, the equation

$$\tfrac{1}{2}(p_1^2 + Zx_1^2)\,\psi(x_1) - \left[\int \psi^*(x_2)\,\tfrac{1}{2}(x_1 - x_2)^2\,\psi(x_2)\,\mathrm{d}x_2\right]\psi(x_1) = \varepsilon\psi(x_1). \tag{3.6}$$

This equation appears integrodifferential, but actually, as $\psi(x_2)$ is of definite parity, the expectation value of the term x_1x_2 vanishes and thus it becomes

$$\tfrac{1}{2}[p_1^2 + (Z - 1)\,x_1^2]\,\psi(x_1) = \varepsilon'\psi(x_1), \tag{3.7 a}$$

where

$$\varepsilon' = \varepsilon + \tfrac{1}{2}\int \psi^*(x_2)\,x_2^2\psi(x_2)\,\mathrm{d}x_2. \tag{3.7 b}$$

Clearly then the HF common potential is again of the harmonic oscillator type, but of frequency $(Z - 1)^{1/2}$ rather than $Z^{1/2}$. The exact hamiltonian can therefore be written as

$$\begin{aligned} H &= H_0'' - \tfrac{1}{2}(x_1 - x_2)^2 + \tfrac{1}{2}(x_1^2 + x_2^2) \\ &= H_0'' - \tfrac{1}{4}(x_1 - x_2)^2 + \tfrac{1}{4}(x_1 + x_2)^2, \end{aligned} \tag{3.8}$$

where H_0'' is the HF hamiltonian of the form

$$H_0'' = \tfrac{1}{2}[p_1^2 + (Z - 1)\,x_1^2] + \tfrac{1}{2}[p_2^2 + (Z - 1)\,x_2^2]. \tag{3.9}$$

As in the previous subsection, we can carry out a scale transformation, but now of the type

$$x_i \rightarrow (Z-1)^{-1/4} x_i \quad i = 1, 2, \tag{3.10}$$

so that the hamiltonian (3.8) becomes

$$H'' = \frac{H}{\sqrt{Z-1}} = H_0 - \delta'' \frac{1}{4}(x_1 - x_2)^2 + \delta'' \frac{1}{4}(x_1 + x_2)^2, \tag{3.11}$$

where H_0 is given by (3.4a) and

$$\delta'' = \frac{1}{Z-1}. \tag{3.12}$$

Our problem now is to find canonical transformations that change the hamiltonian H' of (3.3) and H'' of (3.11) into a separable form similar to the hamiltonian H_0 of (3.4a). The unitary representations of these canonical transformations will allow then to expand the exact solutions of our problem in terms of either non-interacting or HF states.

c Transformation Matrix between the Exact and Non-Interacting, or HF, States. Relation with the Unitary Representation of the Dynamical and Symmetry Groups.

If we carry out the canonical transformation

$$\bar{x}_1 = (1 - 2\delta')^{1/4} \frac{1}{\sqrt{2}} (x_1 - x_2), \quad \bar{p}_1 = (1 - 2\delta')^{-1/4} \frac{1}{\sqrt{2}} (p_1 - p_2),$$
$$\bar{x}_2 = \frac{1}{\sqrt{2}} (x_1 + x_2), \quad \bar{p}_2 = \frac{1}{\sqrt{2}} (p_1 + p_2), \tag{3.13}$$

the hamiltonian H' of (3.3) becomes

$$H' = (1 - 2\delta')^{1/2} \tfrac{1}{2}(\bar{p}_1^2 + \bar{x}_1^2) + \tfrac{1}{2}(\bar{p}_2^2 + \bar{x}_2^2). \tag{3.14}$$

Similarly if we carry out the canonical transformation

$$\hat{x}_1 = (1 - \delta'')^{1/4} \frac{1}{\sqrt{2}} (x_1 - x_2), \quad \hat{p}_1 = (1 - \delta'')^{-1/4} \frac{1}{\sqrt{2}} (p_1 - p_2),$$
$$\hat{x}_2 = (1 + \delta'')^{1/4} \frac{1}{\sqrt{2}} (x_1 + x_2), \quad \hat{p}_2 = (1 + \delta'')^{-1/4} \frac{1}{\sqrt{2}} (p_1 + p_2), \tag{3.15}$$

the hamiltonian H'' of (3.11) becomes

$$H'' = (1 - \delta'')^{1/2}\, \tfrac{1}{2}(\hat{p}_1^2 + \hat{x}_1^2) + (1 + \delta'')^{1/2}\, \tfrac{1}{2}(\hat{p}_2^2 + \hat{x}_2^2). \tag{3.16}$$

If we designate by $\psi_n(x)$ the one dimensional harmonic oscillator states (I.5.2), the eigenstates of H_0, H', H'' become respectively

$$\begin{aligned} &|n_1 n_2\rangle = \psi_{n_1}(x_1)\,\psi_{n_2}(x_2),\ |\bar{n}_1\bar{n}_2) = \psi_{\bar{n}_1}(\bar{x}_1)\,\psi_{\bar{n}_2}(\bar{x}_2), \\ &|\hat{n}_1\hat{n}_2] = \psi_{\hat{n}_1}(\hat{x}_1)\,\psi_{\hat{n}_2}(\hat{x}_2), \end{aligned} \tag{3.17}$$

where we distinguish between the coordinates in terms of which the kets are given by using an angular, round or square bracket. The solution of our simple many body problem is related then with the expansion of the eigenstates of either H' or H'' in terms of those of H_0, i.e. in the determination of the transformation brackets

$$\langle n_1 n_2|\bar{n}_1\bar{n}_2) \quad \text{or} \quad \langle n_1 n_2|\hat{n}_1\hat{n}_2]. \tag{3.18a, b}$$

From the discussion given in sections 2, 3 of chapter I, these transformation brackets are the unitary representations of the canonical transformation (3.13) or (3.15) in a basis in which $H_{0i} = \frac{1}{2}(p_i^2 + x_i^2)$, $i = 1, 2$, are diagonal. These unitary representations can be obtained straightforwardly in a basis in which the coordinates x_1, x_2 are diagonal and then transformed into a basis in which the H_{0i}, $i = 1, 2$, are diagonal by a procedure similar to the one given in (I.5.3). We prefer to use a simpler method based on the fact that both of the canonical transformations (3.13) and (3.15) can be decomposed into rotations by $\pi/4$ and dilatations, e.g. for (3.10)

$$\begin{pmatrix} \bar{x}_1 \\ \bar{x}_2 \\ \bar{p}_1 \\ \bar{p}_2 \end{pmatrix} = \begin{pmatrix} (1-2\delta')^{1/4} & 0 & 0 & 0 \\ 0 & 1 & 0 & 0 \\ 0 & 0 & (1-2\delta')^{-1/4} & 0 \\ 0 & 0 & 0 & 1 \end{pmatrix}$$

$$\times \begin{bmatrix} \frac{1}{\sqrt{2}} & -\frac{1}{\sqrt{2}} & 0 & 0 \\ \frac{1}{\sqrt{2}} & \frac{1}{\sqrt{2}} & 0 & 0 \\ 0 & 0 & \frac{1}{\sqrt{2}} & -\frac{1}{\sqrt{2}} \\ 0 & 0 & \frac{1}{\sqrt{2}} & \frac{1}{\sqrt{2}} \end{bmatrix} \begin{pmatrix} x_1 \\ x_2 \\ p_1 \\ p_2 \end{pmatrix}. \tag{3.19}$$

The rotation implies the same transformation for the creation operators, i.e.

$$\begin{pmatrix}\bar{\eta}_1\\ \bar{\eta}_2\end{pmatrix} = \begin{pmatrix}\frac{1}{\sqrt{2}} & -\frac{1}{\sqrt{2}}\\ \frac{1}{\sqrt{2}} & \frac{1}{\sqrt{2}}\end{pmatrix}\begin{pmatrix}\eta_1\\ \eta_2\end{pmatrix}, \tag{3.20}$$

and thus the state

$$|\bar{n}_1\bar{n}_2) = \frac{\bar{\eta}_1^{\bar{n}_1}\bar{\eta}_2^{\bar{n}_2}}{\sqrt{\bar{n}_1!\,\bar{n}_2!}}|0\rangle = \frac{\left[\frac{1}{\sqrt{2}}(\eta_1 - \eta_2)\right]^{\bar{n}_1}\left[\frac{1}{\sqrt{2}}(\eta_1 + \eta_2)\right]^{\bar{n}_2}}{\sqrt{\bar{n}_1!\,\bar{n}_2!}}|0\rangle$$

$$= \sum_{\substack{n_1,n_2\\ n_1+n_2=\bar{n}_1+\bar{n}_2}} |n_1n_2\rangle\, \mathscr{D}^{\frac{1}{2}(\bar{n}_1+\bar{n}_2)}_{\frac{1}{2}(n_1-n_2),\frac{1}{2}(\bar{n}_1-\bar{n}_2)}\left(0, \frac{\pi}{2}, 0\right), \tag{3.21}$$

where $\mathscr{D}$ is the Wigner $\mathscr{D}$ function[18] for the angles indicated. Thus, from (3.20) the states $|n_1n_2\rangle$ transform as basis for an IR of the symmetry group $\mathscr{U}(2)$ of the two dimensional oscillator. For the dilatation in (3.19) the corresponding matrix elements are given by (I.5.11) in which we replace ω by $(1 - 2\delta')^{1/2}$. Combining then the dilatation and rotation effects, we see that the transformation brackets (3.18a) become

$$\begin{aligned}\langle n_1n_2|\bar{n}_1\bar{n}_2) &= \tfrac{1}{2}[1 + (-1)^{n_1+n_2-\bar{n}_1-\bar{n}_2}]\,[n_1!\,n_2!\,\bar{n}_1!\,\bar{n}_2!\,2^{\bar{n}_2-\bar{n}_1}]^{1/2}\\ &\times (n_1 + n_2 - \bar{n}_2)!\,2^{-n_1-n_2}\\ &\times \left[\sum_{r=0}^{\min(n_1,\bar{n}_2)} \frac{(-1)^{n_2-\bar{n}_2+r}}{(n_1 - r)!\,(\bar{n}_2 - r)!\,(n_2 - \bar{n}_2 + r)!\,r!}\right](1 - 2\delta')^{1/8}\\ &\times \{\tfrac{1}{2}[(1 - 2\delta')^{1/2} + 1]\}^{-1/2} \sum_{s=0}^{\min(\bar{n}_1,n_1+n_2-\bar{n}_2)} \tfrac{1}{2}[1 + (-1)^{\bar{n}_1-s}]\\ &\times \frac{(-1)^{(n_1+n_2-\bar{n}_2-s)/2}}{\left(\frac{\bar{n}_1 - s}{2}\right)!\left(\frac{n_1 + n_2 - \bar{n}_2 - s}{2}\right)!\,s!}\left[\frac{(1 - 2\delta')^{1/2} - 1}{(1 + 2\delta')^{1/2} + 1}\right]^{\frac{n_1+n_2+\bar{n}_1-\bar{n}_2}{2}-s}\\ &\times \left[\frac{4(1 - 2\delta')^{1/4}}{(1 + 2\delta')^{1/2} + 1}\right]^s,\end{aligned} \tag{3.22}$$

where we made use of the explicit formula of the $\mathscr{D}$ functions[18]. In a similar fashion, when we consider the canonical transformation (3.15), we see that the transformation brackets (3.18 b) are given by

$$\langle n_1 n_2 | \hat{n}_1 \hat{n}_2] = \tfrac{1}{2}[1 + (-1)^{n_1+n_2-\hat{n}_1-\hat{n}_2}]\, 2^{-n_1-n_2}[n_1!\, n_2!\, \hat{n}_1!\, \hat{n}_2!\, 2^{-n_1-n_2-\hat{n}_1-\hat{n}_2}]^{1/2}$$

$$\times \sum_{m=0}^{n_1+n_2} \tfrac{1}{2}[1 + (-1)^{\hat{n}_1-m}]\, m!\, (n_1 + n_2 - m)! \left[\sum_{r=0}^{\min(n_1, n_1+n_2-m)} (-1)^{m-n_1+r} \right.$$

$$\left. \times \frac{1}{(n_1 - r)!\, (n_1 + n_2 - m - r)!\, (m - n_1 + r)!\, r!} \right] (1 - \delta'')^{1/8}$$

$$\times \{\tfrac{1}{2}[(1 - \delta'')^{1/2} + 1]\}^{1/2} \left\{ \sum_{s=0}^{\min(\hat{n}_1, m)} \tfrac{1}{2}[1 + (-1)^{\hat{n}_1-s}] \right.$$

$$\times \frac{(-1)^{\frac{m-s}{2}}}{\left(\frac{\hat{n}_1 - s}{2}\right)!\left(\frac{m - s}{2}\right)!\, s!} \left[\frac{(1 - \delta'')^{1/2} - 1}{(1 - \delta'')^{1/2} + 1} \right]^{\frac{\hat{n}_1+m}{2} - s}$$

$$\left. \times \left[\frac{4(1 - \delta'')^{1/4}}{(1 - \delta'')^{1/2} + 1} \right]^{s} \right\} (1 + \delta'')^{1/8}\, \{\tfrac{1}{2}[(1 + \delta'')^{1/2} + 1]\}^{1/2}$$

$$\times \left\{ \sum_{t=0}^{\min(\hat{n}_2, n_1+n_2-m)} \tfrac{1}{2}[1 + (-1)^{\hat{n}_2-t}] \frac{(-1)^{\frac{n_1+n_2-m-t}{2}}}{\left(\frac{\hat{n}_2 - t}{2}\right)!\left(\frac{n_1 + n_2 - m - t}{2}\right)!\, t!} \right.$$

$$\left. \times \left[\frac{(1 + \delta'')^{1/2} - 1}{(1 + \delta'')^{1/2} + 1} \right]^{\frac{n_1+n_2+\hat{n}_2-m}{2} - t} \left[\frac{4(1 + \delta'')^{1/4}}{(1 + \delta'')^{1/2} + 1} \right]^{t} \right\}. \tag{3.23}$$

Thus we see that the transformation brackets that relate the non-interacting, or the HF, states with those of the exact solutions, are given by the unitary representations of canonical transformations. In this elementary problem the brackets were determined in a straightforward fashion. In more complex many body problems they will of course be difficult to obtain. Nevertheless the procedure used in the present case suggests a technique to be followed in general, which we outline in the next section.

4 CANONICAL TRANSFORMATIONS AND THE MANY BODY PROBLEM.

When we are dealing with a many body problem, we are usually interested in the expansion of the set of its eigenstates in terms of the eigenstates of a simpler problem e.g. the problem in which no interactions are present or the one in which we have the particles in a HF common potential. For the simpler problem usually a complete set of integrals of motion is available. If in any way we can determine a similar set for the many body problem, we shall usually also be able to find a canonical transformation that takes us from the integrals of motion of the simpler to those of the full many body problem. We can then look for the unitary representations of this canonical transformation in a basis in which all the integrals of motion of the simpler problem are diagonal. Once we have it we can use the corresponding matrix elements as the coefficients of the expansion of the exact eigenstates in terms of those of the simpler problem.

CONCLUSION

In the present paper we started with the general problem of what is the group of canonical transformations in classical mechanics, and its unitary representation in quantum mechanics. We then particularized our discussion to linear canonical transformations and their representations with applications to systems of oscillators.

It is clear though that the discussions in sections 1 and 2 of chapter I are entirely general. Thus it seems that developments similar to what we carried out here for particles in an harmonic oscillator potential could be extended to other dynamical systems. We plan to explore these possibilities in future publications.

REFERENCES

1. H. Goldstein, *Classical Mechanics* (Addison-Wesley Publishing Co. Reading, Mass. 1950) Chapter VIII
2. E. P. Wigner, *Group Theory* (Academic Press, New York 1959)
3. M. Hamermesh, *Group Theory* (Addison-Wesley Publishing Co., Reading, Mass. 1962)
4. H. Weyl, *The Classical Groups* (Princeton University Press, Princeton, N.J. 1939) Chapter VI
5. P. A. M. Dirac, *Quantum Mechanics* (Third Edition, Oxford Clarendon Press 1947) p. 94 and pp. 103–107
6. S. Goshen and H. Lipkin, *Annals of Physics* (N.Y.) **6**, 301 (1959)
7. A. O. Barut and C. Fronsdal, *Proc. Roy. Soc.* (London) **A 287**, 532 (1965)
   W. J. Holman and L. C. Biedenharn, *Annals of Physics* (N.Y.) **39**, 1 (1966)
8. V. Bargmann and M. Moshinsky, *Nuclear Physics* **18**, 697 (1960)
9. N. N. Bogoliubov, *Nuovo Cimento* **7**, 794 (1958)
   J. G. Valatin, *Nuovo Cimento* **7**, 843 (1958)
10. G. C. Wick, *Phys. Rev.* **80**, 268 (1950)
11. M. Moshinsky, "Group Theory and the Few Nucleon Problem", in *Cargese Lectures in Physics*, Edited by M. Jean (Gordon and Breach, N.Y. 1969) Chapter II, pp. 253 to 262

12. G. Racah, *Group Theory and Spectroscopy* (Lecture Notes of the Institute of Advanced Study, Princeton, N.J. 1951) pp. 16–31
13. B. R. Judd, "Group Theory in Atomic Spectroscopy", in *Group Theory and its Applications*, Edited by E. M. Loebl (Academic Press, N.Y. 1968) pp. 185–187
14. M. Moshinsky, *The Harmonic Oscillator in Modern Physics: From Atoms to Quarks* (Gordon and Breach, N.Y. 1969)
15. P. Kramer and M. Moshinsky, "Group Theory of Harmonic Oscillators and Nuclear Structure" in *Group Theory and Applications*, Edited by E. M. Loebl (Academic Press, N.Y. 1968) pp. 399–402
16. V. Bargmann and M. Moshinsky, *Nuclear Physics* **23**, 177 (1961)
17. Reference 2, p. 115
18. Reference 2, pp. 161–167
19. T. A. Brody and M. Moshinsky, *Tables of Transformation Brackets* (Gordon and Breach, N.Y. 1967)
20. M. Moshinsky, *Nuclear Physics* **13**, 104 (1959)
21. A. Gal, *Annals of Physics* (N.Y.) **49**, 341 (1968)
22. P. Kramer, *Revista Mexicana de Fisica* (In Press)
23. A. de Shalit and I. Talmi, *Nuclear Shell Theory* (Academic Press, N.Y. 1963) Chapter III
24. I. Talmi, *Helv. Phys. Acta* **25**, 185 (1952)
25. K. T. R. Davies, S. J. Krieger, and M. Baranger, *Nuclear Physics* **84**, 545 (1966).
26. M. Moshinsky, "Transformation Brackets for Three and Four Nucleon Systems", in *Clustering Phenomena in Nuclei* (International Atomic Energy Agency, Vienna 1969) pp. 189–196
27. V. C. Aguilera-Navarro, M. Moshinsky, and W. W. Yeh, *Annals of Physics* (N.Y.) **51**, 312 (1969)
28. V. C. Aguilera-Navarro, M. Moshinsky, and P. Kramer, *Annals of Physics* (N.Y.) **54**, 379 (1969)
29. Reference 2, p. 105
30. V. C. Aguilera-Navarro and M. Moshinsky, *Physics Letters* **32B**, 336 (1970)
31. I. R. Afnan and Y. C. Tang, *Phys. Rev.* **175**, 1337 (1968)
32. H. Eikemeir and H. H. Hackenbroich, *Zeits. Phys.* **195**, 412 (1966)
33. R. F. Frosch, J. S. McCarthy, R. E. Rand, and M. R. Yearin, *Phys. Rev.* **160**, 874 (1967)
34. P. Kramer and M. Moshinsky, *Nuclear Physics* **A125**, 321 (1969)
35. M. Moshinsky, *Group Theory and the Many Body Problem* (Gordon and Breach, N.Y. 1967)
36. M. Moshinsky and V. Syamala Devi, *J. Math. Phys.* **10**, 455 (1967)
Y. Akiyama, *Nuclear Data* **2**, 403 (1966)
37. J. Flores and R. Pérez, *Phys. Letters* **26B**, 55 (1967); *Rev. Mexicana Fís.* (In Press)
R. Pérez, Doctoral Dissertation, University of Mexico (1970)
38. W. J. Gerace and D. A. Sparrow, *Physics Letters* **30B**, 71 (1969)
39. M. Moshinsky, O. Novaro, and A. Calles, *J. de Physique* (In Press)
40. P. M. Morse and H. Feshbach, *Methods of Theoretical Physics* (McGraw-Hill Co., New York 1953), Vol. I, pp. 784–787
41. E. Chacón and M. de Llano, *Rev. Mexicana Fís.* **12**, 57 (1963)
42. S. G. Nilsson, *K. Danske Vidensk. Selsk. mat.-fys. Medd.* **29**, N⁰ 16 (1955)

DISCUSSION

L. C. Biedenharn Professor Moshinsky's paper is most interesting and very important in emphasizing the significance of the symplectic group as the group of invariance of the Heisenberg relations.

Sp(2) is the most fundamental example for this structure and I would like to remark further on this. Professor Moshinsky's realization of this group by means of Bloch's projection operators is actually the hard way to realize this dynamical group! One can see this most easily by noting that in phase space x and p are abstractly equivalent, but the use of $\bar{a}\,|0\rangle \equiv 0$ actually forces the operators a and $\bar{a}$ to be inequivalent—and hence destroys the formal similarity of x and p.

Several months ago, Dr. J. D. Louck and I succeeded in giving a more elementary way to realize this dynamical group in a symmetric way. The new idea that is involved is a new inner product for the Hilbert space. One achieve thereby a sort of "Bogoliubov" transformation on the space of one boson. (We called this structure a "symplecton" since it realizes the symplectic group in an elementary way).

This new procedure is significant, it seems, since for example it shows easily that for one particle in three-space—Prof. Moshinsky's example—we see that indeed his structure $SU3 \supset O3 \supset O2$ is actually embedded in $Sp6 \supset SU3$, etc.

We have given the complete analysis of all polynomials in this new space in the form of a theorem. As a bonus, we achieve a new angular-momentum function, $\Delta(abc)$, which (as a coordinate-free function) more properly plays the role of the $(3j)$ function in the series $(3j)$, $(6j)$, ... $(3nj)$ than does the Wigner coefficient ("$(3j)$" symbol) itself. (Wigner, of course, has stressed the coordinate dependence of the currently designated "$3j$" function).

I might conclude by noting that this structure may very well be important abstractly, since it suggests that the concept of a normed ring (which the symplecton structure exhibits quite naturally) may be a more technically amenable concept than that of the von Neumann operator norm.

G. E. Brown I should apologize for trying to inject some physics into the discussion, I guess.

First of all, it's amazing what one can do for the binding energy by expansions in an oscillator basis. Jackson, Laude and Sauer obtain a remarkably good binding energy for the triton. However, I doubt whether this is a good way to get at short-range correlations. Nucleon-nucleon forces are much more violent than those used by Moshinsky. It's very clumsy to build up the rapid variations in the wave function through oscillator functions. Since the problems, when two particles are close together, is essentially a two-body problem—the interaction being so strong that the Pauli exclusion by the other particles has little effect—one should preferably begin by solving for the two-body wave function directly, as one does in a *t*-matrix type approach.

M. Moshinsky The effect that Gerry Brown mentions of short range correlations and their description by the two body equation rather than the many body wave function, is very relevant. It was actually followed in the analysis of Tang himself. It is interesting though to note that our own analysis gives almost as good an approach to the binding energy as does Tang's approach. Thus the construction of many nucleon wave functions with harmonic oscillator states seems to be an effective way to proceed if we go to a sufficiently high number of quanta.

The Nuclear Shell Model in Terms of Pseudo Spin-Orbit Doublets and Pseudo SU(3) Coupling Schemes

K. T. HECHT*

Physics Department, University of Michigan
Ann Arbor, Michigan

1 INTRODUCTION

Strong spin-orbit coupling is one of the basic corner-stones of the nuclear shell model. Shell model calculations in all but the lightest nuclei have therefore been performed in the $j - j$ coupling scheme, intrinsically a much richer coupling scheme than the Russell Saunders scheme of atomic spectroscopy. As a result the types of problems that can be studied by means of a strict shell model approach are severely limited. With only a few neutrons and protons filling a relatively small number of active j-shells, the dimensionality of the shell model space in the $j - j$ scheme blows up very quickly. In order to make progress a search must be made for new coupling schemes. Elliott's $SU(3)$ scheme is the outstanding example of such a scheme. Although it has given much insight as to the nature of rotational spectra in terms of a shell model description, its usefulness for actual shell model calculations is also severely limited (to nuclei up through the first half of the $2s$–$1d$ shell), again because the strong spin-orbit coupling of the shell model destroys the goodness of the $SU(3)$ quantum numbers in heavier deformed nuclei. Nevertheless the periodic table is full of examples of interesting nuclei which should be amenable to understanding by means of a strict shell model approach. In the isotopes of Xe, Cs, Ba, La, Ce, Pr, Nd, Pm, and Sm with 82 neutrons for example, nature has presented us with a family of nuclei which can be understood quite simply in terms of independent particle motion of the protons

* NSF Senior Postdoctoral Fellow, 1970–71.

with $Z > 50$; but a transition to collective behavior is made very quickly with the addition or removal of even a few neutrons. The isotopes of Mo, Tc, Ru, ... with $N = 50$ form another such family. The addition of a relatively small number of neutrons beyond 50 again can lead to strongly collective behavior. In both examples, seemingly only a small number of both active protons and neutrons are involved. In order to give a feasible shell model description even for such "simple' nuclei, however, there is a need for a new intermediate coupling scheme. The pseudo spin-orbit doublet and pseudo $SU(3)$ coupling schemes may serve this purpose. The discovery of the potential usefulness of these coupling schemes has recently been made independently by A. Arima, M. Harvey, and K. Shimizu[1] and by our group at Michigan[2, 3], starting from somewhat different points of view.

Two properties of nuclei are important: (1) The matrix elements of the effective two-body nuclear interaction are functions mainly of the j's and the relative parities of the single particle states, but not the orbital angular momenta l. (It was essentially the similarity between the interactions and spectra for nuclei in the Ni region, ($2p_{3/2}$, $2p_{1/2}$, $1f_{5/2}$), and the nuclei in the $2s_{1/2}$, $1d_{5/2}$, $1d_{3/2}$ shell which led Arima, Harvey, and Shimizu to the discovery of the importance of the pseudo-coupling schemes). (2) The single particle states of the nuclear shell model which are actually nearly degenerate are never the two members of a real spin-orbit doublet, but single particle states such as the doublets $2d_{5/2}1g_{7/2}$, $3s_{1/2}2d_{3/2}$, $2p_{3/2}1f_{5/2}$, or $2f_{7/2}1g_{9/2}$, that is pairs of states of the type l_j, $(l+2)_{j+1}$, which differ by 2 units in l and one unit in j. It is possible to associate to a single nucleon in such a pair of states a pseudo-spin (b spin) of $\frac{1}{2}$ and a pseudo-orbital angular momentum (c spin) with $c = l + 1$, such that the vector coupling $\mathbf{b} + \mathbf{c} = \mathbf{j}$ yields the quantum numbers j of the doublet. So far this is merely a mathematical game. In the same spirit it might be useful to consider the single particle states $2p_{3/2}2p_{1/2}$ $1f_{5/2}$ a nearly degenerate triplet and associate to this triplet single-nucleon pseudo angular momenta $b = 1$, $c = 3/2$. However, such mathematical games will have physical significance only if the many-nucleon total pseudo-spin (B) and total pseudo orbital angular momentum (C) are approximately good quantum numbers in real nuclei. In this sense, the assignment $b = \frac{1}{2}$, $c = l + 1$ to the doublet $l_j(l+2)_{j+1}$ does lead to a physically relevant coupling scheme. On the other hand it will be shown that the assignment $b = 1$, $c = 3/2$ to the triplet $2p_{3/2}2p_{1/2}1f_{5/2}$ does not. The physically relevant bc assignments for this triplet will be: $bc = \frac{1}{2}2$ for ($2p_{3/2}$ $1f_{5/2}$), and $bc = \frac{1}{2}0$ for $2p_{1/2}$; that is, the configuration $2p_{3/2}$ $1f_{5/2}$ $2p_{1/2}$ behaves as a pseudo $d - s$ configuration. (It is interesting to note that an attempt by Vincent[4] to classify nuclei in the real $2s$–$1d$ shell in terms of pseudo angular momenta 1 and 3/2 did not prove fruitful.)

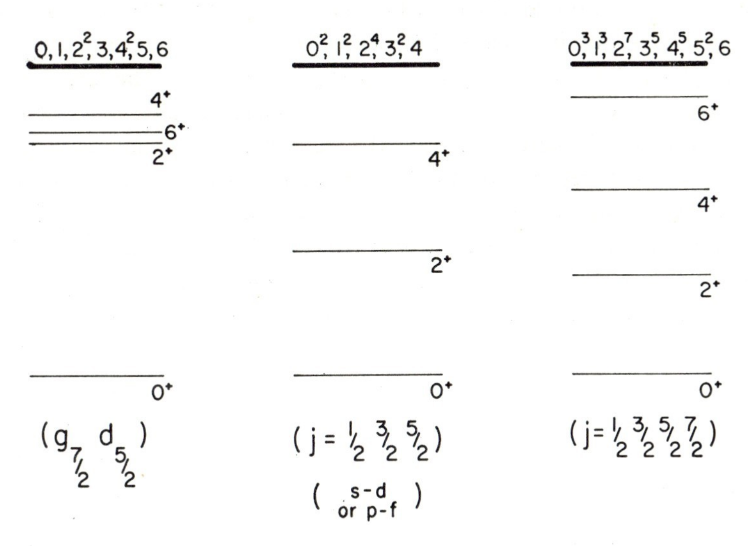

FIGURE 1 SDI Spectra for 2 identical particles. Column (1): degenerate ($g_{7/2}d_{5/2}$) level. Column (2): degenerate $j = 1/2, 3/2, 5/2$ configuration, e.g., 2s-ld shell or $2p_{\frac{1}{2}}2p_{3/2}lf_{5/2}$ shell. Column (3): degenerate $j = 1/2, 3/2, 5/2, 7/2$ configuration. All energies are given in units of $G\Omega$, G = SDI strength, $\Omega = \Sigma_j(j + 1/2)$

In order to investigate the goodness of the many-nucleon pseudo-spin (B) and pseudo orbital angular momentum (C), we must investigate the pseudo-spin and orbital tensor character of the realistic nucleon-nucleon interaction. Since the surface delta interaction[5] (SDI) has, despite its simplicity, proved to be a remarkably good effective interaction for highly truncated shell model spaces, our approach has been to use the SDI as a guide to new nuclear coupling or classification schemes. (The SDI has also been used effectively by C. Quesne of the Université Libre de Bruxelles[6].) Some typical 2-identical particle spectra for the SDI are shwon in Fig. 1 for several examples of configurations of degenerate single particle states. All have one feature in common: Besides the $J = 0$ state of ordinary pairing theory one additional state, corresponding to one specific superposition of 2-particle states, is depressed in energy for each even J value with $J \neq 0$; (if the single particle states of the configuration all have the same parity, these favored pair states have *even* values of J only)* The remaining states are degenerate and are not acted

* In addition, this favored pair state takes up the full strength of the sum rule for the 2^J-pole transition from the ground state.[5]

upon by the SDI. A more realistic interaction will of course give some interaction energy to these levels also; but in the spirit in which the SDI is considered an interaction in which all but the most prominent features of the real effective 2-body interaction have been "turned off," it can serve as an effective guide to the discovery of new coupling schemes. The fact that the SDI favors only one specific superposition of 2-particle states for each value of J is a property it shares with any other separable 2-body interaction. The key question is therefore: Are the favored pair states of the SDI the states actually favored in nature by a realistic nucleon-nucleon interaction? To a very good approximation the answer is yes. This is illustrated in Table 1 by the configuration $j = 1/2, 3/2, 5/2$. The first column shows the j, j' components of the favored $J = 2$ pair state as given by the SDI. Using the 2-body matrix elements of Kuo in a configuration of degenerate $2s_{1/2}$, $1d_{5/2}$, $1d_{3/2}$ single particle states, one $J = 2$ state of the $n = 2$ spectrum is found to be depressed in energy by a considerably greater amount than the rest. The j, j' components for this favored pair state are shown in column 2 and are in remarkably good agreement with those predicted by the SDI. (Some adjustment of phases has been made to put entries in all columns of Table 2 on a common footing[2].) The third column shows that the favored $J = 2$ pair state in a configuration of degenerate $2p_{3/2}$, $2p_{1/2}$, $1f_{5/2}$ single particle states, using realistic 2-body matrix elements, is again very similar. (In this case the two body matrix elements used are the reaction matrix elements

TABLE 1 The jj' components of the favored $J = 2$ pair for the $j = \frac{1}{2}\frac{3}{2}\frac{5}{2}$ configuration

jj'	SDI	Kuo Interaction* Degenerate $2s$–$1d$	Cohen *et al.* (B)** Nickel Region Degenerate $2p_{1/2}, p_{3/2}; 1f_{5/2}$	SU(3) or Pseudo-SU(3) $(\lambda\mu) = (40)$	Model Based on $b = 1$ $c = \frac{3}{2}$
$\frac{5}{2}\frac{5}{2}$	0.447	0.485	0.519	0.326	0.772
$\frac{3}{2}\frac{3}{2}$	0.342	0.331	0.339	0.250	0.168
$\frac{5}{2}\frac{3}{2}$	0.316	0.288	−0.203	0.231	−0.446
$\frac{5}{2}\frac{1}{2}$	0.592	0.592	0.480	0.683	0.188
$\frac{1}{2}\frac{3}{2}$	0.483	0.472	0.586	0.558	0.377

* T. T. S. Kuo, *Nucl. Phys.* **A103** (1967) 71

** S. Cohen, R. D. Lawson, M. H. Macfarlane, S. P. Pandya, and M. Soga, *Phys.Rev.* **160** (1967) 903

calculated by Cohen, Lawson, Macfarlane, Pandya, and Soga from the Hamada Johnston potential using the Kuo-Brown technique, including the perturbative corrections for core excitations.) Similar results for other configurations[7] again support this conclusion. The SDI should therefore serve as a good guide to an understanding of the new pseudo-coupling schemes.

2 THE SINGLE PSEUDO SPIN-ORBIT DOUBLET

To start with the simplest example, we shall examine the spectra for a system of identical nucleons filling the levels of a single pseudo spin-orbit doublet, l_j, $(l + 2)_{j+1}$. Nature has provided us with a number of nuclei of this type. The family of 82-neutron nuclei with $50 \leqq Z \leqq 64$ is a particularly good example[8, 9]: (1) The protons with $Z > 50$ in the family of 82-neutron nuclei are known to fill mainly the $1g_{7/2}$, $2d_{5/2}$ single particle levels. Even in $^{144}_{62}Sm_{82}$, the nucleus which potentially almost fills the $g_{7/2}\, d_{5/2}$ doublet, excitations into the higher $1h_{11/2}$, $2d_{3/2}$, $3s_{1/2}$ orbits account for only about 17% of the ground state wave function[8]. (2) The separation of the $g_{7/2}$ and $d_{5/2}$ levels is relatively small. The best recent estimate of this separation by Wildenthal is 880 keV[11] and shows that this separation is small compared with 2-body interaction energies of about 2.5 MeV in this region, so that the doublet can be considered nearly degenerate in zeroth approximation. (3) The SDI has been shown to be a remarkably good effective interaction for these nuclei by the recent detailed shell model calculations of Wildenthal[10, 11] using the Oak Ridge-Rochester shell model code[12]. The availability of these results also makes it possible to compare rough predictions based on the new coupling scheme with those of a more detailed shell model calculation.

It is interesting to compare the experimental spectra with the predictions of an extreme zeroth order pseudo-spin orbit doublet model. In this model: (1) excitations out of the $g_{7/2}\, d_{5/2}$ orbits are neglected, (2) the $g_{7/2}\, d_{5/2}$ single particle states are taken to be degenerate, and (3) the 2-body interaction is approximated by the SDI. Since the SDI commutes with the total quasi-spin operator[5], the predictions of this extreme zeroth order model are dependent on total seniority number v but independent of proton number n. The experimental information is most extensive for the even nucleus $^{140}_{58}Ce_{82}$ with $n = 8$ so that it will be used as the prime example. The predictions of the extreme zeroth order model are compared with the experimental spectrum for ^{140}Ce in Fig. 2. Except for the $J = 0$ ground state and the favored $v = 2$ states with $J = 2, 4, 6$ the model predicts a degenerate cluster of states with $J = 0, 1, 2^2, 3, 4^2, 5$, and 6 at an excitation energy of (9/7) units above the center of gravity of the favored $v = 2$ states. With the exception of the experimentally observed first excited 0^+ state which is depressed into the region of favored $J \neq 0$ states, and with the possible exception of an as yet missing 5^+ state, all of these states are observed in a cluster between 2.63 and 2.35 MeV which is separated by a gap of 0.3 MeV from the next highest observed positive parity state. In terms of the pseudo-spin orbit coupling scheme states with $v = 2$ for the $g_{7/2}\, d_{5/2}$ doublet are spectroscopically equivalent to the 2-particle states of a pseudo f-shell ($c = 3$) of spin $\frac{1}{2}$ particles. Antisymmetry requirements restrict states with $B = 0$ (pseudo-spin singlets) to those with even C (pseudo orbital angular momentum) and those with $B = 1$ to odd C. It will be shown that the former are the energetically favored states, as indicated in Fig. 2.

Even the electromagnetic information for these nuclei can be understood simply in terms of the extreme zeroth order pseudo-spin orbit doublet model, without the somewhat more laborious calculations which would be required in terms of the equivalent $j - j$ coupling description of these states. The magnetic moment in the first excited 4^+ state of ^{140}Ce has been measured ($\mu = 4.4$ n.m.). It will be shown that the zeroth order pseudo-spin orbit prediction for μ is extremely simple for a pseudo-spin singlet ($B = 0$) state. It is: $\mu = g_l J$, giving $\mu = 4$ n.m. for the 4^+ state, (a result which, perhaps surprisingly, is identical to that for a real spin singlet state and is in good agreement with the observed value, considering the extreme zeroth order nature of the approximation). The observed $E2$ transition rates can also be understood simply (at least qualitatively) in terms of the pseudo-spin orbit doublet model. The observed 0^+ (ground state) $\rightarrow$ 2^+ (1*st* excited state) $E2$ transition probability is strongly enhanced in $^{140}_{58}\mathrm{Ce}_{82}$. (Almost identical enhancement factors are observed in $^{138}_{56}\mathrm{Ba}_{82}$ and $^{142}_{60}\mathrm{Nd}_{82}$). $E2$ transitions among the predominantly $v = 2$ states in ^{140}Ce, however, are hindered, sometimes by large factors. Some of the experimentally known results (expressed in Weisskopf units) are shown in Fig. 2. It will be shown that the electric multipole moment operators are to a very good approximation pseudo-spin scalars ($B = 0$ operators). As a result transitions between $B = 0$ and $B = 1$ states are forbidden. The transition from the $B = 0$ ground state to the energetically favored $J = 2$ (pseudo spin singlet) state takes up the full strength of the sum rule for the quadrupole transition, the transitions to the remaining $J = 2$ states being forbidden. The zeroth order calculations are again very simple and predict

$$B(E2)_{2\rightarrow 0} = \frac{1}{5}\left(\frac{4}{3}\right)\left[\frac{n(14-n)}{6}\right] \text{W. U.} \tag{1}$$

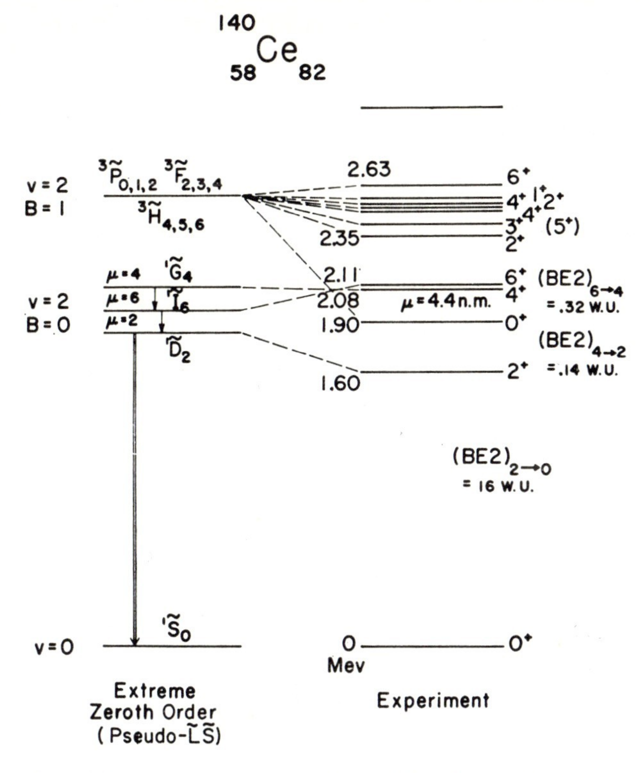

FIGURE 2 ^{140}Ce Spectrum. Comparison of Experimental Spectrum with Extreme Zeroth Order Pseudo Spin-Orbit Coupling Model, assuming degenerate $g_{7/2}d_{5/2}$ levels and surface delta 2-body interaction. Levels are indicated in pseudo Russell Saunders notation (∼). Only the positive parity experimentally observed levels are shown. Some of the electromagnetic information is indicated

where n is the number of protons in the $g_{7/2}\,d_{5/2}$ configuration. The n-dependent factor [ ] gives the cooperative enhancement of the n active protons, whereas the factor 4/3 represents the angular enhancement implicit in the $\Delta B = 0$ selection rule. For ^{140}Ce, ($n = 8$), the above zeroth order prediction becomes $(BE2)_{2\to 0} = 2.1$ W.U., considerably less than the observed value. (The small proton excitations into the $3s_{1/2}$, $2d_{3/2}$, $1h_{11/2}$ can be shown[13] to give an additional enhancement factor of about 1.5. Even with this enhancement an effective proton charge of 2.2 units would be required to match the experimentally observed value). Compared with the predicted enhancement of this transition connecting the 0^+ ground state to the pseudo-spin singlet 2^+ state, the remaining allowed $E2$ transitions connecting the pseudo-spin singlet states are predicted to be hindered in agreement with the observed facts. Compared with the zeroth order prediction $(BE2)_{2\to0} = 2.1$ W.U., the corresponding zeroth order predictions for transitions between the $B = 0$, $v = 2$ states in ^{140}Ce are: $(BE2)_{4\to2} = 0.04$ W.U., and $(BE2)_{6\to4} = 0.01$ W.U. The hindrance arises partly from angular factors and partly from an n-dependent factor which for the $\Delta v = 0$ transitions is proportional to (n-7) and is thus small for the nucleus ^{140}Ce with $n = 8$, near the half-closed position of the pseudo f-shell.

Although the agreement between the zeroth order model and experiment is quite good, the aim here is not to emphasize the goodness of the agreement, (the assumptions inherent in this zeroth order model are after all very extreme); but instead to emphasize the great simplicity of the new coupling scheme. In zeroth approximation the favored $J = 2$ state in ^{140}Ce, for example, can be described as a single state in the pseudo-spin orbit coupling scheme:

$$|(c = 3)^8\ v = 2\ B = 0\ C = 2\rangle = |(\tilde{f})^8\ v = 2,\ ^1\tilde{D}_2\rangle \tag{2a}$$

that is, a state of 8 pseudo f-shell protons coupled to seniority $v = 2$, total pseudo spin $B = 0$, and total pseudo orbital angular momentum $C = 2$. The equivalent description of this *same* state in the conventional $j - j$ coupling scheme would be

$$\begin{aligned}
&\frac{2}{5}\,|(d_{5/2})^2_{J=2}\,(g_{7/2})^6_{J=0}\rangle + \frac{\sqrt{6}}{5}\,|(d_{5/2})^4_{J=2}\,(g_{7/2})^4_{v=0,\,J=0}\rangle \\
&\quad + \frac{1}{3\sqrt{2}}\,|(d_{5/2})^6_{J=0}\,(g_{7/2})^2_{J=2}\rangle + \frac{1}{\sqrt{3}}\,|(d_{5/2})^4_{J=0}\,(g_{7/2})^4_{v=2,\,J=2}\rangle \\
&\quad + \frac{1}{\sqrt{6}}\,|(d_{5/2})^2_{J=0}\,(g_{7/2})^6_{J=2}\rangle - \frac{1}{15}\,|(d_{5/2})^1_{J=5/2}\,(g_{7/2})^7_{J=7/2}\rangle \\
&\quad - \frac{1}{5}\sqrt{\frac{2}{3}}\,|(d_{5/2})^3_{v=1,\,J=5/2}\,(g_{7/2})^5_{v=1,\,J=7/2}\rangle \\
&\quad - \frac{1}{5\sqrt{3}}\,|(d_{5/2})^5_{J=5/2}\,(g_{7/2})^3_{v=1,\,J=7/2}\rangle
\end{aligned} \tag{2b}$$

That is, the state contains 8 almost equally important pieces in the conventional $j - j$ coupling description. Transformation to the pseudo spin-orbit coupling scheme therefore has the potential of reducing the size of the shell model space by a factor of order 8. If a comparable reduction can be achieved for *both* proton and neutron subspaces in nuclei where protons and neutrons are filling different major shells, problems of the type mentioned in the introduction may be amenable to a shell model treatment with the use of present-day computers and shell model codes.

3 THE PSEUDO SPIN-ORBIT COUPLING SCHEME

To show that the energetically favored pair states are the pseudo-spin singlet states and establish the goodness of the quantum numbers B and C let us introduce some of the formalism[2]. To the single nucleon creation operators a^+_{jm} we can associate the operators

$$a^+_{bm_b cm_c} = \sum_j \langle bm_b\, cm_c \,|\, jm\rangle\, a^+_{jm} \tag{3}$$

with $b = \frac{1}{2}$, $c = l + 1$ for a pseudo spin-orbit doublet $l_j(l+2)_{j+1}$. It is possible to define double unit tensor operators in b, c-space

$$[a^+_{bc} \times a_{bc}]^{k_b k_c}_{q_b q_c} = \sum_{m_b m_c} \langle bm_b b - m'_b | k_b q_b\rangle \langle cm_c c - m'_c | k_c q_c\rangle \times a^+_{bm_b cm_c} a_{bm_b' cm_c'} (-1)^{b-m_b'+c-m_c'} \tag{4}$$

The operators with $k_b = 0$, or $k_c = 0$, when expressed in terms of unit tensor operators in conventional $j - j$ coupling language

$$[a^+_j \times a_{j'}]^k_q = \sum \langle jmj' - m' | kq\rangle\, a^+_{jm} a_{j'm'} (-1)^{j'-m'} \tag{5}$$

have the form

$$C^{k_c}_{q_c} = \sum_{jj'} (-1)^{j+b+c+k_c} [(2j+1)(2j'+1)]^{1/2} \begin{Bmatrix} j & c & b \\ c & j' & k_b \end{Bmatrix} [a^+_j \times a_{j'}]^{k_c}_{q_c} \tag{6a}$$

and

$$B^{k_b}_{q_b} = \sum_{jj'} (-1)^{j'+b+c+k_b} [(2j+1)(2j'+1)]^{1/2} \begin{Bmatrix} j & b & c \\ b & j' & k_b \end{Bmatrix} [a^+_j \times a_{j'}]^{k_b}_{q_b} \tag{6b}$$

respectively. In particular, the vector operators with $k_b = 1$ or $k_c = 1$, with suitable normalization factors:

$$\mathbf{B}^{k=1}_q = \sum_{jj'} [\tfrac{1}{3}(2b+1)\, b(b+1)(2j+1)(2j'+1)]^{1/2} (-1)^{j'+b+c+1} \times \begin{Bmatrix} j & b & c \\ b & j' & 1 \end{Bmatrix} [a^+_j \times a_{j'}]^1_q \tag{7a}$$

$$\mathbf{C}^{k=1}_q = \sum_{jj'} [\tfrac{1}{3}(2c+1)\, c(c+1)(2j+1)(2j'+1)]^{1/2} (-1)^{j+b+c+1} \times \begin{Bmatrix} j & c & b \\ c & j' & 1 \end{Bmatrix} [a^+_j \times a_{j'}]^1_q \tag{7b}$$

satisfy the usual angular momentum commutation relations

$$[B_+, B_-] = 2B_0, \quad [B_0, B_\pm] = \pm B_\pm; \qquad [C_+, C_-] = 2C_0, \quad [C_0, C_\pm] = \pm C_\pm \tag{8a}$$

In addition the operators **B** and **C** are commuting angular momentum operators

$$[\mathbf{B}, \mathbf{C}] = 0 \tag{8b}$$

whose sum is the conventional total angular momentum operator

$$\mathbf{J} = \mathbf{B} + \mathbf{C} \tag{8c}$$

That is, the total angular momentum operator **J** has been expressed in terms of a pseudo-spin and pseudo-orbital angular momentum operator. At the same time of course $\mathbf{J} = \mathbf{S} + \mathbf{L}$, where **S** and **L** are the real total spin and real orbital angular momentum operators. However, the last relation is much less useful in the nuclear shell model compared with (8c) since the strong spin-orbit splitting of the shell model may even place the two members of a real spin-orbit doublet into different major nuclear shells.

To further examine the physics implicit in the quantum numbers B and C, the SDI will be transformed into b,c-space. The SDI acts on the favored pair states only: In terms of pair creation operators coupled to total JM

$$A^+_{JM}(jj') = \sum_{m(m')} \langle jmj'm'|JM\rangle \, a^+_{jm} a^+_{j'm'} \tag{9}$$

the favored pairs of the SDI are

$$\mathscr{A}^+_{JM} = \tfrac{1}{2} \sum_{jj'} (-1)^l \, h_J(jj') \, A^+_{JM}(jj') \tag{10a}$$

with

$$h_J(jj') = \left[\frac{(2j+1)(2j'+1)}{(2J+1)}\right]^{1/2} \left\langle j\frac{1}{2}j' - \frac{1}{2}\right| J0 \left.\right\rangle (-1)^{j'+\frac{1}{2}} \tag{10b}$$

For $J = 0$ the favored pair is that of ordinary pairing theory. In particular, the operators $\mathscr{A}^+_{00}$, $\mathscr{A}_{00}$ are the total quasispin operators $\mathscr{S}_+, \mathscr{S}_-$. In terms of $\mathscr{A}^+_{JM}$ the surface delta interaction takes the simple form

$$H_{SDI} = -G \sum_{JM} \mathscr{A}^+_{JM} \mathscr{A}_{JM} \tag{11}$$

If the favored pair operators are transformed to b,c-space, we obtain for the case of a single pseudo spin-orbit doublet

$$\mathscr{A}^+_{JM} = \frac{(2c+1)}{[2(2J+1)]^{1/2}} \langle c0c0|J0\rangle \, A^+_{B=OM_B=O,\, C=JM} \tag{12}$$

That is, the favored pair creation operator is a pair operator coupled to $B = 0$, $C = J$, where

$$A^{+}_{BM_BCM_C} = \sum_{m_b m_c} \langle bm_b bm'_b | BM_B \rangle \langle cm_c cm'_c | CM_C \rangle \, a^{+}_{bm_b cm_c} a^{+}_{bm_b' cm_c'} \tag{13}$$

As a result H_{SDI} is a scalar in both b and c-space, leading to the following important consequences.

1. The SDI is diagonal in B and C. Therefore, if the single particle levels of a pseudo spin-orbit doublet are degenerate, both B and C are good quantum numbers. A pseudo Russell Saunders coupling scheme is a good coupling scheme for nuclei.

2. For a $(B, C)_J$ multiplet with $J = C + B, \ldots, |C - B|$, the eigenvalues of the SDI are independent of J. In the approximation in which the real effective nucleon-nucleon interaction can be replaced by the SDI and the energy separation of the members of the pseudo spin-orbit doublet can be neglected, nuclear states show no fine structure (again using the language of atomic spectroscopy).

Both of these conclusions of course hold for a much wider class of 2-body interactions. The only requirements are that the interaction have only pseudo-angular momentum tensor character $k_b = 0$ and $k_c = 0$, that it be a central interaction in pseudo-space. Arima, Harvey, and Shimizu[1] have made a detailed analysis of the realistic effective interaction in the Ni region. Although the interaction has some relatively large pseudo-spin tensor and vector components ($k_b = 2$ and $k_b = 1$), their matrix elements are in general small compared with those of the $k_b = 0$ part of the interaction.

One other interesting property holds for the SDI (or a more general interaction acting only on $B = 0$ pair states). Just as the operator $\mathscr{A}_{00} = \mathscr{S}_-$ has the property

$$\mathscr{S}_- | n = v \rangle = O \tag{14}$$

the operators $\mathscr{A}_{JM}$ (for any J) have the property

$$\mathscr{A}_{JM} | n = 2B \rangle = O \tag{15}$$

(The operator $\mathscr{A}_{JM}$ cannot change B but it lowers n by 2 units. Since a state of $n - 2$ nucleons cannot have $B = \frac{1}{2}n$, the operator must yield zero when acting on a state with $B = \frac{1}{2}n$.) Just as successive application of the operator $\mathscr{S}_-$ can serve to count the number of nucleons not members of favored $J = 0$-coupled pairs, (i.e., the total seniority number v), the operators $\mathscr{A}_{JM}$ in general count the number of nucleons not members of favored pairs of any J. For a single spin-orbit doublet therefore the number $2B$ serves as a generalized seniority number for the favored $J \neq 0$ pairs.

The pseudo spin-orbit scheme is useful not only in classifying the energy levels but also leads to new selection rules for transition probabilities. The multipole moment operators $\sum_i r_i^k Y_{kq}(\Omega_i)$ might be expected to have complicated tensor character in the pseudo spin-orbit space. However, in the approximation in which the radial part of the matrix elements of r^k are independent of the quantum numbers n, l; i.e., in the approximation in which

$$\int_0^\infty R_{nl} r^k R_{n'l'} r^2 \, dr = \text{constant} = \overline{(r^k)} \tag{16}$$

the multipole moment operators are b-space scalars. (Arima, Harvey, and Shimizu refer to the multipole moment operators with constant radial matrix element as surface multipole operators). In the subspace of a single pseudo spin-orbit doublet, in particular, the kth multipole moment operator has the form

$$\sum_i r_i^k Y_{kq}(\Omega_i) = \overline{(r^k)} \frac{(2c+1)}{[2\pi(2k+1)]^{1/2}} \langle c0c0|k0\rangle (-1)^{c+k} \times [a^+_{\frac{1}{2}c} \times a_{\frac{1}{2}c}]_q^{k_b=0,\, k_c=k} \tag{17}$$

(The phases of the radial functions with $l+2$ are chosen relative to those with l such that the off-diagonal radial matrix elements $nl \rightarrow n'(l+2)$ have the same sign as the diagonal matrix elements.) The approximation (16) is quite good for both harmonic oscillator wave functions[1] and Woods-Saxon wave functions. For the 82-neutron region and Woods-Saxon wave functions for the $1g_{7/2}\; 2d_{5/2}$ doublet, for example, the coefficients of the tensor operators $[a^+_{\frac{1}{2}c} \times a_{\frac{1}{2}c}]_q^{k_b=1, k_c; k}$ are small compared with those with $k_b = 0$ in the exact expansion of the quadrupole operator. In particular

$$\frac{k_b = 1, k_c = 1 \text{ component}}{k_b = 0, k_c = 2 \text{ component}} = 0.1, \quad \frac{k_b = 1, k_c = 3 \text{ component}}{k_b = 0, k_c = 2 \text{ component}} = 0.09$$

A pure $\Delta B = 1$ $E2$ transition must be expected to be of order of 1% of a $\Delta B = 0$ transition.

Since the selection rule $\Delta B = 0$ is dependent on a property of a radial matrix element its validity may be quite different for inelastic scattering processes. In his analysis of inelastic α-scattering on the 82-neutron nuclei ^{140}Ce and ^{139}La, F. T. Baker[14] has calculated radial form factors for protons in the $1g_{7/2}\; 2d_{5/2}$ doublet using the semi-realistic interaction of Morgan and Jackson[15]. These are reproduced in Fig. 3. Although they are strong functions of nl, $n'l'$ in the nuclear interior, they all have approximately the same value in the nuclear surface where the inelastic α-scattering process must be expected to take place. The selection rule $\Delta B = 0$ can therefore be expected to be good for inelastic α-scattering also. Using the pseudo spin-orbit doublet model, Baker[14] finds that the inelastic scattering cross section from a pure pseudo-spin singlet ($B = 0$) ground state in ^{140}Ce to a pure $B = 1$ excited state ($^3\tilde{P}_2$ or $^3\tilde{F}_2$) is 10% and 5%, respectively, of the cross section to a pure $B = 0$, $C = 2$ state (the $^1\tilde{D}_2$ state of Fig. 2).

The magnetic moment operator also has a relatively simple form when expressed in the b, c-scheme

$$\boldsymbol{\mu}_q = g_l \mathbf{J}_q + (g_s - g_l)\left(-\tfrac{1}{3}\,\mathbf{B}_q + \beta[\mathbf{a}^+_{\frac{1}{2}c} \times \mathbf{a}_{\frac{1}{2}c}]_q^{k_b=1,k_c=2;k=1}\right) \tag{18}$$

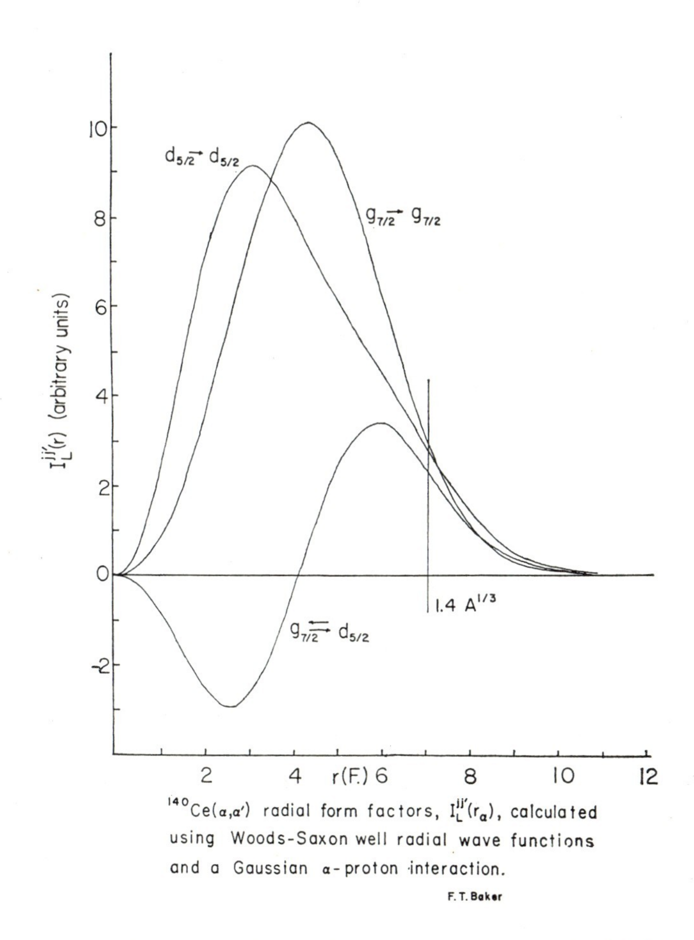

FIGURE 3 Radial Form factors for α-scattering from pseudo f-shell protons, $k = 2$

where the coefficient β depends on the quantum number c. (For the $g_{7/2}$ $d_{5/2}$ doublet $\beta = (4/3)\ [7/15]^{\frac{1}{2}}$.) Since the last two terms can make no contribution to a diagonal matrix element in a $B = 0$ state, the magnetic moment in a pure pseudo-spin singlet state is given by $g_l J$, as already noted in connection with the discussion of ^{140}Ce and Fig. 2.

Introduction of the pseudo angular momentum quantum numbers has reduced the problem of a mixed nuclear configuration such as $g_{7/2}\, d_{5/2}$ to one which is mathematically equivalent to that for a single atomic shell. The $g_{7/2}\, d_{5/2}$ nuclear configuration is equivalent to a pseudo *f*-shell. The spectroscopic classification of the many-particle states and the fractional parentage coefficients needed for detailed calculation have been known since the early work of Racah[16]. Calculations for the $g_{7/2}\, d_{5/2}$ configuration of identical nucleons could in principle have been carried out in the days before the nuclear shell model was born. The classification scheme for such a doublet can be based on the conventional (Racah) group chain

$$U(4c + 2) \supset R_B(3) \times U(2c + 1) \supset R(2c + 1) \supset R(3)$$

(In the case of the pseudo *f*-shell, $c = 3$, the special group[16] G_2 can be sandwiched in between the seniority group $R(2c + 1)$ and the group $R(3)$ generated by the operators **C**. The group G_2, however, seems to have no special physical significance for the nuclear problem).

Alternate group chains starting with $R(8c + 4)$ generated by all operators of the type $[a^+ \times a^+]$, $[aa]$, and $[a^+ \times a]$ may also be useful. There are two possibilities

$$R(8c + 4) \supset R_{\mathscr{S}}(3) \times Sp(4c + 2) \tag{1}$$

where $Sp(4c + 2)$ is generated by the operators $[a^+ \times a]^{k_b k_c}$ with $k_b + k_c =$ odd, and $R_{\mathscr{S}}(3)$ is generated by the 3 components of the quasispin operator $\mathscr{S}$: $[a^+ \times a^+]^{B=0,\, C=0}$, $\frac{1}{2}(N_{\text{op.}} - 2c - 1)$, $[aa]^{B=0,\, C=0}$; or

$$R(8c + 4) \supset R_B(3) \times SP'(4c + 2) \tag{2}$$

where $Sp'(4c + 2)$ is the Helmers group[17] generated by all operators $[a^+ \times a^+]$, $[aa]$, $[a^+ \times a]$ coupled to $B = 0$, and $R_B(3)$ is generated by the 3 components of the pseudo-spin operator **B**.

The more recent classification schemes of Armstrong and Judd[18] based on a quasiparticle factorization of the atomic shells may also be useful, although they may be more difficult to apply to the nuclear case in which the states of *low* pseudo-spin B are the most important.

Calculations in terms of conventional *cfp* expansions are straightforward. The reduced matrix element for the kth multipole moment in the approximation of Eqs. (16) and (17), for example, is given by

$$\langle nv'\alpha'(BC')J'\| \sum_i r_i^k Y_k(\Omega_i)\|nv\alpha(BC)J\rangle = (\overline{r^k}) \sum_{v''\alpha''B''C''} (-1)^{B+C+C'+C''+k+J'}$$

$$\times \left[\frac{(2J+1)(2J'+1)(2C+1)(2C'+1)}{4\pi}\right]^{1/2} (2c+1)\langle c0c0|k0\rangle$$

$$\times \begin{Bmatrix} J & C & B \\ C' & J' & k \end{Bmatrix} \begin{Bmatrix} C & c & C'' \\ c & C' & k \end{Bmatrix} n\langle c^{n-1}v''\alpha''B''C''; \tfrac{1}{2}c|\}$$

$$\times c^n v\alpha BC\rangle \langle c^{n-1}v''\alpha''B''C''; \tfrac{1}{2}c|\} c^n v'\alpha' BC'\rangle \tag{19}$$

where the labels α may be needed for the states with $v > 2$ for which $v(BC)$ do not give a complete classification of the states. For $v, v' \leqq 2$ essentially no *cfp* expansion is needed. The n-dependence of the matrix element can be obtained with quasispin techniques. Operators with $k_b + k_c$ = even are quasispin vectors ($\mathscr{S} = 1$ operators). Wigner coefficients for the quasispin coupling $\overrightarrow{\mathscr{S} = \tfrac{1}{2}(2c+1-v)} \times \vec{1} \rightarrow \overrightarrow{\mathscr{S}' = \tfrac{1}{2}(2c+1-v')}$ lead at once to the n-dependent factors for the $E2$ rates of the 82-neutron nuclei quoted in the discussion of Fig. 2.

Calculations for energy spectra with $v > 2$ are extremely simple in the approximation in which the single particle energy splitting of the pseudo spin-orbit doublet is neglected. Calculations for the $g_{7/2}\, d_{5/2}$ doublet using the SDI have been carried out by Adler[2] and are reproduced in Fig. 4. Since the SDI commutes with the total quasispin operator $\mathscr{S}^2$, the spectra are independent of n and functions only of $v(B, C)$. States with $B = \tfrac{1}{2}v$ and different values of C are all degenerate. This follows from the property given by Eq. (15). Even the states with $B < \tfrac{1}{2}v$ cluster quite closely about their centers of gravity. In the odd A nuclei $^{139}_{57}La_{82}$ and $^{141}_{59}Pr_{82}$ there are two clusters of energy levels in the $v = 3$ region, one centered at roughly 1.5 MeV of excitation appears to be predominantly $B = \tfrac{1}{2}$, while the other between 2 and 2.5 MeV seems to have states with predominant $B = 3/2$ components. Recent inelastic scattering experiments on these nuclei[14, 19] indicate that the transitions from the ground state which is predominantly $B = \tfrac{1}{2}$ excite mainly the positive parity states in the 1.5 MeV region, an example of the selection rule $\Delta B = 0$. An additional simplification has been noted by Baker[14]. The J-dependence of the transition amplitudes is given by a single Racah coefficient [see Eq. (19)]. For a $B = \tfrac{1}{2} \rightarrow B = \tfrac{1}{2}$ transition and relatively large c (for this purpose $c = 3$ can be considered large) transitions implying a pseudospin flip have a much smaller probability compared with the pseudo-spin non-flip transitions. In ^{139}La the ground state has $J = 7/2 = C + \tfrac{1}{2}$. Contributions to the inelastic scattering cross

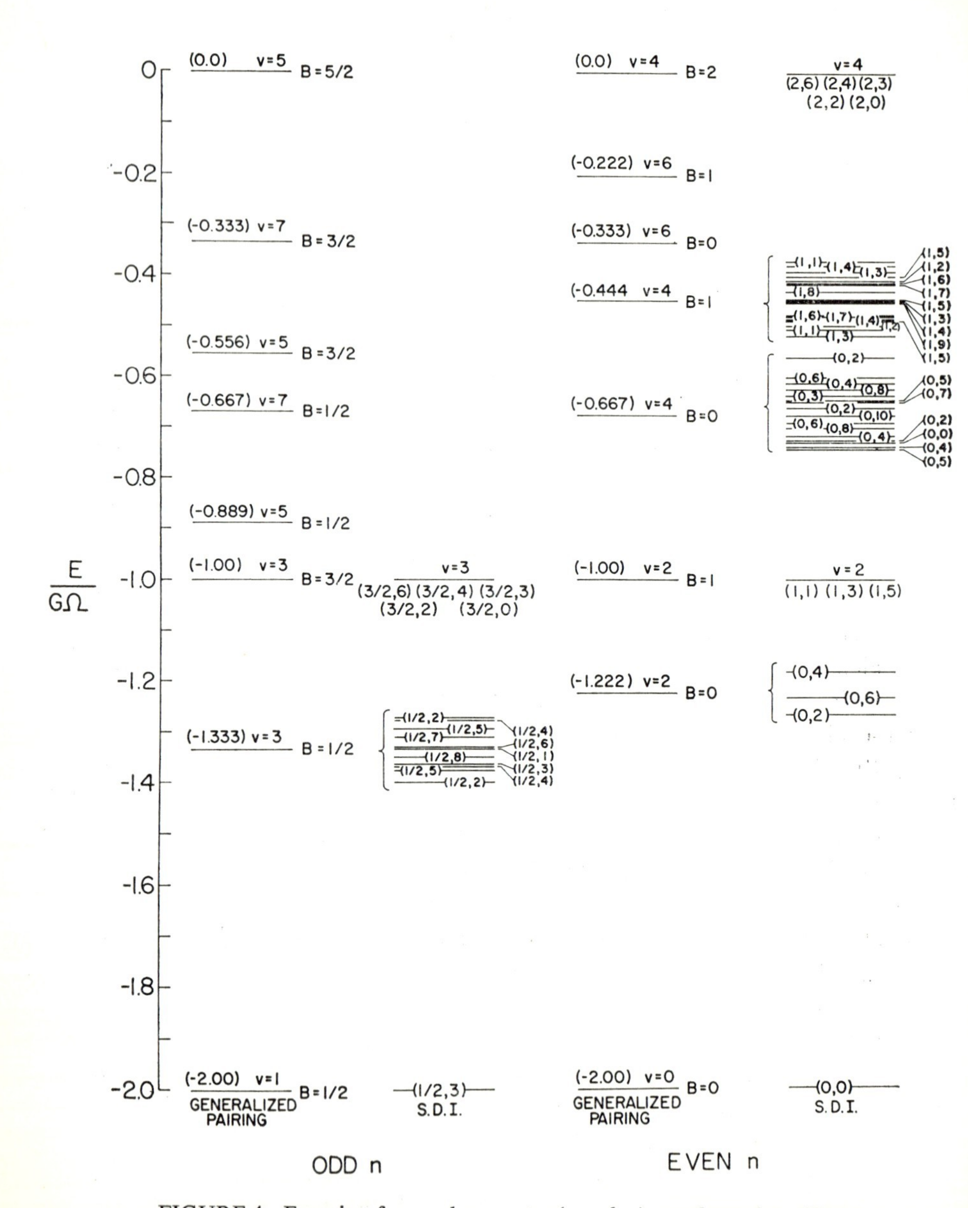

FIGURE 4 Energies for a degenerate $(g_{7/2}d_{5/2})$ configuration. Exact eigenvalues for SDI are shown only for $v \leqq 4$. Levels are labeled by v, (B, C). For the generalized pairing interaction only the low-lying levels of the higher v states are shown

section from the pieces in the final state wave function with $J' = C' + \frac{1}{2}$ are on the average 10 times larger[14] than from those with $J' = C' - \frac{1}{2}$.

Since states of a given B and v cluster quite closely about their center of gravity, the results of the SDI calculation for states with $v \geqq 3$ suggest that the SDI can in zeroth approximation be replaced by an even simpler interaction in which the interaction strength for the favored $J \neq 0$ pairs is taken to be independent of J but differs from the greater pairing strength for $J = 0$, (by a factor of $2/(2c + 3)$ if the interaction strength for $J \neq 0$ pairs is replaced by its average value). This simplified form of the interaction has been named generalized pairing interaction[2], GPI. In terms of the pair creation operators of Eq. (13) it is given by

$$H_{\text{GPI}} = -\frac{1}{2} G(2c+1)\left[A^{+}_{00,00}A_{00,00} + \frac{2}{(2c+3)} \sum_{J>0} \sum_{M} A^{+}_{00,JM}A_{00,JM}\right] \tag{20}$$

It can be expressed in terms of the usual $J = 0$ pairing Hamiltonian and a 2-body pseudo spin-spin interaction

$$H_{\text{GPI}} = -G\frac{(2c+1)}{(2c+3)}\mathscr{S}_{+}\mathscr{S}_{-} + G\frac{(2c+1)}{(2c+3)}\left[2\sum_{i<j}^{n} \mathbf{b}_i \cdot \mathbf{b}_j - \frac{1}{2}\sum_{i<j}^{n} 1\right] \tag{21}$$

and its eigenvalues have the simple form

$$E_{\text{GPI}} = -G\frac{(2c+1)}{(2c+3)}\left[\frac{1}{4}(n-v)(4c+4-n-v) - B(B+1) + \frac{3}{4}n + \frac{1}{4}n(n-1)\right] \tag{22}$$

These are compared in Fig. 4 with the exact eigenvalues of the SDI. For states with $v \geqq 4$, the quantum number B is more important than v in ordering the states as to excitation energy. The $v = 6$ states with $B = 0$ and 1, for example, have come down into the region of $v = 4$ states; while the $v = 5$ states with $B = \frac{1}{2}$ lie very close to the $v = 3$ states with $B = 3/2$. Since states with $v \geqq 4$ are of interest essentially only insofar as they perturb the lower v states (by means of as yet missing parts of the realistic Hamiltonian) precise knowledge of their position may not be important, so that the GPI may be very useful in giving a simple estimate of the approximate position of these higher-lying states.

In the real 82-neutron nuclei the most important part of the realistic Hamiltonian which has been neglected so far is the single particle term. The splitting of the single particle states $1g_{7/2}$ and $2d_{5/2}$ may be as much as 1 MeV, so that it is not a negligible perturbation. For a pseudo spin-orbit doublet with single particle energies ε_{j+1}, ε_j the single particle Hamiltonian can be expressed as

$$H_{\text{s.p.}} = n\left[\frac{1}{2}(\varepsilon_{j+1} + \varepsilon_j) + \frac{1}{2(2c+1)}(\varepsilon_{j+1} - \varepsilon_j)\right] + \frac{2}{2c+1}(\varepsilon_{j+1} - \varepsilon_j)\sum_{i=1}^{r} \mathbf{b}_i \cdot \mathbf{c}_i \tag{23}$$

that is, apart from a constant term proportional to n, the single particle Hamiltonian has the form of a one-body pseudo spin-orbit coupling term. Its matrix elements are subject to the selection rule, $|\Delta B| \leqq 1$, $|\Delta C| \leqq 1$; and it has no diagonal matrix elements in the important lowest-lying $B = 0$ states. In addition, the pseudo spin-orbit coupling term is a quasi-spin vector ($\mathscr{S} = 1$ operator), since its tensor character $k_b + k_c =$ even. The n-dependence of its matrix elements, diagonal in v, is therefore given by the factor $(n - 2c - 1)$. This is zero for the half-full shell, (the nucleus ${}^{139}_{57}La_{82}$ of the 82-neutron family), and relatively small for nuclei near the half-full shell, a fact which helps to contribute to the goodness of the quantum numbers B and C, even in cases where the single particle splitting $\varepsilon_{j+1} - \varepsilon_j$ is relatively large.

Since the pseudo spin-orbit coupling scheme gives an economical description of the shell model space, and since the GPI gives a good approximation to the spectroscopically less important high seniority states, it is interesting to see whether simple calculations, which exploit these facts, can give a good description of the states of the real 82-neutron nuclei. Such calculations have been carried out at Michigan by Jones and Borgman[3]. The aim here was to see whether an approximate calculation involving a very simple treatment of the shell model space can rival the more detailed and precise calculations of Wildenthal[10, 11] involving a shell model space of dimensionality of order of 100. For this purpose the GPI was used to approximate the energies of states with $v \geqq 4$. In the same spirit matrix elements of the single particle Hamiltonian connecting states with $v = 4$ to states with $v' \geqq 4$ were neglected. For all other matrix elements the full single particle Hamiltonian and the SDI (the 2-body interaction of Wildenthal) were used. Results for ${}^{140}Ce$ are shown in Fig. 5, where they are compared with the experimental levels and the calculation of Wildenthal. The calculation described so far, based on a pure $(g_{7/2}\, d_{5/2})^n$ configuration, is shown in column (*c*). Wildenthal has found that excitations into the $s_{1/2}$, $d_{3/2}$ levels are important; and his calculations include the full configurations $(g_{7/2}\, d_{5/2})^n$, $(g_{7/2}\, d_{5/2})^{n-1}$ $s_{1/2}$, $(g_{7/2}\, d_{5/2})^{n-1}\, d_{3/2}$. In the present calculations it is easy to include the excitation of one particle into the $s_{1/2}$ and $d_{3/2}$ orbits (pseudo p-shell); but to retain the simplicity of the treatment it has been assumed that the remaining $n - 1$ nucleons are in their lowest ($v = 1$) seniority state, when coupled to the single pseudo p-shell nucleon. Results[3] are shown in column (*a*) of Fig. 5. It is equally easy to allow, in addition, excitations of one pair of nucleons into either the pseudo p-shell or the $1h_{11/2}$ shell, with the remaining

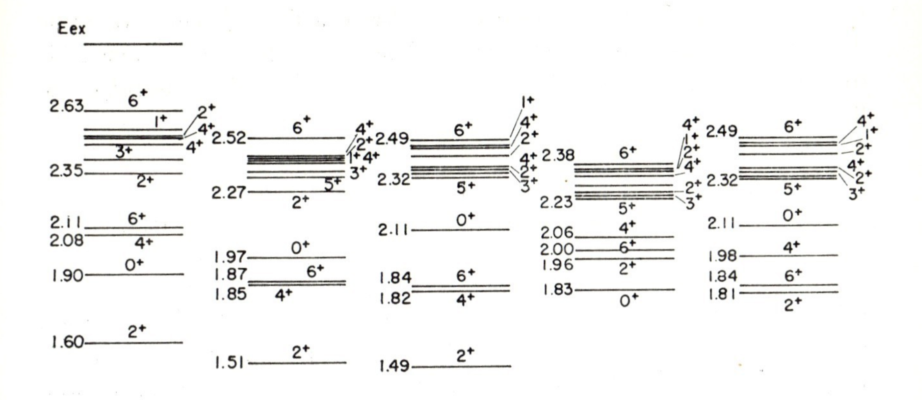

FIGURE 5 ^{140}Ce Spectrum. Comparison of Detailed Shell Model Calculation (Wildenthal) and Simplified Shell Model Calculations based on the Pseudo $\tilde{L}\tilde{S}$-Scheme (Jones and Borgman). The Wildenthal results are taken from ref. 10): SDI strength $G = 0.383$, $\varepsilon_{7/2} - \varepsilon_{5/2} = \Delta = -0.52$ MeV. More recent calculations[11], using $\Delta = -0.88$ give similar results. (a), (b), (c) are the results of Jones and Borgman[3]: (c) Pure $(g_{7/2}d_{5/2})^8$ configuration, $G = 0.38$, $\Delta = -1.0$ MeV. (a) Excitation of one pseudo-p shell proton coupled to pseudo-f shell ($v = 1$) configuration is included; $G = 0.38$, $\Delta_{\tilde{f}} = -1.0$ Mev; $\bar{\varepsilon}_{\tilde{p}} - \bar{\varepsilon}_{\tilde{f}} = 2.5$ Mev, $\Delta_{\tilde{p}} = 0$. (b) Excitation of both one pseudo p-shell proton and one pseudo p-shell or $1h_{11/2}$—pair are included; $G = 0.20$, $\Delta_{\tilde{f}} = -1.0$ Mev, $\bar{\varepsilon}_{\tilde{p}} - \bar{\varepsilon}_{\tilde{f}} = 2.5$ Mev, $\varepsilon_{1h11/2} - \bar{\varepsilon}_{\tilde{f}} = 2.5$ Mev. The ground state wave function for (a), (c) includes: 85% $^1\tilde{S}_0$, 13% $^3\tilde{P}_0$ $v = 2$. First excited 2^+ state wave function for (a) includes: 69% $^1\tilde{D}_2$, 21% $^1[(\tilde{p})^1(\tilde{f})^7]_2$

$(g_{7/2}\, d_{5/2})$ nucleons coupled to $v = 0$. For this shell model space a reduction in the two-body interaction strength is required which results in a somewhat poorer overall description of the spectrum, column (*b*) of Fig. 5; so that, contrary to our expectations, this does not appear to be the best simple truncation of the shell model space. Insofar as the shell model space for column (*a*) most closely approximates that of Wildenthal, it can be seen that there is quite good agreement between the two calculations. This gives some hope that a much more complicated shell model problem may be amenable to treatment by the simplified approach based on the pseudo spin-orbit coupling scheme.

4 MAJOR NUCLEAR SHELLS. THE PSEUDO *SU*(3) SCHEME.

The discussion so far has concentrated on the spectroscopy of a single pseudo spin-orbit doublet. The case of several nearly degenerate single particle states, in particular the case of a major nuclear shell is of special interest. The major nuclear shells can all be considered as made up of pseudo spin-orbit doublets $(bc) = (\frac{1}{2}c_{\max})$, $(\frac{1}{2}c_{\max} - 2)$, ..., $(\frac{1}{2}1)$ or $(\frac{1}{2}0)$, to which is added a single state j_0 of opposite parity with $(bc) = (0 j_0)$, where $j_0 = c_{\max} + 5/2$. The shell $50 \leqq N$ or $Z \leqq 82$ would be classified as

nlj:	$(1g_{7/2}2d_{5/2})$	$(2d_{3/2}3s_{1/2})$	$(1h_{11/2})$
(bc):	$\left(\frac{1}{2}\ 3\right)$	$\left(\frac{1}{2}\ 1\right)$	$\left(0\ \frac{11}{2}\right)$

With these (bc) assignments the favored pair state of the SDI with even J, [Eq. (10)], is a pure $B = 0$ state. It might seem strange that a major shell should include both pseudo-spin $\frac{1}{2}$ and pseudo-spin 0 particles. Since many-particle states of integral and $\frac{1}{2}$-integral spins then correspond to states of opposite parity, however, the assignment is not unnatural. The total **B** operator consists of a sum over the integral c's of terms given by Eq. (7a). The total **C** operator consists of a similar sum, [Eq. (7b)], to which is added the j_0 term: $[\frac{1}{3} j_0(j_0 + 1)(2j_0 + 1)]^{1/2}\, [a^+_{j_0} \times a_{j_0}]^{k=1}_q$. The operators **B** and **C** satisfy all the commutation properties, [Eq. (8)], of pseudo angular momenta. Surface multipole operators (with even k) can be expressed as

$$Q^k_q = \langle r^k \rangle \sum_{cc'} \left[\frac{(2c + 1)(2c' + 1)}{2\pi(2k + 1)} \right]^{1/2} (-1)^c \langle c0c'0|k0\rangle$$
$$\times\, [a^+_{\frac{1}{2}c} \times a_{\frac{1}{2}c'}]^{k_b = 0, k_c = k_q}_q + \frac{(2j_0 + 1)}{[4\pi(2k + 1)]^{1/2}}$$
$$\times\, (-1)^{j_0 - \frac{1}{2}} \langle j_0 \tfrac{1}{2} j_0 - \tfrac{1}{2}|k0\rangle\, [a^+_{j_0} \times a_{j_0}]^k_q \qquad (24)$$

Since the quantum numbers B and C no longer give a complete specification of even the 2-particle states, a search for further quantum numbers must be made. Since single particle states of the same parity behave as pseudo harmonic oscillator states, the possibility of a pseudo $SU(3)$ scheme arises naturally. A single nucleon in a $p_{1/2}\,p_{3/2}\,f_{5/2}$, or pseudo $s-d$ shell, can be assigned $SU(3)$ quantum numbers $(\lambda\mu) = (20)$, while an $s_{1/2}\,d_{3/2}\,d_{5/2}\,g_{7/2}$, or pseudo $p-f$ shell nucleon, would have pseudo $(\lambda\mu)$ quantum numbers (30). Spectra for two identical particles, using a SDI and assuming the degeneracy of such a shell, are reproduced in Fig. 6.

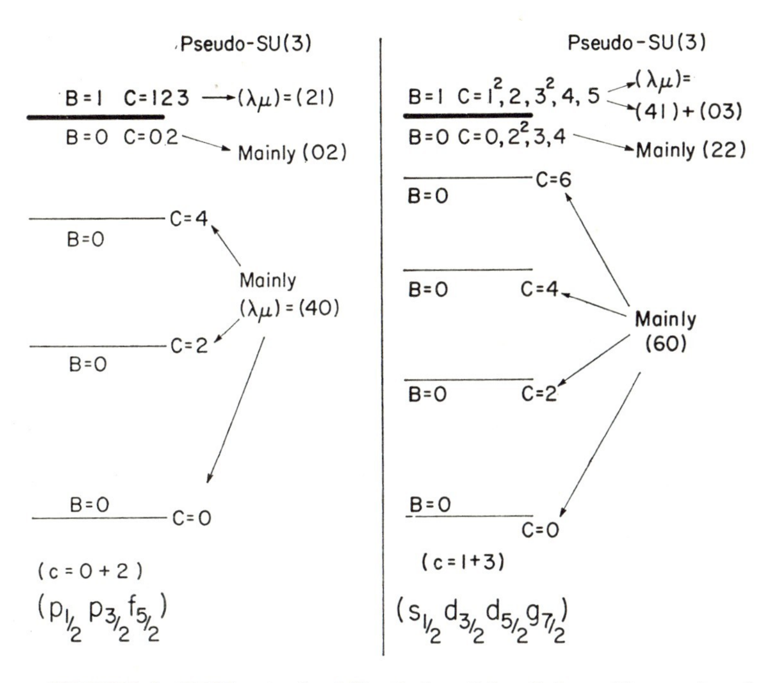

FIGURE 6 SDI Spectra for 2 identical particles. Column (1): pseudo s-d shell. Column (2): Pseudo p-f shell

The specification $B = 0$ is no longer sufficient to characterize the favored pairs. The $B = 0$ two-particle states correspond to the symmetric coupling of $(\lambda 0) \times (\lambda 0)$: for the pseudo $s-d$ shell a combination of (40) and (02);— for the pseudo $p-f$ shell a combination of (60) and (22). Although the favored pair states have as predominant components the pseudo-$SU(3)$ representations $(2\lambda, 0)$, the SDI is far from diagonal in the pseudo $SU(3)$ representation. This is illustrated for the pseudo $s-d$ shell by the 4th column of Table 1. A favored pair state with $J = 2$, based on a pseudo $SU(3)$ model, (pure (40) representation), gives the $s-d$ component of the two-particle wave function too much weight by a factor of $[5/2]^{\frac{1}{2}}$ relative to the d^2 component. For the $J = 0$ pair the factor is 5/2 for the s^2 relative to the d^2 components[2]. Even so, the pseudo $SU(3)$ model gives a much better approximation for the favored pair states than a model based on the $B = 0$ pairs for a single-nucleon pseudo angular momentum assignment $bc = 1\ 3/2$, (column 5 of Table 1), or $bc = 3/2\ 1$, ref. 2.

In a subspace of single particle states of the same parity, the SDI can be written

$$H_{\mathrm{SDI}} = -\frac{G}{2} \sum_{cc'c''c'''} \sum_{JM} F_J(c, c')\, F_J(c'', c''')\, A^{+}_{00,JM}(c, c')\, A_{00,JM}(c'', c''') \tag{25a}$$

where

$$F_J(c, c') = \left[\frac{(2c + 1)(2c' + 1)}{(2J + 1)}\right]^{1/2} \langle c0c'0|J0\rangle \tag{25b}$$

and where the pair creation operators $A^{+}_{BM_BCM_C}(cc')$ are defined in analogy with Eq. (13). When expressed in terms of unit tensor operators, Eq. (4), the interaction has the form

$$H_{\mathrm{SDI}} = -\frac{G}{4} \sum_{cc'c''c'''} \sum_{\substack{k_bk_c \\ q_bq_c}} F_{k_c}(c, c'')\, F_{k_c}(c', c''')\, (-1)^{k_b - q_b + k_c - q_c}$$

$$\times\, [a^{+}_{\frac{1}{2}c} \times a_{\frac{1}{2}c''}]^{k_bk_c}_{q_bq_c}\, [a^{+}_{\frac{1}{2}c'} \times a_{\frac{1}{2}c'''}]^{k_bk_c}_{-q_b,-q_c} \tag{26}$$

Arima, Harvey, and Shimizu point out that a model based on a pure surface quadrupole-quadrupole interaction, i.e., $k_b = 0$, $k_c = 2$ in the notation of Eq. (26), may rival the SDI as a good approximate interaction for the pseudo $s - d$ shell. In such a model of course the $SU(3)$ quantum numbers are good.

Since the generalized pairing interaction gave a very simple yet good approximation for the spectrum of a single pseudo spin-orbit doublet, it is interesting to see whether such an approximation may be useful for the case of two or more doublets. For a single doublet the GPI is based on the approximation

$$F_J(c, c) \simeq \left[\frac{2(2c + 1)}{(2c + 3)}\right]^{1/2} (-1)^{\frac{1}{2}(2c+J)} \quad \text{for all even } J;\ J \neq 0 \tag{27a}$$

in which the $J \neq 0$ interaction strength is replaced by its average value:

$$\frac{\sum_{J>0} (2J + 1)\, (F_J(c, c))^2}{\sum_{\substack{J>0 \\ \text{even}}} (2J + 1)} = \frac{2(2c + 1)}{(2c + 3)} \tag{27b}$$

In the same approximation, with $c' \neq c$, we obtain

$$F_J(c, c') \simeq \left[\frac{2c' + 1}{c' + 1}\right]^{1/2} (-1)^{\frac{1}{2}(c+c'+J)} \text{ with } c' < c. \tag{27c}$$

To make the interaction in the $J \neq 0$ pairs completely independent of pseudo-orbital quantum numbers, it may be advantageous to replace the approximations (27) with

$$\left.\begin{aligned} |F_J(c, c)| &= \sqrt{2} \quad (\text{for } J \neq 0) \\ |F_J(c, c')| &= 1 \qquad\quad (c' \neq c) \end{aligned}\right. \quad J = \text{even only} \tag{28}$$

The approximation (28) is not much more extreme than that implied by Eqs. (27) but is more universal since all dependence on the quantum numbers c has been removed. This interaction will be named the generalized pairing interaction (GPI) for the case of configurations based on more than one pseudo spin-orbit doublet, since it gives equal weight to each c, c' component of the favored $J \neq 0$ pairs. To compare this approximate form of the interaction with the SDI and the $SU(3)$ model, it is interesting to note that the favored $J = 2$ pair state for the pseudo $s - d$ shell (Table 1) has relative intensities for the $s - d$ versus d^2 components given by the ratios: 1 : 1 for GPI, 7 : 5 for SDI, and 7 : 2 for the $SU(3)$ representation (40).

The eigenvalues for the GPI of an even nucleus in a degenerate pseudo $s - d$ shell are plotted in Fig. 7. Since the GPI commutes with the total quasispin operator, the total seniority v is again a good quantum number. The $v = 2$ spectrum is particularly simple. States with $v = 2$, $B = 1$ are all degenerate. Of the states with $v = 2$, $B = 0$ one is depressed in energy for each even value of J. The depression below the center of gravity of the $v = 2$, $B = 0$ states is given in general by: $(\nu - 1)\Delta$, where Δ = separation of the center of gravity of the $v = 2$, $B = 0$ states from the $v = 2$, $B = 1$ states, while ν = number of possible c, c' components in the pair state of a given J. The remaining $(\nu - 1)$ $v = 2$, $B = 0$ states of this J are pushed up in energy and become degenerate with the $v = 2$, $B = 1$ states. The greatest coherence of the favored pair states is thus given to those J values for which the number of possible components c, c' is largest, giving these states the greatest depression in energy. This is a property which the GPI shares, at least qualitatively, with the SDI.

Energies for the higher seniority states are complicated functions of v, B, C, since they depend on the exact content of favored pair strength in each state. This is now not a function of one simple quantum number. The depression of a state C below the center of gravity of the states of fixed v, B is never as great as that of the favored pair states of the $v = 2$ family. In some

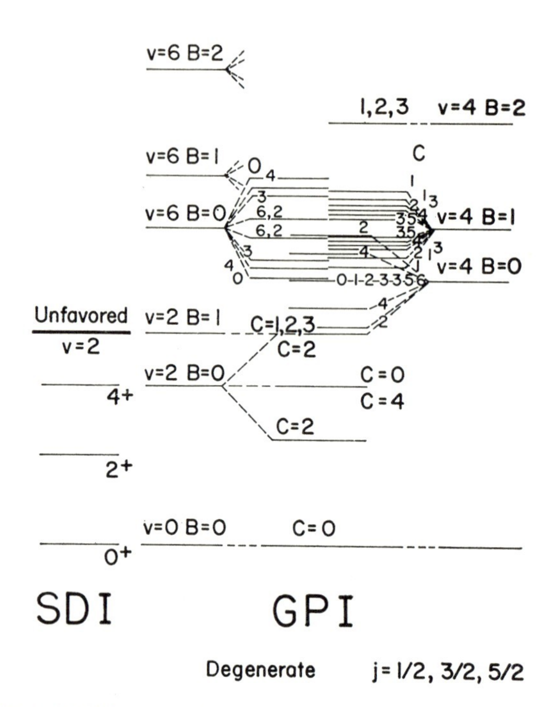

FIGURE 7 GPI Spectrum for an even nucleus,—degenerate pseudo s — d shell. The SDI Spectrum for $v \leqq 2$ is shown for comparison

respects the spectrum is similar to that for the GPI of a single pseudo spin-orbit doublet: the $v = 6$ states with low values of B sink down into the region of $v = 4$ states and may even be in competition with the cluster of higher $v = 2$ states. (A similar property has been noted for the SDI by R. Arvieu[20].)

Unfortunately the eigenvalues of the GPI can no longer be given by a simple general formula as in the case of a single pseudo spin-orbit doublet. Despite its simplicity the GPI does not seem to have any special group theoretical significance so that it does not lead to a new coupling or classification scheme. From this point of view the $SU(3)$ model is much to be preferred. To keep the extreme nature of the approximation in proper perspective, the GPI spectrum of Fig. 7 is to be compared with that of an extreme $SU(3)$ model in Fig. 8. The interaction for this model is taken to be a pure Elliott $Q \cdot Q$ interaction with a pseudo spin-spin interaction to separate states of different spatial symmetry (or B). The spectrum shown is one for

4 identical particles ($T = 2$) in a degenerate pseudo $s - d$ shell. This spectrum is also compared with a "realistic" spectrum for a degenerate $s - d$ shell: the spectrum for 4 identical particles using the Kuo 2-body matrix elements but assuming degenerate single particle states. (I am indebted to J. B. McGrory of Oak Ridge National Laboratory for the calculation of this spectrum.)

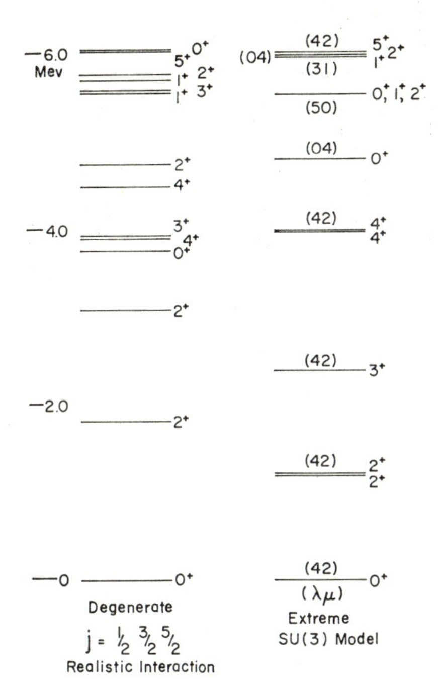

FIGURE 8 Spectra for 4 identical particles in a degenerate s—d shell. Extreme *SU(3)* Model: $E_{(\lambda\mu)BC} = -\alpha[\lambda^2 + \lambda\mu + \mu^2 + 3\lambda + 3\mu - \frac{3}{4}\, C(C+1)] + \beta[B(B+1) - \frac{3}{4}n]$, with $\beta/\alpha \approx 6.7$. The Realistic Spectrum was calculated with the Kuo matrix elements, but assuming degenerate single particle energies (J. B. McGrory, Oak Ridge National Laboratory)

Although the truth seems to lie somewhere between the extremes of the GPI and the $SU(3)$ model, a pseudo $SU(3)$ scheme is to be preferred, partly on physical but in particular on mathematical grounds. So far the discussion has been limited to systems of identical particles, (neutrons only or protons only). It is to be expected that the presence of both protons and neutrons in active shells will improve the goodness of the pseudo $SU(3)$ scheme, even in heavier nuclei where protons and neutrons are filling different major shells. It may even be possible to give a shell model treatment of rare earth nuclei in terms of the pseudo $SU(3)$ coupling scheme. Figure 9 gives a reproduction

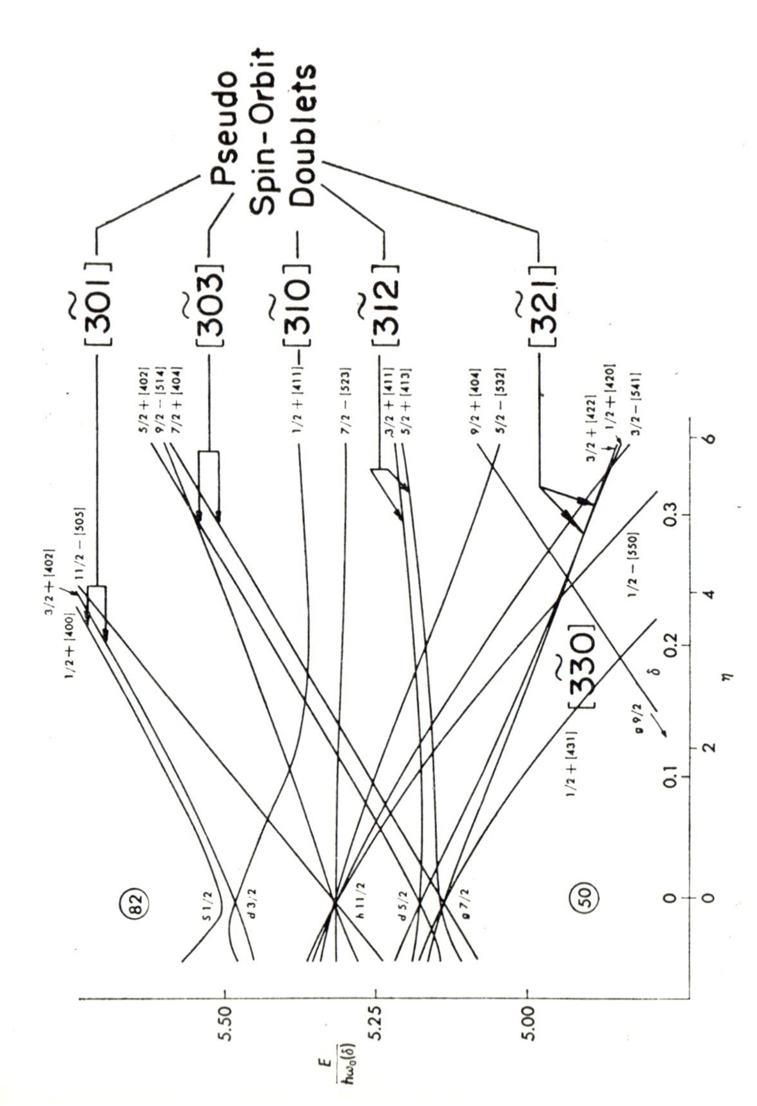

FIGURE 9 Pseudo Spin-Orbit Doublets in the Nilsson Model

of the Nilsson diagram for protons with $50 \leqq Z \leqq 82$ to show that the $g_{7/2}\, d_{5/2}\, d_{3/2}\, s_{\frac{1}{2}}$ or pseudo $p - f$ shell may lead to a good pseudo $SU(3)$ description of the proton configurations in such nuclei. It is possible to assign to these states pseudo-harmonic oscillator quantum numbers $[\widetilde{Nn_z\Lambda}]$ with $\tilde{N} = 3$. Whereas the real spin orbit doublets such as $[404]_{7/2}{}^{+}$, $[404]_{9/2}{}^{+}$are separated by large energies, completely destroying the possibility of describing the rare earth proton configurations in terms of the real $SU(3)$ coupling scheme, the pseudo spin-orbit doublets such as $[\widetilde{312}]$ $3/2^{+}$, $5/2^{+}$ are almost degenerate for all values of the deformation parameter. Just as the Nilsson scheme for the real $2s$–$1d$ shell can be a guide to the real $SU(3)$ representations which are the predominant components of the ground state rotational bands in these nuclei, the Nilsson scheme for the rare earth nuclei may be a guide to the pseudo $SU(3)$ representations which might give a good approximate many-particle shell model description of the proton and neutron configurations in these nuclei. Table 2 shows the leading pseudo $SU(3)$ representations for the odd proton nuclei in the pseudo $p - f$ shell. (The leading representations are those for which $2\lambda + \mu$ has the largest possible value consistent with the Pauli principle.) It is assumed that the systematics of the region based on the Nilsson diagram will be a guide to the odd proton number to be used for a particular nucleus. This assumes that the ground state band proton wave function can be described in terms of a single configuration $(g_{7/2}\, d_{5/2}\, d_{3/2}\, s_{1/2})^{n}\, (h_{11/2})^{Z-50-n}$, whereas in reality there may of course be a large amount of configuration mixing between these and configurations $(g_{7/2}\, d_{5/2}\, d_{3/2}\, s_{1/2})^{n\pm 2}\, (h_{11/2})^{Z-50-n\mp 2}$, ... The systematics of the region based on the Nilsson diagram can also be used to predict the expected K (i.e., K_C) value for the ground state rotational band for each particular value of n. The leading pseudo $SU(3)$ representations always seem to contain the needed K-values, although it is not clear, for example, why $K=2$ should lie lowest in the pseudo $SU(3)$ representation (11,2) for $n = 7$, while $K = 0$ lies lowest in the representation (11,2) for $n = 11$. Similar considerations hold for the neutron configurations of the pseudo $s - d - g$ shell for $82 \leqq N \leqq 126$. The leading pseudo $SU(3)$ representations for the odd neutron nuclei in this region are shown in Table 3.

Although preliminary investigations are under way at Michigan to test such possibilities, simpler calculations by Adler[3] in which protons are assumed to be filling a single pseudo spin-orbit doublet in one major shell while neutrons are filling a different pseudo spin-orbit doublet in another shell indicate that the pseudo Russell Saunders coupling scheme may be a valuable tool in nuclear shell studies. Although calculations to date have been restricted to "pseudonuclei," (using pseudo-coupling schems), it is hoped that they will give some understanding as to the nature of real nuclei.

TABLE 2 Leading Pseudo-$SU(3)$ Representations for Odd Proton Number $50 \leqq Z \leqq 82$

Number of Odd + Parity Protons	1	3	5	7	9	11	13	15	17	19
Leading Pseudo $SU(3)$ Rep. $(\lambda\mu)$	(30)	(71)	(10,1)	(11,2)	(10,4) (11,2)	(11,2)	(9,3)	(47) (55) (63) (71)	(17)	(03)
Expected Ground State K	0	1	1	2	2	0	3	3	1	0

TABLE 3 Leading Pseudo-$SU(3)$ Representations for Odd Neutron Number $82 \leqq N \leqq 126$

Number of Odd-Parity Neutrons	1	3	5	7	9	11	13	15	17	19	21	23	25	27	29
Leading Pseudo $SU(3)$ Rep. $(\lambda\mu)$	(40)	(10,1)	(15,1)	(18,2)	(19,4) (20,2)	(22,2)	(22,3)	(19,7) (20,5) (21,3) (22,1)	(18,7) (19,5) (20,3) (21,1)	(19,3)	(16,4)	(9,10) (10,8) (11,6) $(12,4)^2$ (13,2)	(4,12) (5,10) $(6,8)^2$ $(7,6)^2$ $(8,4)^3$ (9,2) (10,0)	(1,10) (2.8) (3,6) $(4,4)^2$ (5,2)	(04)
Expected Ground State K	0	1	1	2	2	0	3	3	1	1	4	4	2	2	0

REFERENCES AND FOOTNOTES

1. A. Arima, M. Harvey, and K. Shimizu, *Phys. Lett.* **30B** (1969) 517
2. K. T. Hecht and A. Adler, *Nucl. Phys.* **A137** (1969) 129
3. A. Adler, Dissertation, University of Michigan (1970) unpubl.
   W. P. Jones, Dissertation, University of Michigan (1969) unpubl.
   L. W. Borgman, Dissertation, University of Michigan (1969) unpubl.
   W. P. Jones, L. W. Borgman, J. Bardwick, W. C. Parkinson, and K. T. Hecht, *Phys. Rev.* **C4** (1971) 580
4. C. M. Vincent, *Phys. Rev.* **159** (1967) 869
5. I. M. Green and S. A. Moszkowski, *Phys. Rev.* **139B** (1965) 790
   R. Arvieu and S. A. Mozskowski, *Phys. Rev.* **145** (1966) 830
   P. W. M. Glaudemans, P. J. Brussaard, and B. H. Wildenthal, *Nucl. Phys.* **A102** (1967) 593
6. C. Quesne, Thèse, Université Libre de Bruxelles (1968): "Le formalism du quasi-spin et l'interaction delta de surface. Application à la spectroscopie des noyaux impairs à une seule couche fermée."
7. See, e. g., M. Moinester, J. P. Schiffer, and W. P. Alford, *Phys. Rev.* **179** (1969) 984
8. For some recent experimental work, see: B. H. Wildenthal, E. Newman, and R. L. Auble, *Phys. Lett.* **27B** (1968) 628
   Also B. H. Wildenthal, E. Newman, and R. L. Auble, *Phys. Rev.* **C3** (1971) 1199
9. H. W. Baer and J. Bardwick, *Nucl. Phys.* **A129** (1969) 1
   H. W. Baer, J. J. Reidy, and M. L. Wiedenbeck, *Nucl. Phys.* **A113** (1968) 33
   J. C. Hill and M. L. Wiedenbeck, *Nucl. Phys.* **A119** (1968) 53
10. B. H. Wildenthal, *Phys. Rev. Lett.* **22** (1969) 1118; *Phys. Lett.* **29B** (1969) 274
11. B. H. Wildenthal and D. C. Larson, *Phys. Lett.* **37B** (1971) 266
12. J. B. French, E. C. Halbert, J. B. McGrory, and S. S. M. Wong, *Adv. in Nucl. Phys.*, edited by M. Baranger and E. Vogt (Plenum Press 1969)
13. The enhancement factor of 1.5 is obtained with the wave functions corresponding to column (a) of Fig. 5. No such enhancement is obtained with wave functions corresponding to column (b)
14. F. T. Baker, Dissertation, University of Michigan (1970) unpubl.
15. C. G. Morgan and D. F. Jackson, *Phys. Rev.* **188** (1969) 1758
16. G. Racah, *Phys. Rev.* **76** (1949) 1352
17. K. Helmers, *Nucl. Phys.* **23** (1961) 594
18. L. Armstrong and B. R. Judd, *Proc. Roy. Soc.* **A315** (1970) 27
19. H. W. Baer, H. C. Griffin, and W. S. Gray, *Phys. Rev.* C**3** (1971) 1398
20. R. Arvieu, Private communication

DISCUSSION

C. Ripka Charge conjugate nuclei belonging to the first half of the $2s$–$1d$ shell are well described by $L - S$ coupling wave functions and these appear to fail for the description of nuclei belonging to the second half of this shell. The spectra of the latter nuclei are governed by the presence of the ($s_{1/2}$, $d_{3/2}$) doublet and therefore should be described by your pseudo spin-orbit coupling scheme. Can you describe the transition between these two coupling schemes in terms of some continuous deformation, for example?

K. T. Hecht I am afraid that the *transition* from one good coupling scheme to the other looks equally difficult from *either* extreme. The nucleus ^{28}Si looks just as difficult from the point of view of the pseudo coupling schemes as from the point of view of the conventional $L - S$ or $SU(3)$ schemes.

A. Bohr Could you state the general conditions that the interaction must satisfy, in order to leave pseudo-spin invariant. In this connection, what is the estimated strength of the interactions component that is non-central in pseudo-space.

K. T. Hecht The general condition is that the interaction have only components with spherical tensor character $K_b = 0$ in the pseudo-spin space. Arima, Harvey, and Shimizu have made a detailed analysis of the realistic effective interaction for the $2p_{3/2}$, $2p_{1/2}$, $1f_{5/2}$ region (pseudo $s - d$ shell). It does of course have some pseudo-spin tensor ($K_b = 2$) and vector ($K_b = 1$) components. The $K_b = 2$ components, in particular, are perhaps uncomfortably large. However, their matrix elements are minimized by Racah coefficients and are considerably smaller than those of the $K_b = 0$ part of the interaction.

M. Moshinsky You mentioned that the surface delta interaction is almost diagonal in your classification scheme. My question is what are the interactions that are *exactly* diagonal in your scheme and whether they have a physical significance similar, for example, to those that the quadrupole-quadrupole interaction has for the $SU(3)$ scheme.

K. T. Hecht For a single pseudo spin-orbit doublet, the generalized pairing interaction is the interaction which is exactly diagonal. By expressing it as a combination of a two-body pseudo spin-spin interaction, an ordinary ($J = 0$)

pairing interaction, and a two-body unit interaction, we have expressed it in terms of the Casimir invariants of the relevant group chain. Unfortunately, in the case of mixed configurations involving more than one pseudo spin-orbit doublet, we no longer have a sufficient number of Casimir invariants to express a generalized pairing interaction in this simple way.

Shell-model Symmetries

J. P. ELLIOTT
University of Sussex, England

1 INTRODUCTION

The role of symmetry in general and its place within the shell-model of nuclear structure will be discussed. Particular reference will be made to the use of the SU_3 group.

One speaks of symmetry when there exists some group of operators which leaves the Hamiltonian invariant. The most obvious physical consequences of such a symmetry are selection rules in dynamic processes and degeneracies in the energy levels of the static system. Exact symmetries are rare and we shall more often be concerned with approximate, or weak, symmetries.

From the point of view of calculations which attempt to solve the many-body problem $H\psi = E\psi$, the knowledge of a symmetry group means that the Hilbert space L of the wave function φ may be broken up into subspaces $L = \sum_{\alpha, i} L_{\alpha i}$ with the property that H has no matrix elements between spaces with different labels α, i. Here, α labels the distinct irreducible representations of the group while i labels the basis functions of that representation α. Furthermore, the matrix elements of H are independent of i. Nevertheless the matrices of the Hamiltonian are still of infinite dimension. It is only if the Hamiltonian is expressible in terms of the group operators that an exact solution may be deduced using group theory. Such a situation arises only in idealised problems, but even this is useful in understanding some qualitative features of nuclei.

If the Hamiltonian possesses no symmetry, it may still be useful to use a group to obtain a systematic set of basis functions in which to set up the Hamiltonian matrix.

2 STRONG SYMMETRIES.

Rotational symmetry is perhaps the only true symmetry in nuclei, being closely related to Lorentz invariance and leading to the conservation of angular momentum J, the familiar selection rules for multipole operators and to the $(2J + 1)$-dimensional multiplet degeneracies.

There is ample evidence that the strong nuclear forces are, to a high degree, invariant with respect to space-inversions (parity). On the other hand the weak interaction violates parity and leads to a small degree of parity violation in the nuclear wave-functions. This may be detected through the mixing of E.1 and M. 1 modes in γ-decay.

In the same way, nuclear forces are invariant with respect to time-reversal, so far as we can tell, but there is interest in searching for a possible small degree of violation which would appear through a complex mixing ratio δ in a mixed (M. 1 + E. 2) γ-decay.

Within a few percent, nuclear forces are also charge-independent, leading to the concept of conservation of isospin and the isobaric multiplet degeneracies. (See D. H. Wilkinson in this volume). The presence of electromagnetic forces breaks the isospin symmetry. Since they are much stronger than the weak interactions which violate the previous symmetries one should perhaps refer to the isospin as a medium-strong symmetry.

3 WEAK SYMMETRIES AND NUCLEAR FORCES.

For the remainder of this report the symmetries mentioned in § 2 will be regarded as exact. Our attention turns, instead, to the possible existence of further symmetries, albeit very weak symmetries, of the nuclear forces which enable us to devise approximations which make the many-body problem tractable. Consider then the nuclear force which may conveniently be analysed $V = \sum_{k=0,1,2} V(k)$ where $V(k) = (\mathscr{S}^{(k)} \cdot \mathscr{L}^{(k)})$ is a scalar product of a spin tensor $\mathscr{S}^{(k)}$ with an orbital tensor $\mathscr{L}^{(k)}$ the three terms generally being referred to as Central, Spin-orbit and Tensor forces respectively. An analysis of the two-body scattering data shows that each term is present but there are a number of simplifying features in the nuclear problem which make the study of approximate symmetries worthwhile. These are as follows,

(i) The interaction is strongest in the S-states of relative motion and is similar in the ^1S and ^3S states.

(ii) The attractive forces generate an average field which leads to approximately independent particle motion.

(iii) The main effect of the inter-nucleon spin-orbit force seems to be to add an effective one-body spin-orbit force to the average field.

(iv) The tensor force contributes little in first order perturbation theory (from a shell-model as zero order) but gives an appreciable second-order contribution. It is rather slowly converging with increasing energy denominator, but its effect is roughly equivalent to an increase in the effective interaction in relative ^3S states by about 50°/₀.

4 THE SUPERMULTIPLET SCHEME.

The first weak symmetry proposed by Wigner in 1937 follows from feature (i) above. If all but s-state forces are ignored and spin-independence is assumed for the s-states this leads to independance with respect to any unitary transformation in the four-dimensional space of the charge and spin co-ordinates of the nucleon when one takes account also of the charge independence. This SU_4 group is generated by the 15 operators $\Sigma\tau$, $\Sigma\sigma$ and $\Sigma\tau\sigma$. In addition to the isospin and spin labels T and S one has also the label for the irreducible representation of SU_4, a Young diagram with no more than three rows. The reduction $SU_4 \rightarrow SU_2 \times SU_2$ which exhibits the quantum numbers T and S provides a super-multiplet of TS combinations which, if the SU_4 symmetry were exact, would be degenerate. One does not find degeneracies of this kind in nuclei, but the binding energy systematics show some correlations with the expected eigenvalues of the Casimir operator of SU_4 (see L. Radicati in this Volume). It is perhaps important to realise that the SU_4 symmetry does not depend on the assumption of a shell model although it is most often used within such a framework.

With more realistic interactions the supermultiplet symmetry is broken, although at least for light nuclei, the breaking is not large. For example, a recent calculation[1] for the sd-shell nuclei with $A \sim 20$ to 24, using a semi-realistic interaction of spin-dependent central force and one-body spin-orbit force, found the following admixture of supermultiplets given as percentage intensity in the wave function for the ground state,

	[4]	[31]	[22]	[211]	[1111]
Ne^{20}	93.6	6.2	0.1	0.1	0.0

	[42]	[411]	[33]	[321]
Ne^{22}	84.6	10.9	3.2	1.3
Na^{22}	92.4	5.1	1.0	1.5

	[44]	[431]
Mg^{24}	92.3	7.7

The reasons for the rapid convergence are (a) that the one-body spin orbit force can only mix Young diagrams with at most one square different (such as [4] and [31] but not [4] and [22]) and (b) that only the small difference between the ^1S and ^3S state forces contributes to the mixing. Nevertheless the mixing of the next-but-lowest supermultiplet is quite large making any restriction to the lowest supermultiplet of qualitative interest only. Moreover, with the neutron excess in heavier nuclei, and the corresponding increase in isospin T the separation between the lowest supermultiplets decreases, leading to greater mixing as is illustrated above for Na^{22} $(T = 0)$ and Ne^{22} $(T = 1)$.

5 CLASSIFICATION OF MANY-PARTICLE WAVE FUNCTIONS IN THE SHELL MODEL

Let us next consider the symmetries within some shell nl, i.e. we suppose that all lower shells are filled, leaving a number N of nucleons in the configuration $(l)^N$ where $N = 0, 1, \ldots 4(2l + 1)$. We investigate the possible classification schemes with the object of either (i) obtaining an approximate diagonalisation of the Hamiltonian matrix within the shell, or (ii) finding a scheme in which the calculation of the matrix elements is simple. Ideally, one would like to find a scheme which achieves both objects. There are $4(2l + 1)$ states of each single particle implying that the states of the N particle system are the antisymmetric tensors of rank N in this $4(2l + 1)$ dimensional space, in other words that they form the representation [11 ... 1] of the group $U_{4(2l+1)}$. This does not help to classify the states—to do this we must look for *sub-groups* of $U_{4(2l+1)}$. With respect to these sub-groups the representation [11 ... 1] will reduce, introducing the irreducible representation labels of the sub-groups as a means of classification. We look, not for a single sub-group, but for a chain of sub-groups. The more groups in the chain, the more labels will occur and the more complete will be the resulting classification. There are, of course, an immense number of sub-groups but the ones of interest here must contain the isospin and angular momentum groups $SU_2 \times R_3$ as a sub-group. (One may always consider whether it is advantageous to devise a scheme in which certain good quantum numbers like N, T or J are mixed, but for the present, we shall exclude such schemes).

Chains which incorporate the supermultiplet classification will begin with $U_{4(2l+1)} \to SU_4 \times U_{2l+1}$. Because of the simplicity of the representation [11 ... 1] of $U_{4(2l+1)}$, it follows that the representation of U_{2l+1} is uniquely determined by that of SU_4 so that the introduction of U_{2l+1} brings nothing new to the supermultiplet classification. The reductions $SU_4 \to SU_2^T \times SU_2^S$ and $U_{2l+1} \to R_3^L$ bring in the additional labels T, S and L while the final reduction $SU_2^S \times R_3^L \to R_3^J$ introduces J as well. It was long ago pointed out by Racah[2] that the group R_{2l+1} could be included in the chain $U_{2l+1} \to R_{2l+1} \to R_3$. From a physical point of view, this makes good sense since the representation labels of R_{2l+1} provide a measure of the number of pairs of nucleons which are "paired off" to $L = 0$ states of the pair. This comes about because, in the $N = 2$ system, the only effect of the restriction $U_{2l+1} \to R_{2l+1}$ is to separate off the $L = 0$ state which is the R_{2l+1} invariant. With a realistic force, the $L = 0$ state is indeed separated from the others, having the greatest binding, but the lack of degeneracy among the states with $L \neq 0$ will break the R_{2l+1} symmetry.

The main reason why the group R_{2l+1} is of little interest in nuclei, however, is that the nature of the average field does not lead to pure configurations of the kind l^N. In the first place, the spin orbit force splits the l-orbit into the two $j = l \pm \frac{1}{2}$ orbits and secondly, there is usually mixing of configurations with different l. Thus, while there are some mass regions where it makes sense to consider a pure j^N configuration, there is no region, other than the first p-shell, where it is necessary to mix the orbits $j = l \pm \frac{1}{2}$ without also considering other orbits l'. This then brings us to two new problems, the classification in a configuration j^N and the mixing of configurations.

In the j^N configuration we again start from the antisymmetric tensors [11 ... 1] of the group $U_{2(2j+1)}$ with the sub-group $SU_2^T \times U_{2j+1}$ providing the quantum number T which is in one-to-one correspondence with the representation of U_{2j+1}. The SU_2 Young diagram has at most two rows while that of U_{2j+1} is obtained by interchanging rows and columns. The quantum number J corresponds to the sub-group R_3 of U_{2j+1} and again one may interpose[3] a group to form a chain $U_{2j+1} \to Sp_{2j+1} \to R_3$. This time, the new group Sp_{2j+1} is the symplectic group and its role is almost identical to that of R_{2l+1} for the configuration l^N. The two-particle state with zero angular momentum $J = 0$ is now an antisymmetric tensor whereas the $L = 0$ state of l^2 was symmetric. The sub-group of U_{2j+1} with this $J = 0$ state as its invariant is now Sp_{2j+1}. Again the representation labels of Sp_{2j+1} essentially count the number of $J = 0$ pairs. Thus the Sp_{2j+1} classification makes physical sense because with realistic forces the $J = 0$ state of j^2 is invariably the lowest. In a configuration of like particles (e.g. all neutrons) this is particularly good since then $T = \frac{1}{2}N$ and for $N = 2$, J must be even. In this case the state with $J = 0$ usually lies well below those with $J \neq 0$. However, with neutrons and protons, $T = 0$ or 1 for $N = 2$ and the odd J may occur. Then typically, the two states $J = 1$ and $J = 2j$ lie almost as low as that with $J = 0$, thus breaking the symmetry considerably.

Suppose now that two orbits l and l' lie close together in energy. It is then convenient to take them together in providing a space of dimension

$4(2l + 2l' + 2)$ for the single particle, leading to a supermultiplet classification of the states of N particles by the group $SU_4 \times U_{2l+2l'+2}$. There is now much greater scope for sub-groups but the "pairing" classification R_{2l+1} in the pure configuration immediately extends to this mixed configuration through the sub-group $R_{2l+2l'+2}$. The pair-state is now a coherent mixture of the two S-states, $\sqrt{2l+1}\,\Psi(l^2L = 0) + \sqrt{2l'+1}\,\Psi(l'^2L = 0)$.

6 THE SU_3 SCHEME

There is one important type of configuration mixing[4, 5] for which a general discussion may be given. This occurs when the set of mixing orbits is the set of oscillator orbits nl with the same energy $(2n + l + \frac{3}{2})\,\hbar\omega$, such as $0d$, $1s$ or $0f$, $1p$ or $0g$, $1d$, $2s$, etc. In this case it is always possible to include the group SU_3 in the chain $U_{\sum_i(2l_i+1)} \to SU_3 \to R_3$ where i runs over the set of degenerate oscillator orbits. The presence of this group is an immediate consequence of the symmetry of the single particle oscillator Hamiltonian. For the harmonic oscillator in p dimensions it is well known that there is a symmetry group SU_p generated by the infinitesimal operators $A_{ij} = \frac{1}{2}(a_i^\dagger a_j + a_j^\dagger a_i)$ where i and j run over the Cartesian axes 1, 2, ... p and the operators $a_i^\dagger$, a_i create and destroy oscillator quanta in the co-ordinate i. In three dimensions therefore the single particle states with M quanta, i.e. with energy $(M + \frac{3}{2})\,\hbar\omega$, form an irreducible representation of SU_3. It is in fact the simple representation $[M]$ of symmetric tensors with rank M. It follows that, if the single particle functions form a representation of SU_3 then so must the many-particle functions. In the latter case of course the representation will not be irreducible and it is this reduction which helps to classify the many-particle functions. For many purposes it is more convenient to choose operators which transform irreducibly under rotations, taking linear combinations of the A_{ij} to form a scalar H_0 (the oscillator Hamiltonian), a vector L_q^1 (the angular momentum) and a second rank tensor Q_q^2. The extension from one particle to many is exactly analogous to that for angular momentum. In the N-particle system the infinitesimal operators of SU_3 are therefore $L_q^1 = \sum_{k=1}^{N} L_q^1(k)$ and $Q_q^2 = \sum_{k=1}^{N} Q_q^2(k)$ with the former being simply the total orbital angular momentum. The quadrupole operators Q_q^2 have the explicit form $(r \wedge r)_q^2 + b^4(p \wedge p)_q^2$, which follows directly from the relation $a_x^\dagger = (x - ib^2p_x)/b\sqrt{2}$.

It is easy to see the physical significance of the SU_3 classification from the presence of this quadrupole operator. After the removal of single particle terms and the square of the total angular momentum, the Casimir operator for SU_3 is simply the two body "quadrupole force" $\sum_{k<k'} (Q^2(k) \cdot Q^2(k'))$. The presence of the momentum in Q simply ensures that there is no mixing of different oscillator configurations. Within a configuration, the equivalence between r^2 and p^2 in the oscillator means that the quadrupole force may be written in the more familiar form

$$\sum_{k<k'} r_k^2 r_{k'}^2 \, P_2(\cos \theta_{kk'}).$$

where P_2 is a Legendre polynomial and $\theta_{kk'}$ the angle between r_k and $r_{k'}$. It is dangerous to ignore completely the single-particle terms contained in the Casimir operator in this argument. To produce SU_3 eigenfunctions, the two-body quadrupole force needs to be accompanied by a one-body term in which the large l are higher in energy. Empirically, the spectra of ^{17}O and ^{41}Ca show that such a term is present in the *sd*-shell but not in the *pf*-shell. In the expansion of an arbitrary force in multipoles the short range components are slowly convergent while the long range parts occur mainly in the low order multipoles. Thus the monopole term is largely responsible for the spherical average field while the next term of even parity will tend to produce a quadrupole term in the field. Since the group operators cannot mix oscillator orbits, the only way in which a deformation may be generated is in the arrangement of the valance nucleons. This is precisely what happens in the SU_3 classification as may be seen in the structure of the wave functions.

There is a very simple technique for constructing the eigenfunctions of the quadrupole force. In the system of N particles in an oscillator shell M, using a Cartesian basis for the oscillator, first maximise the number M_z of quanta in the z-direction and, if any freedom remains, maximise the number of remaining quanta in the x-direction. This choice of z and x is of course arbitrary and it is the Pauli principle which prevents all quanta from going into the z-direction. The differences $M_z - M_x = \lambda$, $M_x - M_y = \mu$ are a measure of the deformation of the function χ so constructed. It then follows that the set of functions

$$\psi(KLM) \sim \int D^L_{KM}(\Omega) \, \mathscr{R}(\Omega) \, \chi \, d\Omega$$

obtained by projecting all possible values of KLM from χ spans an SU_3 multiplet. Moreover, this multiplet with the greatest deformation lies lowest in energy with an attractive quadrupole force. Furthermore, the eigenvalues of the quadrupole force within the multiplet form a set of rotational bands with energies proportional to $L(L+1)$. In terms of the Young diagram labels for the representations of SU_3, this multiplet would have two rows of length $\lambda + \mu$ and μ and it is usual to refer to this as the $(\lambda\mu)$ representation.

The SU_3 classification thus provides an exact microscopic model in which a quadrupole force leads to a ground state with high deformation, (large quadrupole moment) and the spectrum and transition probabilities have rotational characteristics. It differs from the rotational model in one important way, namely that the rotational bands terminate at a comparatively low angular momentum, such as $J = 8$ in ^{20}Ne and $J = 12$ in ^{24}Mg. The reason for this is simply that, with the single particles restricted to the *sd*-shell, the intrinsic function χ contains a restricted range of L-values. The termination is thus a direct result of the assumption of a single oscillator configuration. One then asks "Do the bands terminate in the experimental data?" but no decisive answer has been given. To show that they do not terminate one must identify, in ^{20}Ne, a $J = 10^+$ level at about 16 MeV which belongs to the ground state band, i.e. which γ-decays with a transition probability consistent with others in the band. From a theoretical point of view one's first reaction is that the bands need *not* terminate in this way because, when configuration mixing is allowed the more realistic intrinsic function χ will contain unlimited angular momenta. In other words, although it may make sense to restrict to the lowest configuration when discussing the ground state and the first 4 MeV of excitation, it is not reasonable to so restrict the configurations at 16 MeV. The spectrum of ^{16}O tells us that the next even-parity configuration begins at 6 MeV. It seems to me that there are two extreme possibilities, either

(i) that with increasing J the rotational levels gradually contain less of the lowest configuration and more of the excited configurations in a smooth way so that for example at $J = 10^+$ in ^{20}Ne the amplitude of the lowest configurations has dropped to zero, or

(ii) that there will be a major discontinuity in the band properties at the termination point. Personally, I would be surprised if there were no discontinuity at this point but the extent of the discontinuity will surely differ from one nucleus to another. We earlier showed results which, in the (*sd*) shell, suggested that the convergence in a supermultiplet classification was quite rapid. The group SU_3 provides a sub-classification of this since, strictly, we use the group $SU_4 \times SU_3$. The physical significance given to the SU_3 scheme through the quadrupole force suggests that it would provide a useful additional means of truncation. (Remember that Mg^{24} has matrices of the order $2{,}000 \times 2{,}000$ even within the *sd*-shell.) The work of Akiyama *et al.*[1] investigates this question in the nuclei $18 < A < 24$. They conclude, for example, that a selection of about 28 out of a full basis of 525 states with $J = 2$, $T = 1$ in Ne^{22} will provide about 99.5% of the *sd*-shell wave function. This is impressive. It means that much more attention should be paid to the mixing of excited major shells than to the remaining 0.5% in the *sd*-

shell! For present day computers, the size 28 is small whereas 525 is close to the limit of practicality. The truncation suggested by Akiyama *et al.* is based partly on theoretical argument and partly on practical experience. One starts with the leading representation $(\lambda\mu)$ with greatest deformation and with the lowest supermultiplet. The admixtures into this are dominated by two mechanisms,

(i) The (s. l) force which, under SU_3, transfroms according to the representation (11). This immediately identifies those $(\lambda'\mu')$ which mix directly with the leading state.

(ii) The non-quadrupole components of the central force which is found to mix states $(\lambda'\mu')$ for which $\lambda' + 2\mu' = \lambda + 2\mu$ and also those with $\lambda' = \lambda - 2, \mu' = \mu - 2$, the latter being identified with a pairing component.

Calculations with the complete basis of *sd*-shell states have been made by the Oak Ridge group[6] for the nuclei $18 < A < 22$. Unfortunately, they were made with slightly different two-body forces so that a precise test of the truncation may not be made but the results of the two calculations look very similar. One should here mention also the calculation of Flores and Perez[7] which showed that convergence in the supermultiplet scheme was more rapid than in $j - j$ coupling, allowing particles to be excited from $j = l + \frac{1}{2}$ to $j = l - \frac{1}{2}$.

7 EXTENSIONS OF THE SU_3 SCHEME

Let us now talk about improvements or extensions of the SU_3 scheme. The structure of the lowest SU_3 wave function, being a projection from an intrinsic function suggests an obvious improvement, namely the use of a more realistic intrinsic function. For the SU_3 function, the intrinsic state is the Slater determinant built on single particle functions appropriate to the potential $\alpha H_x + \beta H_y + \gamma H_z$ where H_x etc., are one-dimensional oscillator Hamiltonians with the same length parameter and with $\gamma > \alpha > \beta$. In other words, the single particle functions may be regarded as Hartree-Fock functions in which

(i) there is no spin-orbit force and

(ii) the departure from sphericity allows mixing only within an oscillator shell. These two features ensure that the wave function does not depend on the deformation parameters α, β, γ. Many calculations have been made[8,9] taking into account the first of these features. They are referred to as restricted Hartree-Fock calculations and the angular momentum eigenstates are again constructed by an angular momentum projection from the Slater determinant. The function space of such a calculation lies still within the *sd*-shell, for example but now the projected state is a mixture of different super-multiplets. This method yields results[9] in ^{21}Ne, for example, which are very similar to those of the large matrix diagonalisations carried out by the Oak Ridge group. Improvements of the second kind (ii) have been carried out by the Belgian group[10] at Mol. They have been interested in the relative positions of levels belonging to different major shells like, for example the excitation energy of the 0^+ level at 6 MeV in ^{16}O. The SU_3 scheme may be equally well applied to excited configurations so that for the two particle-two hole and four particle-four hole configurations of greatest symmetry, the lowest SU_3 representations would be (42) and (84) respectively. Using a simplified[11] force fitted to the binding energies of He^4 and nuclear matter they found the excitation energies to be 32.0 and 47.2 MeV respectively. They next allowed the single particle functions to take on extra freedom of the Hartree-Fock kind such as $|0s\rangle \rightarrow |0s\rangle + a\,|1s\rangle + b\,|0d\rangle$. Minimising against the parameters a and b produces excitation energies of 22.3 and 8.5 MeV respectively. The dramatic decrease of excitation energy of the four particle-four hole state from 47.2 MeV to 8.5 MeV brings it close to the observed level at 6 MeV. This is almost entirely due to a binding energy gain of about 41 MeV for the four particle-four hole state and only 2 MeV for the ground state. The two particle-two hole state remains high, at 22 MeV excitation.

Within the framework of group theory there has been an attempt to describe the mixing of major shells in terms of the group $SL(3, R)$. (See the contribution of L. C. Biedenharn in this volume.) The essence of this approach is to replace the quadrupole operator $Q = (r \wedge r)^2 + b^4(p \wedge p)^2$ by the operator $\tilde{Q} = (r \wedge p)^2$. The link with major-shell mixing becomes apparent if we define $Q^{\pm} = ((r \pm ib^2p) \wedge (r \pm ib^2p))^2$. The operators $Q^{\pm}$ respectively raise and lower the number of oscillator quanta by two units of $\hbar\omega$ and simply $\tilde{Q} = (Q^+ - Q^-)/4ib^2$ .The physically important, momentum independent, quadrupole operator is given by $(r \wedge r)^2 = (2Q + Q^+ + Q^-)/4$. From this point of view the operator $\tilde{Q}$ is less related to physical reality than Q and consequently, although $SL(3, R)$ leads to configuration mixing it is not obviously of any physical significance in nuclei.

The possibility of using group theory to describe the mixing of two adjacent major shells, like 0*p* with 0*d*, 1*s*, has been studied by Sato *et al.*[12] Here it is possible to use the chain $U_9 \rightarrow SU_3^1 \times SU_3^2 \rightarrow SU_3 \rightarrow R_3$ where the final SU_3 group is the usual one described earlier but where the SU_3 factors in the product group are quite unusual. Their operators do not have definite parity, for example the vector operators are $\frac{1}{2}(L^1 \pm iW^1)$ where W^1 is an odd-parity vector operator coupling 0*p* to 0*d* and 1*s*. Although the basis functions in the product group have mixed parity, like $(d_1 + ip_1)/\sqrt{2}$, states of definite parity are recovered in the final reduction to SU_3. In the publication[9] only one of the factors SU_3^1 was identified but it is clear that there is a much richer structure to this system. In fact, for any pair of adjacent oscillator shells say with energy $(M + \frac{3}{2})\,\hbar\omega$ and $(M + 1 + \frac{3}{2})\,\hbar\omega$ there exists a product group classification according to $SU_{M+2} \times SU_{M+2}$. So far as I know, little work has been done on this group. It may be useful in describing the mixing of major shells but at first sight it appears rather formidable, classifying together, for example, the states of p^4, $p^3(ds) \ldots (ds)^4$.

8 THE MIXING OF PARTICLE NUMBER.

Throughout this article, all group operations have conserved the particle number. Considerable light is thrown on some of the group structures by introducing[13,14] groups which transform between states with different numbers of particles. The role of the symplectic group in the configuration j^n is a case in point. The group U_{4j+2} has infinitesimal operators $(a^+a)^{rk}$ where the tensor operator labels r and k refer to isospin and angular momentum respectively. Since the states of given particle number belong to the same irreducible representation of this group it may seem pointless to construct a larger group but that is not entirely true. The inclusion of the operators $(a^+a^+)^{rk}$ and $(aa)^{rk}$ generate the group R_{8j+4} and all states in the entire shell are contained in the simple representations $(\frac{1}{2}\,\frac{1}{2} \cdots \pm\frac{1}{2})$. From R_{8j+4} there is now a chain of sub-groups which bypasses U_{4j+2}, namely $R_{8j+4} \to Sp_4 \times Sp_{2j+1}$, where the second factor is the symplectic group met before with operators $(a^+a)^{ok}$ with odd k while the group Sp_4 contains the isospin operators $(a^+a)^{10}$, the number operator $(a^+a)^{00}$ and the pair creation and destruction operators $(a^+a^+)^{01}$, $(aa)^{01}$ for the $J = 0$, $T = 1$ pair. The simplicity of the representation of R_{8j+4} implies that there is a unique correspondence between the representations of Sp_{2j+1} and Sp_4. Notice that an irreducible representation of Sp_4 will reduce $Sp_4 \to SU_2^T \times U_1$ giving the pair of labels TN. Thus each multiplet of Sp_{2j+1} will appear for a range of values of the particle number and isospin. Thus, through the introduction of the larger group R_{8j+4}, states of different particle number are brought together under a single representation of Sp_4. This throws new light on the usual symplectic classification[3], but leads to no new classification. The quantum numbers of the group Sp_4 correspond to the seniority and reduced isospin of the symplectic group classification. For maximum isospin the group Sp_4 contracts to an SU_2 group leading to the "quasi-spin" quantum number.

Following work by Armstrong and Judd,[15] a new chain of sub-groups of R_{8j+4} has been found[16] which does indeed introduce a new classification but at the expense of having the number operator non-diagonal. There are, in fact, two slightly different chains,

$$R_{8j+4} \rightarrow R^{\lambda}_{4j+2} \times R^{\mu}_{4j+2} \rightarrow (SU^{\lambda}_{2} \times Sp^{\lambda}_{2j+1}) \times (SU^{\mu}_{2} \times Sp^{\mu}_{2j+1}) \rightarrow SU_{2} \times Sp_{2j+1} \rightarrow SU_{2} \times R_{3}$$

$$\searrow (SU^{\lambda}_{2} \times R^{\lambda}_{3}) \times (SU^{\mu}_{2} \times R^{\mu}_{3}) \nearrow$$

The idea rests on a factorisation which is achieved by defining new quasi-particle operators λ^{+} and μ^{+} containing equal weights of particle and hole. The reduction $R^{\lambda}_{4j+2} \rightarrow SU^{\lambda}_{2} \times Sp^{\lambda}_{2j+1}$ is identical with the usual quasi-spin treatment of like particles. However, the operators of SU^{λ}_{2} and SU^{μ}_{2} are combinations of the pair operators and the isospin operators, both of which transform like $T = 1$, $J = 0$. Thus the labels $T_{\lambda}(T_{\mu})$ for the representations of these two groups may equally well be regarded either as quasi-spin or isospin for the $\lambda(\mu)$ space. The fascinating feature is that the physical isospin is the sum of these operators so that T is obtained by standard vector coupling of T_{λ} and T_{μ}. In the same way, vectorially, $J = J_{\lambda} + J_{\mu}$ where $J_{\lambda}(J_{\mu})$ refers to the group $R^{\lambda}_{3}(R^{\mu}_{3})$ with operators $(\lambda^{+}\lambda)^{01}$. One therefore has four labels T_{λ}, T_{μ}, J_{λ}, J_{μ} in addition to J and T. The advantage of the scheme is

(i) that the labelling is more complete. In fact no additional quantum numbers are needed until $j \geqq 9/2$.

(ii) the calculation of matrix elements of two body operators may be carried out in a formally very simple manner[15,17] without the use of coefficients of fractional parentage. Nothing more than angular momentum coupling is involved. The disadvantage is that a state labelled by $|J_{\lambda}T_{\lambda}J_{\mu}T_{\mu}, JT\rangle$ does not generally have definite particle number, although by symmetrising with respect to λ and μ, mixing occurs only between states for which the number difference is a multiple of four. One must either project out the particle number or work with a large basis which runs over all particle numbers in the shell. The particle conserving Hamiltonian will, of course, produce eigenfunctions with definite particle numbers.

In conclusion, you will see that there is a host of possible weak symmetries but that their usefulness depends on the balance between their effectiveness in making a reliable truncation and the extra complication which sometimes accompanies a sophisticated classification.

REFERENCES

1. Y. Akiyama, A. Arima, and T. Sebe, *Nucl. Phys.* **A 138** (1969) 273
2. G. Racah, *Phys. Rev.* **76** (1949) 1352
3. B. H. Flowers, *Proc. Roy. Soc.* **A 215** (1952) 120

4. J. P. Elliott, *Proc. Roy. Soc.* **A 245** (1958) 128
5. M. Harvey, *Advances in Nuclear Physics*, Vol. 1, (1968) Plenum Press
6. E. C. Halbert, J. B. McGrory, B. H. Wildenthal, and S. P. Pandya, *Advances in Nuclear Physics*, Vol. 4, (1970) Plenum Press
7. J. Flores and R. Pérez, *Phys. Lett.* **26B** (1967) 55
8. W. H. Bassichis and G. Ripka, *Phys. Lett.* **15** (1965) 320
9. I. P. Johnstone and H. G. Benson, *Nucl. Phys.* **A 134** (1969) 68
10. M. Bouten, M. C. Bouten, H. Depuydt, and L. Schotsmans, *Nucl. Phys.* **A 158** (1970) 217
11. D. M. Brink and E. Boeker, *Nucl. Phys.* **A 91** (1967) 1
12. T. Ishidzu, H. Kawarada, and M. Sato, *J. Math. Phys.* **10** (1969) 683
13. H. J. Lipkin, *Lie Groups for Pedestrians* (1965) North-Holland
14. K. T. Hecht, *Phys. Rev.* **139** (1965) 794
15. L. Armstrong and B. R. Judd, *Proc. Roy. Soc.* **A 315** (1970) 27
16. J. P. Elliott and A. J. Evans, *Phys. Lett.* **31B** (1970) 157
17. K. T. Hecht and S. Szpikowski, *Nucl. Phys.* **A 158** (1970) 449

DISCUSSION

A. Bohr The group theoretical classification of many-particle states has played a very significant role in nuclear structure studies, by exhibiting various important correlation effects of a general nature. However, the coupling schemes characterized in this manner are of a somewhat schematic character, because of the assumptions involved concerning the single particle energy spectra and the restrictions on the configurations included.

It is therefore of importance that, for the major correlation effects recognized in the nucleonic motion, it is possible to give descriptions that generalize the group theoretical classification. Thus, the properties of the U_3 model appear as a special case of the rotational description based on a deformed single particle potential. In a similar manner, the classification in terms of the symplectic group, which is related to the nuclear pairing effect, is generalized to arbitrary configuration mixings by the description in terms of a deformation of the pair density. In both these cases, the correlation effect can thus be viewed in terms of a deformation of a generalized one-particle density with an associated collective motion of rotational structure.

J. P. Elliott One must regard the coupling schemes in two ways. The first is in the sense of providing a simple schematic scheme to exhibit qualitative features. At the same time, detailed calculations with realistic forces must incorporate many different qualitative features. To do this we must choose a basis in which to work and for this purpose it is often advantageous to choose one which corresponds to one of the simple schemes.

M. Moshinsky You discussed a model in which you consider two adjoining harmonic oscillator shells and the chains of groups that describe them. I would like to point out that a simpler picture exists that includes even more shells. This picture was introduced by Kadkika, Parik etc. and it has to do with a chain than includes a $U(4)$ group whose given irreducible representation (I.R.) includes all $U(3)$ I.R. up to a given level. My question concerns the possibility of using this $U(4)$ classification scheme to take into account the contribution of the excitation of the core to the rotational bands.

J. P. Elliott There are indeed a great many groups which may be used in classifying states when there is mixing of configurations. It is important however to look for groups which have some physical significance. The group of Ishidzu *et al.* which I mentioned has the advantage that it mixes adjacent oscillator shells which is, of course, the most likely configuration mixing.

L. C. Biedenharn I would just like to make a remark á propos of the two remarks made: one by Bohr and the other by Moshinsky. The point at issue is very closely related to the deformation of the field. If, for example, the deformation is caused by a quadrupole effect, then the *os* orbit would mix with the *d* orbit which will then mix in other orbits. If you pursue this to the limit, saying that you can mix everything coupled by the quadrupole, you have nothing other than $SL(3R)$ again.

I think perhaps I was too emphatic this morning. $SL(3R)$ has many aspects and the one that I chose to emphasize is only one of many you can take. For example, Cusson shows how one can indeed generate deformed fields, which allow one to define effective quadrupole moments in terms of the deformations. The physical point is very clear; if you make a deformed field approximation, in a sense you have unlimitedly many possibilities at your disposal, because you can deform continuously. Equivalently you must have a non-compact group—this is merely another language for describing it.

H. J. Lipkin What are the relative advantages and disadvantages of this new scheme which does not conserve particle number with the following old-fashioned scheme. In treating the j^n configuration with neutrons and protons, the group $R_5 \times Sp(2j + 1)$ is used. The $R5$ group contains the subgroup $R_3 \times R_2$ where generators are isospin and the particle number and difficulties arise in this classification because there is no group between R_5 and $R_3 \times R_2$. However, if we are willing to abandon particle number, the R_5 classification becomes trivial, via the chain $R_5 \supset R_4 \supset R_3$ and a classification scheme is obtained with many additional quantum numbers.

J. P. Elliott The new scheme has two advantages. It is a more complete classification in the sense that no arbitrary labels are required to distinguish states until $j \geqq 9/2$. Furthermore, the structure of the wave functions involves only the coupling coefficients of the group R_3 which are well known. It remains to be seen whether the classification has physical significance or whether it is only a simple mathematical device for building up the energy matrix.

K. T. Hecht From the practical point of view, classification by means of the R_5 group is particularly simple for low seniority states. Classification by means of the new $(\lambda\mu)$-quasi-particle scheme is equally useful for high and intermediate seniorities.

W. Heisenberg May I make a remark concerning the various symmetries, which have been treated more or less on the same level in this report: SU_3 as a weak "symmetry", SU_2 as much better, P and T as very strong symmetries, the Lorentz group as an almost perfect symmetry. I think one should make a clear distinction between exact symmetries and approximate symmetries. An exact symmetry may appear as broken by an asymmetry of the ground state. In this case there must, according to Goldstone's theorem, exist long range forces or particles of mass zero responsible for the breaking, this can be checked experimentally. Therefore, the experiments allow us to consider the Lorentz group and SU_2 as exact groups (broken by the ground-state in connexion with gravitation or Coulomb forces), while SU_3 cannot be an exact group.

This distinction may seem unimportant for nuclear physics. But the fact e.g., that in weak interaction $SU_2 \times SU_2$ algebra for the current gives very good results, while $SU_3 \times SU_3$ does not, may be connected with the fact, that SU_2 is an exact group, while SU_3 is not.

J. P. Elliott It is of course only a matter of degree but I would certainly agree that the supermultiplet group SU_4 and the nuclear SU_3 group are only very weak symmetries. It is right therefore that they should be clearly distinguished from the strong symmetries. Nevertheless from the mathematical point of view, they are symmetries in the same sense.

D. H. Wilkinson I should like to comment on the experimental situation concerning the termination or otherwise of the rotational bands and the associated problem of whether or not the $B(E2)$ values decline as one approaches the point where the band would terminate in the simple shell model or $SU(3)$ description. It is not yet possible to examine this problem in the heavier nuclei but it can be done and has been done, if only so far in a tentative way, in several light nuclei of which I will quote just ^{8}Be and ^{19}F as examples. In ^{19}F, where the simple $SU(3)$ expectation is that the band should terminate at $J = 13/2$, states of higher spin have indeed as far eluded detection near the energies expected for them if the band in question were to continue semi-classically. Furthermore the $B(E2)$ values do seem to decline somewhat relative to the semi-classical expectation, so one moves towards the top of the band. In ^{8}Be, on the contrary, the ground state band appears to continue beyond its "shell model" top of $J = 4$ as states of $J = 6,8$ and possibly even higher have been reported. Unfortunately no $B(E2)$ values at all are available within this band. I should like to emphasize that, in my view, neither the breaking-off of the ^{19}F band nor the continuation of the ^{8}Be band can yet be regarded as established unequivocally but the results are highly suggestive.

I should now like to comment on the theoretical $SU(3)$ expectation that the rotational bands will terminate. I do not believe that we should expect such a termination, or at least an abrupt termination, for the following reason. When $SU(3)$ calculations are carried out in a spherical basis we may gain a good description of the excitation energies of the members of the band but we always fall short of the experimental $B(E2)$ and must supplement the calculation by the traditional device of effective charges. This is equivalent to saying that we must leave the spherical basis and adopt a deformed basis or, what is the same thing, introduce excited configurations. But if this is done we are ipso facto making available further single-particle orbitals and so providing the material by which the band can extend itself beyond the point at which it would otherwise have to terminate. Of course when calculations are performed in a deformed basis, such as those of Bouten *et al.* in the $1p$-shell, we do not expect the termination as found in the spherical basis and it is a matter of extreme satisfaction in these calculations that the experimental $B(E2)$, which are very large indeed in the case of ^{9}Be, are well reproduced without the necessity of introducing effective charges. We may confidently anticipate that such calculations will generate bands extending beyond those belonging to the spherical basis.

A final, and perhaps rather graceless, experimental comment is that the most dramatically-enhanced $E2$ transitions anywhere in the periodic table, in relation to the Z-value involved, are found in ^{10}B where no significant band structure in the conventional sense is to be discerned.

B. R. Mottelson I feel that I must differ with Prof. Biedenharn when he describes the $SL(3R)$ approach and the usual rotational description as essentially the same. The crucial question is the definition of the collective variables that describe the rotational motion. In the version of $SL(3R)$ presented this morning, the orientation of the nucleus is described in terms of its quadrupole mass tensor. The collective variables described by Prof. Bohr are much more complicated functions of both the particle positions and momenta and lead to rather different results for many matrix elements. The crucial question here is the proper definition of what goes around when a nucleus rotates.

L. C. Biedenharn Once again Prof. Mottelson has brought light on what was a murky statement by myself. He is perfectly correct that $SL(3R)$ is a mathematical group and the physics is done by identifying the operators of the group with physical operators. In one case if you say that the mass quadrupole is the electric quadrupole (as we did this morning for the transition symmetry) then you are stuck. The model is completely defined and it will either agree or desagree with nature. On the other hand you can take a more parametric point of view and argue that $SL(3R)$ is a group of deformations and rotations and parametrize this in terms of unknown collective operators which you are seeking to elucidate. This I think is your point; and I certainly would not say that I disagree.

M. Bouten About the connection between enhanced $E2$-transitions and cut-off of rotational bands, I want to report that results of PHF calculations do not seem to confirm this. On the one hand, we found that for p-shell nuclei, $E2$ transitions between low-lying states were quite accurately reproduced by the PHF-functions without any effective charge. Unfortunately, so far, we did not calculate high L-values in the p-shell, e.g. $L = 6$ or $L = 8$ in ^{8}Be, but we did calculate the $L = 10$ level in ^{20}Ne, and found it considerably higher than expected from the $L(L + 1)$ law.

Concluding Remarks

by E. P. WIGNER

THIS IS the third conference in nuclear physics that I have attended within a year. I have here been learning very, very much again and this is nice, but the fact that one has learned so much on three successive occasions gives one the impression that he still knows an awfully small part of all that he would like to know. This is somewhat discouraging.

Dr. Amaldi was so kind as to ask me to summarize what I have learned and experienced during the conference. I will follow the tradition of these conferences and be very open, voicing my reservations or lack of understanding clearly.

In this spirit, let me begin with a general remark. For much of the time, up to this morning, I had the impression that we did not form a fully homogeneous group, that we were, more or less, divided into three groups, none with a full appreciation for the ideas and conclusions of the members of the other groups. This, I fear, interfered to a certain extent with the ultimate purpose of the conference: to acquaint all of us with all that is worth while to know about the symmetry relations' role in nuclear physics. Today, however, all this has changed. We all participated with keen interest in what we heard presented. Today's sessions were very much in the spirit of the Solvay Congresses, as their purpose was originally visualized by the founders of this institution.

Perhaps I am turning in the wrong direction when I mention, next, two subjects which are very important in nuclear physics and about which we heard very little, at least up to this morning. Perhaps, as a result of the rapidity with which information now accumulates, we had too many and not too few subjects under discussion. Let me mention, nevertheless, the two subjects about which it would have been nice to hear a bit more. The first of these is the $j - j$ coupling shell model. It would have been good to learn a little more about the present status of this model, its recent successes, and its difficulties. It would have been good, also, to hear about the interface of this and the collective model from the viewpoint of the shell model.

The other subject which was, until this morning, sort of pushed under the table is the giant resonance phenomenon. In some cases, single levels exhibit some very marked properties. Thus, from any level of an isotopic spin multiplet, the Fermi β transition leads, at low excitation, preponderantly to a single level. In other cases, a conglomerate of energy levels shares a similar property. The original example for this is the γ ray dipole absorption. This is a giant resonance because it does not lead to a single but to a multiplicity of levels all situated, though, within a relatively narrow energy range of a few MeV. Another example is the one which I happened to mention this afternoon: the dissolution of the analog state in the many states, with lower T, in its neighborhood. Actually, this last subject did not go unheeded at our conference—Dr. French discussed it in some detail this morning. In fact, he arrived at conclusions which I found quite surprising.

Let me now review the three subjects which were discussed during the first few days.

THE COLLECTIVE MODEL

Let me first tell you what was difficult for me to accept of the discussion of this subject. There were two such things: the first one is that it uses a classical picture a little too freely. We all admire Niels Bohr's original picture of the hydrogen atom, how he explained the Balmer series with such ingenuity. Nevertheless, today we do not use his picture any more and we talk in terms of the wave functions. It seems to me that the collective model is too often represented in terms of a classical picture which is just as far from or just as close to reality (to the wave functions) as are Bohr's orbits. Thinking in terms of Bohr orbits, I would be led to believe that I can put the H atom into rotation, just as I can any nucleus. This, of course, is impossible. However, thinking in classical terms may lead to similar difficulties in nuclear physics. A case in which it leads to such a difficulty as actually recognized by Dr. Bohr a long time ago. If the deformation of the nucleus is small, then a rotation of the nucleus and a shape vibration are represented by exactly the same wave function. In this case the appellations, rotational state or vibrational state, are therefore misleading; in some ways they merge. One concludes that even at higher deformations the two are not as distinct as the terminology indicates or as the picture which it creates in one's mind. They are not as different as a ball put into a kind of oscillation and a ball which is spun around.

The second difficulty which I have—and it is my fear that this difficulty is shared by many others—is that it is rarely clear to me whether a type of motion that is described illustrates a stationary state or a giant resonance. Surely, those in the field would not confuse the two situations, but the lack of specification is confusing many of us.

Let me say at this point that no one surpasses me in my admiration for the collective model and the ingenuity and intuition that went into its development. But this is not always the case for the language that is used. Before discussing the new achievements of the model, permit me now to compliment Dr. Bohr's analogy between translational and rotational motion. He pointed out that if we have a nucleus, let us say at rest, we can put it into motion. This is a rigorous concept and the stationary nature and the exact state vector of the moving nucleus follow from Galilean or Lorentz invariance. Similarly, he said we can put it into rotation, though this is not a rigorously defined process, because rotating coordinate systems are not equivalent to nonrotating coordinate systems. Nevertheless, the analogy is very appealing.

Let us now turn to the new conclusions about which we learned from Drs. Bohr and Mottelson. One of the very interesting observations is that, if we express in a rotational band the energy as function of ω, it is a simpler function than as function of the angular momentum. Of course, initially the energy is given in terms of the angular momentum I but, as Dr. Bohr explained, if we make a plot of the energy as function of angular momentum or as function of $I(I + 1)$ (which gives a somewhat simpler plot) and if we take the derivative of this and denote it by ω and then plot the energy as function of ω, the functional dependence is much simpler than in terms of I. In particular (if we look at it as a power series) this converges very much faster. I do not know whether we fully know what this means physically, but it is one of the interesting facts.

Let me insert next something which I did not learn here. This is a result of Y. Sharon. He considered the rotational spectra from the point of view which was used also by Drs. Bohr and Mottelson but was first proposed, I believe, by Peierls and Yoccoz. Sharon assumed a rigid intrinsic state. We heard today from Dr. Bouten about similar calculations. The change in kinetic energy is, as we know, in first approximation proportional to $I(I + 1)$. In second approximation there is a term proportional to the square of this. However, the coefficient of this square is positive definite, and that means that if one plots the energy as function of $I(I + 1)$, it is convex from below. The experimental curve is concave from below, the coefficient of the square of $I(I + 1)$ is negative. Sharon then calculated the change of the potential energy as function of I and found that it decreases and can render the coefficient of the square of $I(I + 1)$ negative. I am bringing this out because it may be important to realize that, in calculating the rotational spectra of nuclei, not only the change in the kinetic but also that in the potential energy has to be taken into account. In this regard, there is a great difference between molecular rotation and nuclear rotation. In the case of molecular rotation, the potential energy is, in the approximation of a rigid molecule (which is, of course, an approximation) independent of the angular momentum I and equal to the potential energy of the state with fixed nuclei. This plays the role of the intrinsic state in the theory of molecular spectra. The reason for the independence of the potential energy on I is that, no matter how little the intrinsic state is rotated, the resulting state is orthogonal to the initial one. Since the operator of the potential energy does not change the position of the nuclei, the potential energy can be calculated for every position of the intrinsic state separately, and since all these potential energies are equal to that of the intrinsic state, the total potential energy becomes equal to that of the intrinsic state, independent of I.

This is not so in the case of nuclei because a small rotation of the intrinsic state does not render its state vector orthogonal to the state vector of the unrotated intrinsic state. The amount of rotation needed to render the rotated intrinsic state essentially orthogonal to the initial intrinsic state depends, of course, on the deformation but is surprisingly large even for a significantly deformed intrinsic state. The lack of orthogonality makes the calculation of the potential energy as function of the angular momentum I more difficult and, at any rate, the potential energy is no longer independent of I.

Next, we learned from both Dr. Bohr and Dr. Mottelson about the possibility of a number of new phenomena. Dr. Bohr, in particular, mentioned the possibility of a phase transformation, at a definite I. This would be a very interesting phenomenon. I presume it would mean that the energy, as function of I, or of ω, would not be smooth but would show a kink or discontinuity of some sort. This, however, was not the only new phenomenon anticipated and Dr. Mottelson, in particular, spoke about vibrations with very high amplitudes and about the interaction of these vibrations with single particle motion. The phenomena foreseen by Dr. Mottelson are numerous and would be most interesting to observe. To be sure, we all heard before about the interaction between collective and single particle motion but never as concisely as this time.

Let me mention next a problem presented by the collective theory which always puzzled me. It is due to the fact that, after all, the nucleus does not consist of a very large number of constituents as does a liquid drop or a solid body. How closely can its shape be defined? I once made a calculation of the definition of the shape presented by a number of particles in a sphere and randomly distributed therein. The average value of the deformation, that is the difference between the maximum and minimum moments of inertia, is much larger than one would expect. This means that the statistical fluctuations of the shape are not negligible as compared with the departure of most intrinsic states from spherical. Now the statistical fluctuations of the shape, calculated this way, may not give an entirely fair picture of the actual fluctuations because the nuclear matter does have properties similar to liquid matter and the fluctuations in density and shape are smaller than if the particles moved independently of each other. There is a repulsive core to begin with and the exclusion principle also has an effect. Nevertheless, the shape fluctuations are surely real and obscure the picture of the nuclear shape. It would be good to clarify it.

Basically, the same question arises in another context and it arose also in today's discussion. If the wave function of the $J = 0$ state is given, to what extent does this determine what we call the nuclear shape? If many photographs of the nucleus are taken, and these superposed on each other, the resulting shape is, of course, spherical. The individual pictures may be so elongated that this is at once evident but they are, on the average, less elongated than the intrinsic state. The question remains: is there a real definition of the deformation of the nucleus? Naturally, many intrinsic shapes give the same $I = 0$ state and may give even the same $I = 2$ state, $I = 4$ state, and so on. Since we talk so much about the shape of the nucleus, it would be good to have a really solid definition of what we do mean by that shape.

The next problem which I wish to raise is: how far do the rotational spectra go? We heard today that in beryllium, according to calculations, levels up to $I = 8$ exist and we heard about experimental data indicating rotational levels up to $I = 10$. The experimental data and the calculations do not quite agree: according to the calculations there is no break in the I dependence of the energy at $I = 6$, which is the last angular momentum permitted by shell theory. On the other hand, if I understood Dr. Wilkinson correctly, the experimental data do show a considerable jump in the energy between the $I = 6$ and the $I = 8$ states. Since I believe in a certain validity of the shell model, I would not be surprised by such a jump. Dr. Wilkinson told me though that the experimental data are not yet definitely established.

GROUP THEORY AS TOOL FOR CALCULATIONS

This subject is more closely related to the subject which was, originally, meant to dominate our conference. We heard about two types of calculations. Both types were concerned with the mathematics of the shell model. The first type was directed toward an improved classification of the states that particles within a shell or within two shells, perhaps, can produce. Dr. Judd's discussion was, I think, exclusively concerned with this and he wanted to develop a scheme in which the vanishing of so many matrix elements appears natural. Dr. Moshinsky's talk on the group theory of the oscillators of which we have the 73-page manuscript was also concerned with this but went further toward actual calculations. The usefulness of his model always surprises me.

The density distribution of the nucleons, from which his calculations start, is very different from the density distribution of actual nuclei. The density distribution, in the limiting case of very large number of nucleons, is, to begin with, $\varrho(r) = (R^2 - r^2)^{3/2}$, R being the nuclear radius. This is, essentially, the density distribution of the nuclei in an oscillator potential. This distribution is, of course, modified by the admixtures of excited states, caused by the interactions between the nucleons. The admixtures unquestionably modify the $(R^2 - r^2)^{3/2}$ density and the purpose of the calculation, which makes such far-reaching use of group theoretical theorems, is just to calculate the admixtures. It is surprising that this can be done adequately when the final density differs as much from the initial one as it seems to do in this case. I happen to have a paper by Dr. Heisenberg's son here, showing the density of the protons in the Nd^{142} nucleus, as measured by the Stanford group of which he is a member. The density, as measured, assumes its maximum around $\frac{3}{4}R$ where the model's initial density is less than a third of the maximum density assumed at the center of the nucleus.

Nevertheless, as we know, the theory reproduces the experimental data in many cases very accurately—surprisingly so. We must hope that this is not the manifestation of something which, at one time or another, bothered all of us: that occasionally theories which are quite far from reality give good agreement. Lawson pointed this out most eloquently by means of his pseudonium theory. I do not believe, though, that this fear is justified in this case.

The last remark brings us naturally to Dr. Hecht's report. The theory he reported on is truly surprising. It starts by replacing, in a particular instance, the $d_{5/2}$ and $g_{7/2}$ orbits with $f_{5/2}$ and $f_{7/2}$ orbits. For some quantities, this surely gives entirely incorrect results. Thus, the density of a particle in the $d_{5/2}$ orbit goes to zero at the center as r^2, for the g orbits as r^4. If there are particles in both orbits, the density near the center is proportional to r^2. The replacement of these orbits by the f orbits would give an r^3 density dependence. The density at the origin is only one particular quantity for which the replacement leads to incorrect conclusions. However, as far as the energy levels are concerned, the model has amazing usefulness.

Before leaving this subject, let me return for a minute to Dr. Judd's report and be more specific concerning the significance of the groups he used. Dr. Judd was concerned with the filling of 14 single particle states. Since each of these states can be filled or empty, we have 2^{14} orthogonal states, and the transformations of these are given by $U(2^{14})$. This is, of course, an enormous group, in fact clearly unnecessarily large; it can be broken up into 15 subgroups, the kth of which is $U\binom{14}{k}$ and refers to the case of k particles and $14 - k$ holes. Even these are extremely large groups, the largest one being $U(3432)$. What Dr. Judd showed is that it is not necessary to use these groups in order to specify the possible states. Instead of the transformations of the 2^{14} states of the system, he considered the transformations of the bilinear forms of the creation and annihilation operators. The operator $a_\varkappa^+ a_\lambda$ annihilates the orbit (single particle state) λ and fills the state $\varkappa$. There are 14^2 such operators. The operators $a_k^+ a_\lambda^+$ fill both orbits $\varkappa$ and λ but, because of $(a_\varkappa^+)^2 = 0$, $a_\varkappa^+ a_\lambda^+ = -a_\lambda^+ a_\varkappa^\lambda$, there are only $14 \times 13/2$ such operators. Similarly, there are $14 \times 13/2$ operators $a_\varkappa a_\lambda$. We have, altogether, $14 \times (14 + 13) = 28 \times 27/2$ operators. Their transformations form the group $0(28)$—a much smaller group than most of those mentioned before. What Dr. Judd found is that all the 2^{14} states of the 14 orbit problem can be uniquely characterized by means of this group and one can explain by means of it the vanishing of all the matrix elements which do vanish. Why this is true is a mathematical question and a mathematical question of some depth to which I, at least, do not know the answer.

Let me now turn to the last subject—the old-fashioned application of group theory to derive the consequences of symmetries and invariances, accurate and approximate.

CONSEQUENCES OF THE ACTUAL SYMMETRIES —ACCURATE AND APPROXIMATE

The difference between the two types of applications of group theory was brought out in our discussions most clearly by Dr. Heisenberg. What gives our present subject a new life is the much improved knowledge of the forces between nucleons. This was brought out, perhaps most dramatically, by Dr. Wilkinson who compared the three nucleon-nucleon interactions. He said that the $p - p$ and $n - n$ interactions—responsible for charge symmetry—are now believed to be equal within a couple of tenths of a per cent. I believe he will be the first to admit that it is very difficult to be absolutely sure of this because the $n - n$ interaction can be measured, at present, only as final state interaction. Even the measured direct interactions $p - p$ and $p - n$, are difficult to interpret in terms of an interaction Hamiltonian, and the difficulty is much greater if the information comes from final state interaction measurements. However, a good case can be made for the equality of $p - p$ and $n - n$ interactions also from other data.

Similarly, Dr. Wilkinson had a measure for the difference between $p - n$ and $p - p$ interactions. It was long suspected that the former is stronger but the accurate measure of the difference, about 2.8 per cent, is certainly quite new.

Isotopic Spin-SU_2 Symmetry was Dr. Wilkinson's next subject. This symmetry is, of course, a consequence of the equality of the three nucleon-nucleon interactions. The first of the consequences considered was the second order dependence of the masses of the members of isotopic spin multiplets on T_z (which is half of the difference between neutron and proton numbers). Dr. Wilkinson said that the coefficient of the third order term, T_z^3, is only about 10 keV—very small indeed.

It is not easy to draw inferences from this on the equality of the three kinds of nucleon-nucleon interaction because any interaction that has vector character in the isotopic spin space gives (in first and second approximation) a similar T_z dependence of the masses and the same is true in first approximation even of a tensor operator in that space. (An operator of scalar nature—we hope that the overwhelming part of the interaction is of this nature—does not give rise to any difference between the masses of the members of the isospin multiplet.) Hence, what the second order dependence of the masses in T_z proves is only that the third approximation of a vector-like operator in the isotopic spin space, and the second approximation of any tensor-like operator, do not play a significant role in the Hamiltonian. Something that does give information and evidence for the equality (and hence scalar nature) of the three nucleon-nucleon interactions is the β decay of spin zero angular momentum ($J = 0$) nuclei into their isotopic spin partners. To these transitions, the Gamow-Teller matrix element does not contribute because it has vector character in ordinary space. Wilkinson pointed out that the so-called ft values of all these transitions are very nearly the same (about 3050), thus giving evidence for the equality of the aforementioned three interactions. As a slight exception, Dr. Wilkinson mentioned Al^{26}, decaying into Mg^{26}, with an ft value apparently somewhat below the others. However, this decreased ft value is not absolutely sure and, second, it concerns a degree of accuracy which is beyond that of ordinary nuclear physics. We are usually satisfied if two transition probabilities, which should be equal according to theory, agree within about 15 per cent and the apparent difference in this case is much less than that.

I would like to mention something here that I learned from Dr. J. D. McCullen of the University of Arizona. He investigated the β decay of Sc_{21}^{42}, a $J = 0$ nucleus. The decay product is Ca_{20}^{42}, with a $J = 0$ normal state, the isotopic spin partner of the Sc^{42} normal state. The second excited state of Ca^{42} also is a $J = 0$ state but it is not the isotopic spin partner of the Sc^{42} normal state. Hence, the β transition thereto is forbidden by the isotopic spin rules and it is, in fact, 10,000 times weaker than the transition to the partner, that is the normal state. It is, naturally, possible that other circumstances also play a role in this much decreased transition probability but that two such propabilities should differ by a factor 10,000 remains remarkable.

The regularities which one can derive from the validity of the isotopic spin concept for nuclear reaction cross sections and polarizations are not equally well confirmed. Dr. Wilkinson discussed the polarization resulting from two nuclear reactions which should yield equal polarizations. The two polarizations, as he showed them to us, did look alike. However, his data were confined to small angles. At larger angles, the polarizations resulting from $H^2 + H^2 \rightarrow He^3 + n$, and from $H^2 + H^2 \rightarrow H^3 + H^1$ are quite different. So are, I understand, the cross sections of the $He^3(p, p)$ and of the $H^3(n, n)$ reactions. The conflict with the validity of the isotopic spin concept may be more apparent than real because, occasionally, very small deviations give very large effects. Nevertheless, it is my impression that we yet have to learn a great deal about the consequences of the very close equality of the three nucleon-nucleon interactions, when are these consequences valid, and when not.

I should mention, perhaps, the limitation of the validity of the rules derived from the isotopic spin concept: this validity is restricted, except for the lightest nuclei, to low energies. Dr. Wilkinson mentioned this many times. I happened to read, a few days ago, about the $N^{14} + \alpha$ reaction, going through the F^{18} compound nucleus. The yields of the different reaction products obtained make it evident that, for excitation energies of 15 Mev or so, the isotopic spin of the levels is no longer pure. This will always be the case if levels with different isotopic spins are close to each other—in such a case even a relatively small interaction, such as the electrostatic one, causes a significant amount of mixing. We know, in fact, that for heavier nuclei the states with isotopic spin $T = T_\xi + 1$, i.e., exceeding that of the normal state, are dissolved into giant resonances.

Of the not strictly valid symmetries, that of charge independence is most accurate in nuclei. When I say this, I consider the rotational (and of course the translational) invariance to be strictly valid, which is, as far as we know, truly correct. I also count the reflection invariances among the strictly valid ones, as they are in practice valid as far as nuclear physics is concerned. Nevertheless, I should recall some of the beautiful experiments about which we heard and which exhibit tiny violations of the parity concept. May I add, to what we heard in this connection, a reference to Lobashov's beautiful work on the circular polarization of yet another transition, the $5/2^+ \rightarrow 7/2^+$ transition in Ta^{181}. It is of the order of 10^{-6}, in consonance with the usual assumption that the parity-violating forces play a subordinate role in the parts of nuclear physics, other than β decay.

The SU_3 symmetry, my next subject, deals with a symmetry which is less closely valid than the isotopic spin's SU_2 symmetry. Dr. Elliott was the first to recognize the usefulness of the model, based on the assumption of the validity of this symmetry, and it was discussed at our meeting by Dr. Elliott himself. Dr. Elliott also referred to the SU_4 model and showed that this has, in the cases considered by him, an accuracy of about 90 per cent, that is, 90 per cent of the total wave function belongs to a single representation of SU_4. The SU_3 symmetry is a great deal better yet and appears to have, in many cases, a truly surprising validity. There would be little point in my repeating what Dr. Elliott said on this subject—his presentation was lucid and complete.

There is one reservation, though, on the confirmation of this and partly also of the SU_4 model which will be discussed later. The reservation is that most of the confirmation is derived by comparison with other calculations, and only to a small degree by comparison with experimental data. This is a little bothersome. The agreement between two calculations is not as significant as the agreement with experiment and we were only too often reminded of this in the course of our discussions. It is, of course, much easier to calculate something on the basis of a more or less confirmed theory than to persuade an experimental physicist to measure something which may be very difficult to measure. Nevertheless, there is no real substitute for experimental confirmation and if the agreement of a model only with theory, to be sure a more elaborate theory, can be established, the model becomes a calculational tool, similar to those discussed before.

The Supermultiplet (or SU_4) Theory. Our next subject concerns the symmetry which is least accurate: the SU_4 symmetry. This would be strictly valid if the forces between nucleons were independent of both isotopic and ordinary spin. The independence of isotopic spin was discussed in detail before; it is a reasonable approximation. The dependence on ordinary spin is, on the other hand, much greater. We know this ever since Rabi's measurement of the quadrupole moment of the deuteron. One can well be surprised, therefore, by the degree of validity of this model, as discussed particularly by Dr. Radicati. Let me discuss only that part of the evidence which is based on comparison with experimental information. This comparison will show a remarkable similarity with the comparison of the SU_2 theory with experiment, inasmuch as it will deal with the same two types of data. It will be less convincing.

The mass formula of the supermultiplet model has a much wider applicability than that of the isotopic spin theory: it extends to the normal states of all nuclei. However, neither its derivation from the basic postulates of the theory, nor its confirmation by Radicati, are equally clear. I am sure Dr. Radicati will agree with this statement.

As to β decay, it does follow from supermultiplet theory that only the transitions between members of the same supermultiplet are allowed. The Fermi, that is $\Delta J = 0$, transitions are, as we saw, allowed only between members of the same T multiplet, the Gamow-Teller ones also allowed between members of those T multiplets which are united into the same supermultiplet. It follows that the ft values of β transitions between members of different supermultiplets should be much larger than those within the same supermultiplet, i.e., should be well above 3000. Actually, they are, as Dr. Radicati told us, as a rule about 100 times larger. Again, this could perhaps be attributed to many other cases and, as I believe I pointed out in the course of the discussion of Dr. Radicati's paper, it can be deduced from postulates somewhat less far-reaching than the total SU_4 symmetry. It remains true that this conclusion of the supermultiplet theory seems to be confirmed.

Incidentally, the large ft values, that is the small values of the corresponding Gamow-Teller β decay matrix elements, are in many cases in sharp conflict with the $j - j$ coupling shell model. This is very confusing because this model gives, *at least* for the energy levels, regularities—what I call Talmi rules—which are, in many cases amazingly accurate. The fact that the β decay probabilities are in many cases so much lower than this model would postulate, is very puzzling. The magnetic moments, as I learned from Dr. Talmi, are not given by the model as well as the energy levels but not as poorly as the β decay matrix elements.

The low decay rate of C^{14} was brought up in the discussion as to be in disagreement with the postulates of supermultiplet theory. It was assumed that the normal state of the daughter nucleus, of N^{14}, is in the same supermultiplet as that of C^{14} so that the transition should be rapid. A possible explanation of this discrepancy is that the normal state of N^{14} is a $^3D_1$ state and that it is, therefore, not part of the supermultiplet which contains the $^1S_1$ normal state of C^{14}. Dr. Bertsch called my attention to a paper by Mangelson, Harvey, and Glendenning which seems to confirm this assumption. Nevertheless, I admit that, even if we accept this reasoning, the very low transition rate remains puzzling. It appears to indicate a validity of the supermultiplet model which is much closer than one would expect, a surprisingly low transition rate between the S and D supermultiplets. One would expect the ^3S and ^3D states to mix to a reasonable extent—in fact, the $j - j$ model postulates such a mixing.

CLUSTER STRUCTURE

The discussion of the cluster model was very interesting and, again, the enormous range of applicability, and the visualizability of its pictures of the nuclei are amazing. Naturally, one worries about the fact that the state vectors given by these models correspond to $L - S$ coupling, just as those of the supermultiplet theory. Furthermore, it appears to be more difficult to incorporate the departures from $L - S$ coupling into this model than into supermultiplet theory.

My final observation draws attention to the new tools of nuclear physics about which we heard, from several speakers, usually somewhat incidentally. I am referring to the tools furnished by particle physics, μ decay data, hypernuclei of various kinds, probing of the nucleus not only with electrons but also with muons, pions, and other particles. These new tools promise to furnish new insights into nuclear structure, and perhaps also into the Symmetry Properties of Nuclei which was our original subject.

We had an interesting meeting and we'll all return home enriched in knowledge of nuclear physics. As the last speaker, I wish to express the appreciation of all of us to the organizers of the meeting, to the group here for its constant help and hospitality, and last but not least to those who stand behind them.